**JOHN R. HOLUM**
*Augsburg College*

W9-AZS-005

STUDY GUIDE
WITH ANSWERS TO EXERCISES
TO ACCOMPANY

# Elements of General and Biological Chemistry
### Eighth Edition

## John Wiley and Sons, Inc.
*New York    Chichester    Brisbane    Toronto    Singapore*

Copyright © 1991 by John Wiley & Sons, Inc.

All rights reserved.

Reproduction or translation of any part of
this work beyond that permitted by Sections
107 or 108 of the 1976 United States Copyright
Act without the permission of the copyright
owner is unlawful. Requests for permission or
further information should be addressed to the
Permissions Department, John Wiley & Sons, Inc.

ISBN 0 471 53498-6
Printed in the United States of America

10 9 8 7 6 5 4 3

Printed and bound by Port City Press, Inc.

# PREFACE

Each chapter in this Study Guide matches a chapter in *Elements of General and Biological Chemistry*, eighth edition. Six features are included.

1. A discussion of the purpose of the chapter and a list of specific objectives.
2. A glossary of the important terms in the chapter.
3. Special help units. These may include aids in learning how to work problems, practice in writing formulas or structures, or lists of the important properties of functional groups. Several special sets of drill exercises, together with their answers, are also included.
4. Self-testing questions. Both completion and multiple-choice questions are provided.
5. Answers to the exercises and the self-testing questions in this Study Guide.
6. Answers to all exercises in the text.

John R. Holum
Augsburg College
Minneapolis, MN 55454

# CONTENTS

# CHAPTER 1

# GOALS, METHODS, AND MEASUREMENTS

## STRATEGY FOR STUDY

Mastering a body of knowledge comes in two main steps. Neither can be omitted. The first must occur before the second will work.

1. Study the material until it is *understood*, until you can say to yourself "that makes sense," or "now I get it," or something similar. With easy material, one reading will get you this far. With the more difficult, you'll need the text, class lectures and discussions, and sometimes extra help.

2. Drill yourself on the understood material until you *know* it. In some cases, this means memorizing the definitions of new words. In others, you should work problems of a purely drill nature as well as questions that require thought.

Don't make the mistake of confusing *understanding* with *knowing*. Understanding is only the first step, although an essential one, to knowing.

What will count from this course throughout your professional life, including later courses and in-service training, is *remembered chemistry*. "Understood chemistry" is of itself almost never "remembered chemistry." "Understood chemistry" taken alone is chemistry very soon forgotten; it is chemistry that can seldom, if ever, be applied in new situations. It's like a house stopped at the foundation. On the other hand, "rote-memorized chemistry" is almost never "remembered chemistry," either. It is like a house without a foundation and will soon collapse. A complete house requires both a foundation and a superstructure.

This *Study Guide* is designed to help you get from the understanding stage to the knowing stage. The text, lectures, and outside help are your chief aids to understanding. The *Study Guide* can help you by providing examples of worked problems. Sometimes alternative methods are presented. Definitions of the important terms used in each chapter of the text are provided. A number of drill exercises, together with all their answers are given. You are also given an opportunity to test yourself by two kinds of questions - completion and multiple choice. As a final check, you can go to the list of objectives for the chapter being studied and see if you can do each of them.

The two main steps for mastering a body of knowledge can be broken down into a series of smaller steps, as follows:

Step 1.    Read the chapter in the text. If at all possible read it once before the class discussion. Use a pencil - not a pen - to place a check in the margin near any part you have trouble understanding. You don't help yourself much by underlining what you do understand. Discover, instead, what you do not understand.[1]

Step 2.    Use the lecture period to clear up as many of the puzzles as possible. Erase each check mark as understanding moves in and puzzlement moves out. If the lecture leaves any points still unclear, see your instructor or a teaching assistant about them as soon as you can. Often another classmate can help you.

Step 3.    After reaching the understanding stage, begin the knowing stage. Learn the definitions of the important terms. They appear in boldface in the text and are defined in the glossaries of each chapter in this *Study Guide*. These same terms are organized at the end of the *text* into a book-length glossary.

Step 4.    Where the material involves calculations, rework the examples. Try all of the problems for which answers are provided in order to test yourself. The *Study Guide* has many additional problems for you to solve.

Step 5.    When you are quite sure you have done sufficient studying, test yourself with the Self-Testing Questions in this *Study Guide*. All of the answers are given and, in many instances, explanations are also provided. If you do not do very well on these questions, take it as a sign that you need to do more studying.

----

[1] I owe the formulation of this step and the next to S. Paul Steed of American River College, Sacramento, California. He calls this approach the EYI study technique - "eliminate your ignorance" - and it makes great good sense. (*Journal of Chemical Education*, December 1976, page 745.)

Step 6.    As a final test, check the list of objectives for the chapter as given in this *Study Guide*. Be sure you can *write out* a correct response for each objective. Sometimes working in pairs helps.

## OBJECTIVES

After you have studied this chapter and have worked the exercises in it, you should be able to do the following. (Other objectives may also be assigned.)

1.   *Distinguish between a number and a physical quantity.*

2.   *Distinguish between a base unit and a standard of measurement and between a base unit and a derived unit.*

3.   *Give the present name (in the English translation) and the official abbreviation for the successor to the metric system.*

4.   *Name, define, and give the symbols of the SI base units of length, mass, temperature, and time and the derived unit for volume.*

5.   *Describe the advantage of the use of base-10 in the SI.*

6.   *Use the conventions for writing figures and symbols for SI units.*

7.   *Use the prefixes kilo-, centi-, milli-, and micro- with SI units.*

8.   *Convert back and forth between kilograms, grams, milligrams, and micrograms.*

9.   *Convert back and forth between liters, milliliters, and microliters.*

10.   *Use conversion factors to change from one system to another.*

11.   *Tell how many significant figures a number has.*

12.   *Correctly round off numbers obtained by arithmetic operations on physical quantities.*

13.   *State the difference between "precision" and accuracy," and tell what is meant by an exact number.*

14.   *Convert degrees Fahrenheit to degrees Celsius (or vice versa).*

15.   *Do calculations involving density.*

16.   *State what we mean by "calorie" and "kilocalorie."*

17.   *Define each term listed in the Glossary.*

## GLOSSARY

**Accuracy.**  In science, the degree of conformity to some accepted standard or reference; freedom from error or mistake; correctness.

**Base Quantity.**  A fundamental quantity of physical measurement such as mass, length, and time; a quantity used to define derived quantities such as mass/volume for density.

**Base Unit.**  A fundamental unit of measurement for a base quantity - such as the kilogram for mass, the meter for length, the second for time, the kelvin for temperature degree, and the mole for quantity of chemical substance; units to which derived units of measurement are related.

**Calorie.**  The amount of heat that must be transferred to 1 g of water at 14.5 °C to raise its temperature 1° Celsius to 15.5 °C.

**Centimeter (cm).**  A length equal to one-hundredth of a meter.  1 cm = 0.01 meter = 0.394 inch.

**Chemical Energy.**  Energy possessed by a substance by virtue of its ability to release other forms of energy as it undergoes a chemical change.

**Chemical Property.**  The ability of a substance to undergo a change into a different substance.

**Chemical Reaction.**  A change in which substances are converted into different substances.

**Chemistry.**  The study of the composition, structures, and properties of substances.

**Conversion Factor.**  A fraction that expresses a relationship between quantities that have different units; for example, 2.54 cm/1 in.

**Degree Celsius.**  One-hundredth of the interval on a thermometer between the freezing point of water and the boiling point of water.

**Degree Fahrenheit.**  One-one hundred and eightieth (1/180) of the interval on a thermometer between the freezing point and the boiling point of water.

**Density.**  The ratio of the mass of an object to its volume; the mass per unit volume with the usual units being grams per milliliter.  D = mass/volume.

**Derived Quantity.**  A quantity based on a relationship involving one or more base units of measurement; for example, volume (length)$^3$ or density (mass/volume).

**Derived Unit.**  A unit of a derived quantity such as g/mL (density).

**Energy.**  A capacity to cause a change that, in principle, can be harnessed for useful work.

**Factor-Label Method.**  A strategy for solving a problem involving physical quantities in which the proper cancellation of units is used as an aid in setting up the solution.

**Gram (g).** A mass equal to one-thousandth of the kilogram mass, the SI reference standard. 1 g = 0.001 kilogram = 1000 milligrams; 1 pound = 454 g.

**Heat.** The form of energy transferred between two objects that initially have different temperatures.

**International System of Units (SI).** The successor to the metric system of measurements that retains the units of the metric system and their decimal relationships but employs new reference standards.

**Kelvin Degree (K).** Identical in size to the degree Celsius but labeled so that K = °C + 273.

**Kilocalorie (kcal).** 1 kcal = 1000 cal

**Kilogram (kg).** A mass equal to that of a reference standard made of a platinum-iridium alloy and kept in France; originally the mass of 1000 cubic centimeters of water at 4 °C. 1 kg = 1000 grams = 2.20 pounds (avoirdupois) = 37.27 ounces (avoirdupois).

**Kinetic Energy (K.E.).** Energy of motion; K.E. = $1/2\ mv^2$.

**Length.** The base quantity for expressing distances or how long a thing is.

**Liter (L).** A volume equal to 1000 cubic centimeters. 1 liter = 1000 milliliters = 1.057 liquid quarts.

**Mass.** A quantitative measure of inertia based on an artifact at Sévres, France, called the kilogram mass; mass is the measure of the amount of matter in an object relative to that reference.

**Measurement.** An operation that compares an unknown physical quantity with known standards.

**Meter (m).** The base unit of length in the International System (SI). 1 m = 39.270 inches = 3.281 feet = 1.094 yards = 100 centimeters.

**Microgram (μg).** A mass equal to one-thousandth of a milligram. 1 μg = 0.001 mg = 0.000 001 g. (Its symbol is sometimes given as mcg or as γ, gamma, in pharmaceutical work.)

**Microliter (μL).** A volume equal to one-thousandths of a milliliter, 1 μL = 0.001 mL = 1 x $10^{-6}$ L

**Milligram (mg).** A mass equal to one-thousandth of a gram. 1 mg = 0.001 g; 1000 mg = 1 g; 1 grain = 64.8 mg.

**Milliliter (mL).** A volume equal to one-thousandth of a liter. 1 mL = 0.001 L = 16.23 minim = 1 cubic centimeter; 1 liquid ounce = 29.57 mL; 1 liquid quart = 946.4 mL.

**Millimeter (mm).** A length equal to one-thousandth of a meter. 1 mm= 0.0394 inch = 0.001 meter.

**Physical Properties.**   Characteristics of substances that can be observed and measured without causing changes into different substances.

**Physical Quantity.**   A property to which we assign both a numerical value and a unit; physical quantity = number x unit.

**Potential Energy.**   Stored or inactive energy.

**Precision.**   In science, the fineness of the measurement or the degree to which successive measurements agree with each other when several are taken one after the other. (Compare with **Accuracy**.)

**Property.**   A characteristic of something by means of which we may identify it.

**Second (s).**   The SI unit of time; the sixtieth part of a minute.

**Scientific Notation.**   The method of writing a number as a product of two numbers, one being $10^x$, where x is some positive or negative whole number.

**SI.**   See **International System of Units**.

**Significant Figures.**   The number of digits in a numerical measurement or result that are known with certainty to be accurate plus the first digit whose value is uncertain.

**Specific Gravity.**   The ratio of the density of an object to the density of water.

**Specific Heat.**   The ability of one gram of a substance to absorb heat; specifically, the number of calories of heat that one gram of a substance can absorb for every degree Celsius change in its temperature.

**Standard, Reference.**   A physical description or embodiment of a base unit.

**Temperature.**   The measure of the hotness or coldness of an object.

**Time.**   A period during which something endures, exists, or continues.

# MULTIPLES AND SUBMULTIPLES BY MOVING A DECIMAL POINT

The purpose here is to become skilled in changing a number by units of 10, 100, 1000, or more or by 0.1, 0.01, 0.001, etc., simply by moving the decimal point. We will practice applying this skill to a set of problems that involves specific quantities stated in metric or SI units.

The most common metric or SI conversions are made by moving the decimal point either three places or multiples of three places. The prefix used indicates the type of change to be made. Consider those prefixes most commonly encountered in the life sciences and in chemistry:

| Prefix | Exponential Equivalent |
|---|---|
| kilo- | $10^3$ |
| REFERENCE | 1 (or $10^0$) |
| milli- | $10^{-3}$ |
| micro- | $10^{-6}$ |

To change a quantity expressed in the unit of the reference (e.g., liters) to the equivalent expressed by the next smaller unit in this list (e.g., milliliter), simply move the decimal three places to the right.

**Example:**     3 liters = 3000 milliliters

3.0 liters becomes 3.0,0,0,0 milliliters

right $\longrightarrow$   1 2 3

Remember, to go from a larger to a smaller unit we always move the decimal point to the right.

To change a physical quantity from a smaller to a larger unit, we move the decimal point to the left.

**Example:**     2000 millimeters = 2 meters

2000.0 millimeters becomes 2.0.0.0.0 meters

3 2 1 $\longleftarrow$ left

Note that "milli-" means "a thousandth"; thus, the meter is a larger unit than the millimeter.

We can summarize the most often used conversions by the following table.

| The Conversion | The Number of Places to Move the Decimal Point | Examples |
|---|---|---|
| Reference unit to milli- | 3 to the right | 0.001 g = 1 mg<br>0.002 L = 2 mL<br>0.0008 m = 0.8 mm |
| Milli- to micro- | 3 to the right | 0.015 mL = 15 μL<br>0.462 mg = 462 μm |
| Kilo- to reference unit[a] | 3 to the right | 1.452 kg = 1452 g |
| Reference unit to centi-[b] | 2 to the right | 0.560 m = 56.0 cm |
| Milli- to reference unit | 3 to the left | 156 mg = 0.156 g<br>4325 mL = 4.325 L<br>25 mm = 0.025 m |
| Micro- to milli- | 3 to the left | 146 μL = 0.146 mL<br>36.5 μg = 0.0365 mg |
| Milli- to deci-[c] | 2 to the left | 150 mL = 1.50 dL |
| Reference unit to kilo-[a] | 3 to the left | 465 g = 0.465 kg |

[a] Used almost exclusively in connection with masses.
[b] Used almost exclusively in connection with lengths.
[c] Used almost exclusively in connection with volumes.


## DRILL EXERCISES

### I. EXERCISES IN METRIC UNITS

The purpose of these exercises is to help you become very skillful in converting from a quantity to one of its multiples or submultiples. Every conversion in these exercises involves nothing more than moving the decimal point. Fill in the blanks of the second column with a number that, together with its unit, is equivalent to the quantity in the first column. Another purpose of the exercises is to learn the accepted abbreviations of the various SI units (consult the table in the text for these).

| | Quantity | Equivalent | |
|---|---|---|---|
| **Examples:** | 0.001 m | 1 | mm |
| | 1000 m | 1.000 | km |
| | 1654 g | 1.654 | kg |

| | | | | | | |
|---|---|---|---|---|---|---|
| 1. | 0.01 m | _____ cm | | 11. | 0.527 mg | _____ μg |
| 2. | 0.10 m | _____ cm | | 12. | 0.120 mL | _____ μL |
| 3. | 2985 m | _____ km | | 13. | 15 mm | _____ cm |
| 4. | 1564 mg | _____ g | | 14. | 15 cm | _____ m |
| 5. | 156.4 mg | _____ g | | 15. | 0.98 m | _____ cm |
| 6. | 15.64 mg | _____ g | | 16. | 1.620 g | _____ mg |
| 7. | 1640 mL | _____ L | | 17. | 0.101 L | _____ mL |
| 8. | 0.002 L | _____ mL | | 18. | 0.067 g | _____ mg |
| 9. | 454 g | _____ kg | | 19. | 250 mL | _____ L |
| 10. | 150 mL | _____ L | | 20. | 0.0001 mL | _____ μL |

## II.  EXERCISES IN SCIENTIFIC NOTATION

The purpose of these problems is to help you become skilled in converting back and forth between numbers expressed in the usual way and their equivalents in scientific notation.

**SET A.**   Change each number to the form it has in scientific notation.

1.    1,062,457        _____

2.    0.00543          _____

3.    111.6            _____

4.    0.00000521       _____

5.    5.025            _____

**SET B.**   Change each number given in scientific notation to its equivalent stated in the usual way.

1.    $6.150 \times 10^3$       _____

2.    $5.362 \times 10^2$       _____

3.    $2.35 \times 10^{-2}$       _____

4.    $8.79 \times 10^{-5}$       _____

5.    6,542,001        _____

## III. EXERCISES IN SIGNIFICANT FIGURES

Rule 1.   Zeros are always significant if nonzero numbers occur on both their right and their left.

Rule 2.   Zeros are never counted as significant if they are used to set off the decimal point on their left.

Rule 3.   Trailing zeros to the right of the decimal point are always significant.

Rule 4.   Trailing zeros to the left of the decimal point are significant only if the author somewhere says or implies so.  (We will assume so.)

Rule 5.   Exact numbers are treated as having an infinite number of significant figures.

**Examples:**    4.0054    _____5_____    (Rule 1)

4540    _____4_____    (Rule 4 as further noted in the parentheses)

0.000454    _____3_____    (Rule 2)

4.5400    _____5_____    (Rule 3)

| | | | | | | |
|---|---|---|---|---|---|---|
| 1. | 60,500 | _____ | | 6. | 0.0006050 | _____ |
| 2. | 6.0500 | _____ | | 7. | $1.98 \times 10^{10}$ | _____ |
| 3. | 0.605 | _____ | | 8. | $1.9800 \times 10^{10}$ | _____ |
| 4. | 0.6050 | _____ | | 9. | 123.50 | _____ |
| 5. | 0.000605 | _____ | | 10. | 12,350 | _____ |

## IV.    EXERCISES IN THE FACTOR-LABEL METHOD FOR CONVERTING FROM ONE SYSTEM OF UNITS TO ANOTHER

The key idea:  Get the final unit you want fixed in your mind and work toward your answer being in that unit.

The key operation:  Multiply and divide (cancel) units like numbers.

The key strategy:  Set up multiplications or divisions to get all unwanted units to cancel.  If they don't, you know you have set up the solution to the problem incorrectly.

We shall study and learn to do this by means of examples.

**Sample Problem 1:**   Change 10.0 inches into centimeters.

A table of conversion factors (e.g., Table 1.1 in the text) tells us that there are 2.54 cm in 1 inch.

We may write this basic relation in either of two equivalent ways.  Both state the same relation between the inch and the centimeter.  One is simply the inverse, or the reciprocal, of the other.

$$\frac{2.54 \text{ cm}}{1 \text{ in.}} \quad \text{or} \quad \frac{1 \text{ in.}}{2.54 \text{ cm}}$$

The divisor line is read as "per." The first ratio says "2.54 cm per 1 in." The second says "1 in. per 2.54 cm."

We want our answer to be in centimeters. Therefore, we must use the ratio that will give us that answer. That is the "key idea" for this problem.

Step 1.     Write down the given (be sure
to include the unit.)                    10.0 in.

Step 2.     Multiply the given by the
right conversion factor.

Do we write it as        (A)     $10.0 \text{ in.} \times \dfrac{2.54 \text{ cm}}{1 \text{ in.}}$ ?

or

Do we write it as        (B)     $10.0 \text{ in} \times \dfrac{1 \text{ in.}}{2.54 \text{ cm}}$ ?

In other words, which way do we use the relationship between inches and centimeters? In (B), the units will not cancel to leave the answer solely in "cm". In (A), the units cancel correctly.

(A)     $10.0 \text{ in.} \times \dfrac{2.54 \text{ cm}}{1 \text{ in}}$

Step 3.     Do the arithmetic. Do whatever multiplications or divisions of numbers that there are left to do.

$$10.0 \times 2.54 \text{ cm} = 25.4 \text{ cm}$$

**SET A.**  Before continuing, work enough of the following until you are sure you understand and can set up conversion factors in either of two ways.

1.   There are 5280 feet in a mile.
2.   There are 2.20 pounds in a kilogram.
3.   There are 454 grams in a pound.
4.   There are 36.4 inches in a yard.
5.   There are 1000 milligrams in a gram.
6.   There are 1000 meters in a kilometer.
7.   There are 29.6 milliliters in a fluid ounce.
8.   There are 7000 grains in 454 grams.
9.   There are 1000 kilograms in a metric ton.
10.  There are 1000 grams in a kilogram.

**Sample Problem 2:**   How many inches are there in 127 centimeters?

Step 1.    Write down the given.                        127 cm

Step 2.    Multiply the given by                         Do we write it
           the right conversion
           factor.                                       (C)    127 cm    x    $\dfrac{2.54 \text{ cm}}{1 \text{ in.}}$ ?

                                                          or do we write it

                                                          (D)    127 cm    x    $\dfrac{1 \text{ in.}}{2.54 \text{ cm}}$ ?

           Only (D) lets us cancel the
           units in a way that leaves
           the answer in the unit
           called for in the problem.
           The units for (C) would be

           $\dfrac{(cm)(cm)}{in.}$   or   $\dfrac{cm^2}{in.}$   !

Step 3.    Do the arithmetic.    127  x  $\dfrac{1 \text{ in.}}{2.54}$ = $\dfrac{127}{2.54}$  in.  =  50.0 in.

**SET B.**   Work the following problems using the conversion factors stated in Set A.

1.    How many feet are there in 0.5000 mile?

2.    How many miles are there in 15,000 feet?

3.    How many yards are there in 100 inches?

4.    How many milligrams are there in 0.100 gram?

5.    How many grams are there in 5.0 grains?

These problems were actually quite simple.  The real power of the factor-label method comes when you don't have a single conversion factor to use to work a problem and you have to improvise from two or more factors.  Using the conversion factors given in Set A, let us see how this works by studying examples.

**Sample Problem 3:**    How many metric tons are there in 1000 pounds?

We don't have a factor to convert pounds to metric tons directly.  We have to improvise.  We do have a factor relating metric tons to kilograms (No. 9 in Set A).

$$\dfrac{1000 \text{ kilograms}}{1 \text{ metric ton}} \quad \text{or} \quad \dfrac{1 \text{ metric ton}}{1000 \text{ kilograms}}$$

We also have a factor relating kilograms to pounds (No. 2 in Set A).

$$\dfrac{2.20 \text{ pounds}}{1 \text{ kilogram}} \quad \text{or} \quad \dfrac{1 \text{ kilogram}}{2.20 \text{ pounds}}$$

Following our steps, we have:

Step 1.    Write down the given.                      1000 pounds

Step 2.    Pick a factor that allows
           you to cancel "pounds."
           Multiply all this by            1000 ~~pounds~~ x $\frac{1 \text{(kilogram)}}{2.20 \text{ ~~pounds~~}}$
           another factor that allows
           you to cancel kilograms.
                                  1000 ~~pounds~~ x $\frac{1 \text{ ~~kilogram~~}}{2.20 \text{ ~~pounds~~}}$ x $\frac{1 \text{ (metric ton)}}{1000 \text{ ~~kilograms~~}}$

(In principle, you would keep doing this, picking additional conversion
factors that let you cancel unwanted units, until only the desired unit(s)
remains. As you can see, we are through using conversion factors in this
sample problem.)

Step 3.    Do the arithmetic.

$$1000 \times \frac{1}{2.20} \times \frac{1}{1000} \text{ metric tons} = 0.455 \text{ metric ton}$$

The advantage of leaving all the arithmetic to the end is that one can often
cancel some of the numbers too.

**Sample Problem 4:**   How many milligrams are there in 1.00 grain? (The needed
units are given in Set A.) Here is how the final solution will look before the arithmetic
is done.

$$1.00 \text{ grain} \quad \times \quad \frac{454 \text{ grams}}{7000 \text{ grains}} \quad \times \quad \frac{1000 \text{ milligram}}{1 \text{ gram}}$$

Final Answer: 64.9 milligrams per grain

**SET C.**   Practice what you have learned from Sample Problems 3 and 4 by working
the following.

1.    How many grains are there in 1.00 kilogram?

2.    How many pounds are there in 2.0 metric tons?

3.    How many milligrams are there in 0.5 kilogram?

## SELF-TESTING QUESTIONS

**Completion.**   Fill in the blanks with the words or phrases that best complete each
statement or answer the question.

1.    The successor to the metric system is the _____.

2.    The physical embodiment of a base unit is called a reference _____.

3.   The base unit of length in the metric system is the _____.

4.   The base unit of length in the successor to the metric system is the

_____.

5.   The reference standard for the measurement of length in the metric system is

_____.

6.   Is this also the reference standard for the measurement of length in the successor to the metric system? _____

7.   If not, how do the two differ in length? _____

8.   The inertia of an object is its ability to _____.

9.   The quantitative measure of an object's inertia is called the _____.

10.  Which of these are base units and which are derived?

   (a)   volume _____        (c)   area _____

   (b)   mass _____          (d)   length _____

11.  The base unit of mass in the metric system is the _____.

12.  The base unit of mass in the successor to the metric system is _____.

13.  Scientists would like to have a reference standard for mass that is not an artifact (not a manufactured object) because _____

_____.

14.  The base unit of time in the SI is the _____.

15.  The base unit for the temperature degree in the SI is the _____.

16.  The degree _____ was formerly called the degree Centigrade.

17.  There are _____ degree divisions between the freezing point and the boiling point of water on the Fahrenheit scale.

18.  The meter is a little longer than _____ feet.

19.  Half an inch would be _____ than 1 centimeter.
                        (shorter or longer)

20. In 1.0 milliliter of water there are about 16 drops.  One drop is therefore
    _____ microliters.

21. One kilogram of water occupies one _____ of volume (in SI terms),
    and this volume is roughly one _____ in the "English" system.

22. To change 0.00056 to scientific notation we move the decimal point 4 places to
    the _____ and use an exponent of
         (left or right)
    _____ for the 10 part of the expression; e.g., $5.6 \times 10^{\underline{\quad}}$

23. Express these numbers in scientific notation in which the first part of the number
    is between 1 and 10.

    (a)  156                      _____

    (b)  4,360,890,000,000        _____

    (c)  0.00000043               _____

    (d)  0.10004                  _____

24. Complete the following conversions of the first numbers given to alternative
    expressions in scientific notation in which the exponents are number divisible
    by 3.

    (a)  $94,500,000 = 94.5 \times 10^{\underline{\quad}}$

    (b)  $0.000896 = 896 \times 10^{\underline{\quad}}$ or $0.896 \times 10^{\underline{\quad}}$

25. Write the accepted SI abbreviations for each unit.

    (a)  milligram    _____      (c)  deciliter   _____

    (b)  microliter   _____      (d)  milliliter  _____

26. Reexpress these quantities using the SI prefix supplied.

    (a)  1500 g = _____ kg

    (b)  0.0000080 L = _____ μL

    (c)  0.0045 mL = _____ μL

(d)    4502 mg = _____ g

(e)    0.015 kg = _____ g

27.    If the correct value of the mass of an object is 14.5068 g but someone reported it as 24.5068 g, we'd say that the reported measurement isn't very

_____.

   (precise or accurate)

28.    If someone reported that the volume of a liquid is 12 mL and it is known to have a volume of 12.478 mL, the report can be described as not very

_____.

   (precise or accurate)

29.    State how many significant figures are in each physical quantity.

   (a)        1.00050 gram        _____

   (b)        $1.000 \times 10^4$ L        _____

   (c)        0.0105 mL        _____

30.    The inch is legally defined as 2.54 cm, exactly. In doing calculations, how many significant figures can be assumed are in 2.54 cm when it occurs in the conversion factor:

$\dfrac{2.54 \text{ cm}}{1 \text{ in.}}$ ?        _____

31.    When we add 24.567 mL to 482.4 mL, how do we round off the sum and how is the sum correctly expressed?

(The calculator answer is 457.833.)        _____

32.    If we subtract 0.458 m from 362 m, how is the difference correctly expressed?

_____ (The calculator answer is 361.542.)

33.    In multiplying 2.3 in. times $\dfrac{2.54 \text{ cm}}{1 \text{ in.}}$ , how is the product

correctly expressed? _____ (The calculator answer is 5.842.)

34.    If we multiply 3.678 cm by the conversion factor $\dfrac{1 \text{ in.}}{2.54 \text{ cm}}$ ,

the answer is correctly expressed as _____ (The calculator answer is 1.4480314.)

35.  If we multiply 13.97 cm by the conversion factor $\dfrac{1 \text{ in.}}{2.54 \text{ cm}}$ , the answer is

correctly expressed as _____ (The calculator answer is 5.5)

36.  The relationship: 1 liter = 1.057 quart can be expressed by means of what two

ratios that we can use as conversion factors?  _____

_____

37.  If 1 gram = 15.43 grain and 1 dram = 60 grains (exactly), then how many drams

are in 425 gram? _____

38.  What is 75 °F in °C? _____ In K? _____

39.  Energy is a capacity for causing _____.

40.  The energy possessed by an object in motion that can be attributed to the motion

itself is called the object's _____, and it equals half the

product of the _____ and the _____.

41.  The kind of energy that an object is said to have solely because of its location
relative to the ground (such as possessed by a rock perched on the edge of a

window ledge high above the sidewalk) is called the _____

of the object.

42.  The kind of energy that a fuel such as wood or coal has that permits us to obtain

heat from it by its reaction with oxygen is called the _____ of the fuel.

43.  One calorie of energy will raise the temperature of _____

(a mass)

of _____ by _____ on the _____

(a substance)            (a temperature change)

scale, specifically between the readings of _____ and _____ °C.

44.  The specific heat of iron is 0.10 cal/g °C.  The specific heat of water is

_____.  If 1 g of iron at 25 °C is given the same energy that

changes the temperature of 1.0 g of water by 5 degrees Celsius, then the new

temperature of the iron sample will be _____.

45. The mass of an object divided by its volume equals its _____.

46. The density of mercury at room temperature is (rounded) 13.5 g/mL. The density of water at room temperature (again rounded) is 1.0 g/mL. What is the specific gravity of mercury at room temperature? _____

47. At the same temperature, object A has a density of 1.6 g/mL and object B has a density of 1.1 g/mL. Suppose a sample of A and a sample of B have the same mass. Which sample would occupy a larger volume? _____

**Multiple-Choice.** For each of the following select all of the correct answers found among the choices (in some questions two or more choices will correctly answer the question).

1. An example of a base unit in the SI has the symbol
   (a)  km          (b)  kg          (c)  kcal          (d)  ks

2. When properly expressed in scientific notation, the quantity 0.01205 liter becomes
   (a)  12.05 mL                    (c)  $1.205 \times 10^3$ L
   (b)  $12.05 \times 10^3$ L       (d)  $1.205 \times 10^{-2}$ L

3. A distance of 100 cm is the same as
   (a)  1 meter    (b)  10 cc    (c)  1000 mm  (d)  0.001 kilometer

4. A temperature of 60 °C is the same as
   (a)  15.6 °F    (b)  10 cc    (c)  108 °F    (d)  140 °F

5. A length of 25.4 mm is the same as
   (  a)  1 in.      (b)  10 in.    (c)  2.54 cm    (d)  0.0254 meter

6. A measurement reported as $1.5050 \times 10^{-3}$g has how many significant figures?
   (a)  4          (b)  5          (c)  7          (d)  8

7. A substance that is denser than water might have
   (a)  a density of 1000 grams per liter
   (b)  a specific gravity of 0.5
   (c)  a density of 18 micrograms per microliter
   (d)  a specific gravity of 1

8. One milliliter is the same as
   (a)  0.001 deciliter              (b)  1000 microliters
   (c)  1 mm                         (d)  0.001 liter

9. Which unit is closest in size to 500 grams?
   (a)  1 oz        (b)  1 dr      (c)  0.5 kg      (d)  0.5 lb

10. A temperature of 27 °C is the same as
    (a)  246 K     (b)  -27 K     (c)  300 K     (d)  27 K

11. A volume of 145 mL is the same as
    (a)  0.145 L     (b)  0.145 kL  (c)  1450 μL  (d)  1.45 L

12. A mass of 0.203 g is the same as
    (a)  2.03 kg     (b)  0.0203 kg (c)  20.3 mg   (d)  203 mg

13. A mass of 5,000,000 μg is the same as
    (a)  5000 mg (b)  5 kg        (c)  5 g         (d)  0.5 g

14. How many meters are in 1.00 bolt of cloth, if
        1 bolt = 120 feet (exactly), and
        1 foot = 0.3048 meter (exactly) ?
    (a)  36.6 m     (b)  394 m     (c)  0.00254 m     (d)  40.0 m

15. How many furlongs are in 5.00 mile if
        1 furlong = 220 yards (exactly)
          1 yard = 3 feet (exactly)
      and 1 mile = 5280 feet (exactly) ?
    (a)  360 furlong              (c)  $1.29 \times 10^{-4}$ furlong
    (b)  $1.94 \times 10^{6}$ furlong         (d)  40.0 furlong

16. To raise the temperature of 250 mL of water 10 °C would take a minimum of how much heat?
    (a)  2500 cal (b)  250 cal     (c)  2.5 kcal    (d)  2500 kcal

17. A liter of water at 90 °C was cooled to 50 °C.  How much heat was removed?
    (a)  4 kcal     (b)  40 kcal     (c)  40,000 cal (d)  400 kcal

18. The density of liquid water at 20°C may be accurately given in any one of these ways.  Which has the greatest precision?
    (a)  1 g/mL     (b)  1.0 g/mL     (c)  0.999 g/mL     (d)  0.99862 g/mL

19. A sample of a mineral has a volume of 4.50 mL and a mass of 13.5 grams.  What is its density?
    (a)  3.00 g/mL (b)  0.33 mL/g   (c)  0.333 g/mL     (d)  3.00 mL/g

20. The density of gem quality diamond is 3.513 g/ml at 20 °C.  To three significant figures what is its specific gravity? (See question 18 for the density of water at 20 °C.)
    (a)  3.5165     (b)  3.52        (c)  3.52 g/mL     (d)  3.52 mL/g

21. Benzene boils at 80 °C.  What is its boiling point in degrees F?
    (a)  27 °F       (b)  176 °F     (c)  353 °F     (d)  160 °F

## ANSWERS

### ANSWERS TO DRILL EXERCISES

**I.    Exercises in Metric Units**

1. 1 cm
2. 10 cm
3. 2.985 km
4. 1.564 g
5. 0.1564 g
6. 0.01564 g
7. 1.640 L
8. 2 mL
9. 0.454 kg
10. 0.150 L
11. 527 µg
12. 120 µL
13. 1.5 cm
14. 0.15 m
15. 98 cm
16. 1620 mg
17. 101 mL
18. 67 mg
19. 0.250 L
20. 0.1 µL

**II.    Exercises in Exponential Notation**

Set A.
1. $1.062457 \times 10^6$
2. $5.43 \times 10^{-3}$
3. $1.116 \times 10^2$
4. $5.21 \times 10^{-6}$
5. 5.025 (not 5.025 x $10^0$; this is not done.)

Set B.
1. 6150
2. 536.2
3. 0.0235
4. 0.0000879
5. $6.542001 \times 10^6$

**III.    Exercises in Significant Figures**

1. 5 (Rule 1, Rule 4)
2. 5 (Rules 1 and 4)
3. 3 (Rule 1)
4. 4 (Rules 1 and 3)
5. 3 (Rules 1 and 2)
6. 4 (Rules 1, 2, and 3)
7. 3 (Rule 3)
8. 5 (Rule 3)
9. 5 (Rule 3)
10. 5 (Rule 4)

## IV. Exercises in the Factor-Label Method for Converting from One System of Units to Another

Set A

1. $\dfrac{5280 \text{ feet}}{1 \text{ mile}}$ or $\dfrac{1 \text{ mile}}{5280 \text{ feet}}$

2. $\dfrac{2.20 \text{ pounds}}{1 \text{ kilogram}}$ or $\dfrac{1 \text{ kilogram}}{2.20 \text{ pounds}}$

3. $\dfrac{454 \text{ grams}}{1 \text{ pound}}$ or $\dfrac{1 \text{ pound}}{454 \text{ grams}}$

4. $\dfrac{36.4 \text{ inches}}{1 \text{ yard}}$ or $\dfrac{1 \text{ yard}}{36.4 \text{ inches}}$

5. $\dfrac{1000 \text{ mg}}{1 \text{ g}}$ or $\dfrac{1 \text{ g}}{1000 \text{ mg}}$

6. $\dfrac{1000 \text{ m}}{1 \text{ km}}$ or $\dfrac{1 \text{ km}}{1000 \text{ m}}$

7. $\dfrac{29.6 \text{ mL}}{1 \text{ fl oz}}$ or $\dfrac{1 \text{ fl oz}}{29.6 \text{ mL}}$

8. $\dfrac{7000 \text{ grains}}{454 \text{ grams}}$ or $\dfrac{454 \text{ grams}}{7000 \text{ grains}}$

9. $\dfrac{1000 \text{ kg}}{1 \text{ ton}}$ or $\dfrac{1 \text{ ton}}{1000 \text{ kg}}$

10. $\dfrac{1000 \text{ g}}{1 \text{ kg}}$ or $\dfrac{1 \text{ kg}}{1000 \text{ g}}$

Set B

1. $0.5000 \text{ mile} \times \dfrac{5280 \text{ feet}}{1 \text{ mile}} = 2640 \text{ feet}$

2. $15000 \text{ feet} \times \dfrac{1 \text{ mile}}{5280 \text{ feet}} = 2.841 \text{ miles}$

3. $100 \text{ inches} \times \dfrac{1 \text{ yard}}{36.4 \text{ inches}} = 2.75 \text{ yards}$

4. $0.1 \text{ gram} \times \dfrac{1000 \text{ mg}}{1 \text{ g}} = 100 \text{ mg}$

5. $5.0 \text{ grains} \times \dfrac{454 \text{ grams}}{7000 \text{ grains}} = 0.32 \text{ gram}$

Set C

1. $1.00 \text{ kg} \times \dfrac{1000 \text{ grams}}{1 \text{ kg}} \times \dfrac{7000 \text{ grains}}{454 \text{ grams}} = 1.54 \times 10^4 \text{ grains}$

(The pocket calculator result, 15,418.5022, has to be rounded to 3 significant figures.)

2.    $2.0 \text{ metric ton} \times \dfrac{1000 \text{ kilogram}}{1 \text{ metric ton}} \times \dfrac{2.20 \text{ pounds}}{1 \text{ kilogram}} = 4.4 \times 10^3 \text{ pounds}$

3.    $0.5 \text{ kg} \times \dfrac{1000 \text{ grams}}{1 \text{ kg}} \times \dfrac{1000 \text{ milligrams}}{1 \text{ gram}} = 5 \times 10^5 \text{ milligrams}$

## ANSWERS TO SELF-TESTING QUESTIONS

### Completion

1.    SI (International System of Units)
2.    standard
3.    meter
4.    meter
5.    the distance between two scratches on a bar of platinum-iridium alloy kept at Sévres, France
6.    No
7.    They do not differ at all in length.
8.    resist a change in motion
9.    mass
10.   (a)  derived      (b)  base       (c)  derived      (d)  base
11.   kilogram mass (at Sévres, France)
12.   the same kilogram mass
13.   manufactured artifacts are subject to corrosion and possibly to theft or loss through natural disaster
14.   second
15.   kelvin
16.   Celsius
17.   180
18.   3
19.   longer
20.   63 microliters.  Using the factor-label method it works out like this:

$$\dfrac{1.0 \text{ m}\cancel{L}}{16 \text{ drops}} \times \dfrac{1000 \text{ microliters}}{1 \text{ m}\cancel{L}} = 63 \dfrac{\text{microliters}}{\text{drop}}$$

21.   liter, quart, (liquid, U.S.)
22.   right; -4; $5.6 \times 10^{-4}$
23.   (a)   $1.56 \times 10^2$
      (b)   $4.36089 \times 10^{12}$
      (c)   $4.3 \times 10^{-7}$
      (d)   $1.0004 \times 10^{-1}$

24. (a)  6            (b)  -6;  -3
25. (a)  mg           (b)  μL         (c)  dL          (d)  mL
26. (a)  1.500 kg  (b)  8.0 μL     (c)  4.5 μL      (d)  4.502 g
    (e)  15 g
27. accurate
28. precise
29. (a)  6            (b)  4            (c)  3
30. an infinite number
31. 457.8 mL
32. 362 m
33. 5.8 cm
34. 1.448 in.
35. 5.500 in.
36. $\dfrac{1 \text{ liter}}{1.057 \text{ quart}}$ or $\dfrac{1.057 \text{ quart}}{1 \text{ liter}}$
37. 109 dram
38. 24 °C; 297 K
39. change
40. kinetic energy; mass, velocity squared
41. potential energy
42. chemical energy
43. 1 g; water; one degree; Celsius; 14.5, 15.5
44. 1.0 cal/g °C; 75 °C
45. density
46. 13.5
47. Object B.  Being less dense, it needs more room to have the same mass as the more dense A.

**Multiple-Choice**

1.  b              8.  b and d        15.  d
2.  d              9.  c              16.  a and c
3.  a, c, and d   10.  c              17.  b and c
4.  d             11.  a              18.  d
5.  a, c, and d   12.  d              19.  a
6.  b             13.  a and c        20.  b
7.  c             14.  a              21.  b

# CHAPTER 2

# THE NATURE OF MATTER: THE ATOMIC THEORY

## OBJECTIVES

While this chapter may seem remote from our theme - the molecular basis of life - mastering its contents will be of vital importance in understanding the information presented in the rest of the book.

The specific objectives of this chapter can be classified under three main objectives.

1.  *Continue to accumulate the basic vocabulary of chemistry. Most of the terms in the glossary will be used again and again throughout the course.*

2.  *Learn how to write the electron configuration for an atom of any element with atomic number from 1 to 20. We need a thorough knowledge of electron configurations in order to be able to understand how these elements form various kinds of chemical bonds.*

3.  *Learn how the elements can be sorted into families having somewhat similar properties. The Periodic Table will be a helpful aid in correlating the properties of the elements as we proceed to later chapters.*

After studying this chapter and working all of the assigned exercises, you should be able to do the following.

1. *Give the names of the three kinds of matter and describe how they are different.*

2. *Name the three states of matter and tell how they are different.*

3. *Give the names of the two kinds of changes that can take place among substances and describe how they can be distinguished.*

4. *Use the general technical terms that apply to the substances involved in a chemical reaction, both before and after the change.*

5. *Use the technical terms that describe if heat is released or absorbed during a chemical reaction.*

6. *State the law of conservation of mass and the law of definite proportions.*

7. *Describe how Dalton used the laws of chemical combination to support the view that atoms are real.*

8. *List and describe the properties of the three subatomic particles.*

9. *Use a mass number and an atomic number to work out the electron configuration and the composition of the nucleus of any atom from number 1-number 20 according to principal energy levels.*

10. *Repeat objective 9 for an isotope of the atom used to test objective 9.*

11. *To the extent of our introductory treatment of this chapter, tell the difference between (a) an atom and an element, (b) an element and a compound, and (c) a compound and a mixture.*

12. *Given the name of any element in Table 2.4, write its atomic symbol.*

13. *Given the atomic symbol of any element in Table 2.4, write its name.*

14. *Explain how the concepts of mass number and atomic weight differ.*

15. *Tell from an electron configuration if an element is a metal or a nonmetal.*

16. *State the Periodic Law and give one property that illustrates it (in general terms).*

17. *Use the Periodic Table to tell which elements are in families, which are metals, which are nonmetals and, for a specific element, find its atomic number and atomic weight.*

18. *Given an atomic number, find the element on the Periodic Table.*

19. *Use the Periodic Table to determine how many electrons are in the outer energy levels of the atoms of any of the representative elements.*

20. *Define the terms in the Glossary.*

## GLOSSARY

**Alkali Metals.**   Those in Group IA of the Periodic Table; lithium, sodium, potassium, rubidium, and cesium.

**Alkaline Earth Metals.**   Those in Group IIA of the Periodic Table; beryllium, magnesium, calcium, strontium, barium, and radium.

**Atom.**   An exceedingly small particle having only one nucleus and no electrical charge; it is the smallest particle of a particular element that can be said to be completely unique for that element.

**Atomic Mass Number.**   See Mass Number.

**Atomic Mass Unit (amu).**   A unit of mass equal to $1.67 \times 10^{-24}$g (rounded) and very close to the mass of the proton or the neutron.

**Atomic Number.**   The electrical charge on an atom's nucleus and, therefore, the number of protons possessed by that atom as well as the number of its electrons.

**Atomic Weight.**   The numerical average of the relative masses of the atoms of all of the isotopes of an element as they occur in their natural mixture in nature.

**Carbon Family.**   The elements in Group IVA of the Periodic Table; carbon, silicon, germanium, tin, and lead.

**Chemical Property.**   Any one of the chemical reactions that a particular substance is known to undergo.

**Chemical Reaction.**    Any event in which substances change into new substances.

**Compound.**    In chemistry, any substance made by a chemical combination of at least two, but usually more, elements and in which the elements have combined in a definite proportion by weight and by atoms.

**Dalton's Theory.**    An atomic theory of matter proposed by John Dalton to account for the laws of chemical combination. Its postulates are: that matter consists of atoms; that atoms are indestructible; and that all atoms of one element are identical in mass and other properties; that the atoms of different elements are different in mass and other properties; and that when a compound forms, a small but definite number of atoms of each of the elements forming the compound join to make up the fundamental particle of the compound.

**Electron.**    A subatomic particle bearing one unit of negative charge and having a small fraction (1/1837) of the mass of a proton or a neutron.

**Electron Configuration.**    The most stable arrangement (i.e., the one having the lowest energy) of the electrons of an atom, ion, or molecule.

**Element.**    A substance that cannot be broken down into anything that is both stable and simpler; a substance in which all of the atoms have the same atomic number and the same electron configurations.

**Endothermic Change.**    A change that requires the use of heat to start the event and to sustain it.

**Energy Level.**    A general region of space near an atomic nucleus where an atom's electrons exist. A "low energy level" is one close to the nucleus and a "high energy level" is one farther away from the nucleus.

**Exothermic Change.**    A change accompanied by the release of heat.

**Family, Chemical.**    A vertical column in the Periodic Table; a group.

**Gaseous State.**    A state of matter characterized by an indefinite shape and volume.

**Group.**  See **Family, Chemical**.

**Halogens.**    The elements in Group VIIA of the Periodic Table; fluorine, chlorine, bromine, iodine, and astatine.

**Inner Transition Elements.**    Elements 58-71 (lanthanide series) and 90-103 (actinide series).

**Isotope.**    A substance in which all of the atoms are identical in atomic number, mass number, and electron configurations.

**Law of Conservation of Mass**.    Matter is neither created nor destroyed in chemical reactions; the masses of all the products equals the masses of all the reactants.

**Law of Definite Proportions.**    The elements that make up a compound occur in it in definite proportions by weight and, therefore, in definite proportions by atoms.

**Liquid State.**    A state of matter characterized by a definite volume but an indefinite shape.

**Mass Number.**    The sum of the protons and neutrons in the nucleus of an atom of an isotope.

**Matter.**    Anything that occupies space and has mass. The three states of matter are solid, liquid, and gas.

**Metal.**    Any element that is shiny, that conducts electricity well, and can be hammered into sheets and drawn into wires.

**Metalloid.**    An element having both metallic and nonmetallic properties.

**Mixture.**    Together with elements and compounds, one of the three kinds of matter; a substance made up of two or more elements or compounds combined physically in no particular proportion by weight and separable into its component parts by physical means.

**Neutron.**    An electrically neutral subatomic particle having a mass of 1 amu (atomic mass unit) and always found, when in atoms, within their nuclei.

**Nitrogen Family.**    Group VA in the Periodic Table; nitrogen, phosphorus, arsenic, antimony, and bismuth.

**Noble Gases.**    Group 0 in the Periodic Table; helium, neon, argon, krypton, xenon, and radon.

**Nonmetal.**    An element that has little or no malleability, ductility, ability to conduct electricity; an element whose atoms have 4, 5, 6, 7, or 8 outside level electrons.

**Nucleus.**    The particle at the center of the atom and that contains all of the atom's neutrons and protons (and, therefore, all of the positive charge within the atom).

**Outside Level.**    An atom's highest energy level that contains at least one electron.

**Oxygen Family.**    Group VIA in the Periodic Table; oxygen, sulfur, selenium, tellurium, and polonium.

**Period.**    In chemistry, a horizontal row in the Periodic Table.

**Periodic Law.**    Many properties of the elements are periodic functions of their atomic numbers.

**Periodic Table.**    A display of the symbols, atomic numbers, and atomic weights of the elements in their order of increasing atomic number and that groups elements that are in the same family in vertical columns called groups or families.

**Physical Change.**    Any event that is not chemical.

**Physical Property.**  . Any observable characteristic of a substance other than a chemical property. Physical properties include: color; density; melting point; boiling point; temperature; the ability to reflect, transmit, or block light; the ability to conduct electricity (although some substances undergo chemical changes if they are made to conduct an electric current); and a number of other "observables."

**Product.**    A substance that forms from a chemical reaction.

**Property.**    Any feature of something that we can use to recognize it again.

**Proton.**    A subatomic particle bearing one unit of positive charge. On the atomic mass (or weight) scale, each proton has a mass of 1 amu (atomic mass unit).

**Reactant.**    A substance that undergoes chemical change.

**Representative Element.**    Any element in the groups of the Periodic Table numbered as A-groups; IA - VIIA plus Group 0.

**Solid State.**    A state of matter characterized by both definite shape and definite volume.

**States of Matter.**    The three possible physical conditions of aggregation of matter - solid, liquid, or gas.

**Subatomic Particle.**    An electron, a proton, or a neutron. Sometimes the atomic nucleus is called a subatomic particle.

**Transition Element.**    An element occurring between Group IIA and Group IIIA in the long periods of the Periodic Table; a metallic element other than one in Group IA or IIA and other than an actinide or a lanthanide.

## SELF-TESTING QUESTIONS

### Completion

1.  The general name we give to anything that occupies space and has mass is

    _____.

2.  The three *states* of matter are _____, _____, and _____.

3.  The three *kinds* of matter are _____, _____, and _____.

4.  Which two kinds of matter, by definition, obey the law of definite proportions?

    _____ and _____

5.  When heat is evolved as a result of a chemical reaction, the reaction is said to be

    _____.

6.  Element A and element B, both colorless, odorless gases, are mixed to give a colorless, odorless gas that, when cooled sufficiently, changes to a liquid identified as the liquid form of A with a gas above it identified as B. The mixing of A and B is an example of a _____ change.

7.  Element X and element Y, both colorless gases that can be kept in glass containers, are mixed. A great deal of heat is generated and, after cooling, the mixture, still a colorless gas, slowly dissolves the glass container. The mixing of X and Y is an example of a _____ change.

8.  If the separate masses of elements X and Y (question 7) are 2 g each, then according to the law of _____ the mass of the result of mixing X and Y would be _____.

9.  The main postulates of Dalton's atomic theory are

    (a) _____

    (b) _____

    (c) _____

(d) _____

(e) _____

10.   If atoms are indestructible, then the ratios of the atoms present in a compound can be expressed only in _____ numbers.

11.   If atoms are present in a compound in a definite ratio by atoms, and if atoms are indestructible, then the atoms must also be present in a definite ratio by

_____.

12.   The general term for the pieces released when an atom is smashed is

_____, and the three that are of prime interest in chemistry are the

_____, the _____, and the _____.

13.   Positively charged particles with a mass of 1 amu in an atom are called

_____.

14.   A positively charged particle in an atom that can have in many instances a mass several times 1 amu is called the _____ of an atom.

15.   Contributing mass to an atom but no electric charge is a particle called the

_____.

16.   Contributing hardly any mass to an atom but furnishing units of negative charge are particles called _____.

17.   An atom is an electrically _____ particle having only one _____ and a net electrical charge of _____; it is the smallest representative sample of a kind of matter called an _____.

18.   If nuclei from two or more elements are present in a sample of matter, the sample may be an example of a _____ or of a mixture.

19.   There are _____ different elements.

20.   The number of positive charges on a nucleus is called its _____.

21.   The sum of the protons and the neutrons in an atom is called its _____.

22. Isotopes are atoms related by having identical _____ but different

    _____.

23. The most stable arrangement of the electrons about the nucleus of an atom is called the atom's _____.

24. The "solar system" picture of the atom derives from a theory advanced by

    _____.

25. Of more lasting importance, this scientist was the first to postulate that the electrons in an atom were restricted to definite energy _____, also called energy states.

26. Bohr also postulated that an atom can neither lose nor gain _____ as long as its electrons never changed energy levels.

27. We number the levels from the lowest energy level and out beginning with number _____.

28. An electron in the third level could reach a state of lower energy by going to the _____ level, assuming there is room in it.

29. If the electron (problem 28) made this transition, the atom in which it occurred would _____ energy.
    (emit or absorb)

30. The maximum number of electrons that can be held by each level is, for

    level-1, _____;    for level-2, _____;

    level-3, _____,

31. How many electrons are there in the outside levels of each of the following elements, identified by their atomic numbers? (You should be able to deduce the answers without referring to any tables or charts.)

    (a)  18 _____          (d)  1 _____

    (b)  13 _____          (e)  5 _____

    (c)  7 _____           (f)  4 _____

32. Using only the atomic numbers given and deducing the correct electronic configurations in each case (main levels only), predict what each would be - metal or a nonmetal.

(a)  15 _____          (d)  14 _____

(b)  10 _____          (e)  13 _____

(c)  11 _____

33. Nearly all naturally occurring elements consist of mixtures of a small number of different kinds of atoms called _____.

34. Two isotopes make up element X. One has atomic number 35 and a mass number of 79. The mass number of the other isotope is 81 and its atomic number is _____.

35. The ratio of the numbers of atoms of each of the two isotopes described in the previous question is 1:1. Therefore, the atomic weight of element 35 is _____.

36. Of the two terms, atom and element, which describes a particle? _____. Which describes a substance consisting of particles? _____.

37. Write the atomic symbol of each of the following elements.

(a)  mercury    _____        (i)  carbon    _____

(b)  potassium  _____        (j)  copper    _____

(c)  phosphorus _____        (k)  sodium    _____

(d)  sulfur     _____        (l)  lead      _____

(e)  oxygen     _____        (m)  neon      _____

(f)  nitrogen   _____        (n)  iron      _____

(g)  chlorine   _____        (o)  silver    _____

(h)  calcium    _____

38. Write the name of each of the following elements.

    (a) Cu    _____        (i) Zn    _____

    (b) Mg    _____        (j) Ba    _____

    (c) Li    _____        (k) P    _____

    (d) K    _____         (l) Na    _____

    (e) Mn    _____        (m) Br    _____

    (f) Pt    _____        (n) F    _____

    (g) Al    _____        (o) Cl    _____

    (h) Pb    _____

39. The isotope of which element has an atomic number equal to its mass number?

    _____

40. According to the _____ law, many properties of elements are _____ functions of their atomic _____.

41. Vertical columns in the Periodic Table (as displayed in the inside front cover of the text) are called _____.

42. Horizontal rows in the Periodic Table are called _____.

43. All the atoms of the elements in Group VIIA have _____ electrons in their outside levels; those in Group IIA have _____.

44. The transition elements are generally _____
    (metals or nonmetals)
    and their atoms may have from _____ to _____ electrons in their outside levels.

45. As one crosses from left to right in the second period of the Periodic Table, energy level number _____ is filling.

**Multiple-Choice**

1.  One of the states of matter is
    (a) mixture      (b) liquid      (c) compound      (d) atom

2.  One of the kinds of matter is
    (a) mixture      (b) gas      (c) nucleus      (d) atom

3.  One of the "building blocks" of matter is the
    (a) mixture      (b) liquid      (c) crystal      (d) atom

4.  An example of a purely physical change is
    (a) digesting      (b) cooking      (c) raining      (d) rusting

5.  An example of a chemical change is
    (a) subliming      (b) evaporating      (c) melting      (d) bleaching

6.  Dalton's theory proposed that matter is made of particles called
    (a) elements      (b) atoms      (c) mixtures      (d) compounds

7.  If atoms of A and of B chemically combine in a ratio of 1:1 by atoms but a ratio of 2:1 by mass, then compared to the atoms of B, those of A are
    (a) half as heavy                (c) twice as heavy
    (b) equal in weight              (d) three times as heavy

8.  Lord Rutherford's contribution to our understanding of atomic structure was that atoms
    (a) are like solar systems       (c) have dense inner cores
    (b) contain electrons            (d) consist of isotopes

9.  Bohr postulated that electrons occur in atoms
    (a) in discrete isotopes
    (b) in discrete energy states
    (c) in a condition of constant shifting between states
    (d) in hard, dense, inner cores

10. The correct symbol for the element potassium is
    (a) P      (b) Po      (c) Pt      (d) K

11. The correct symbol for the element sulfur is
    (a) S      (b) Su      (c) SU      (d) Sf

12. An atom with 20 protons would have
    (a) a nuclear charge of 20+      (c) 20 electrons
    (b) a nuclear charge of 20-      (d) 40 neutrons

13. An element whose atoms have 10 protons, 10 neutrons, and 10 electrons would have an atomic number of

(a)  10          (b)  20          (c)  30          (d)  40

14. All of the atoms of any naturally occurring element will have identical numbers of

(a)  protons     (b)  nuclei      (c)  neutrons     (d)  electrons

Problems 15 to 19 should be answered without reference to tables or charts.

15. An element of atomic number 19 consists of atoms having how many electrons in their outside levels?

(a)  eight       (b)  nine        (c)  one          (d)  19

16. An element of atomic number 14 is probably

(a)  a metal                      (c)  a transition element

(b)  a nonmetal                   (d)  a noble gas

17. Standing immediately to the left of the element of atomic number 17 in the Periodic Table is an element with

(a)  a charge of 16+ on its nuclei

(b)  a charge of 18+ on its nuclei

(c)  a charge of 17+ on its nuclei

(d)  a charge of 9+ on its nuclei

18. Standing immediately below the element of atomic number 17 in the Periodic Table is an element with

(a)  17 electrons in its outside level

(b)  7 electrons in its outside level

(c)  17 protons in its nuclei

(d)  7 protons in its nuclei

19. If element X is a gas, then the element immediately above it in the Periodic Table is most likely a

(a)  metal       (b)  nonmetal  (c)  gas          (d)  transition element

20. An element whose atoms have one, two, or three electrons in their outside levels is probably

(a)  a nonmetal  (b)  a gas      (c)  a liquid     (d)  a metal

21. An element whose atoms have seven or eight electrons in their outside levels is probably

(a)  a nonmetal                   (c)  a transition element

(b)  a solid                      (d)  a metal

22. If an element of atomic number 20 had two isotopes and one isotope had 20 neutrons while the other had 22 neutrons, and if these isotopes were present in a 50:50 ratio, what would be the atomic weight of the element?

    (a)  20.5        (b)  21        (c)  40.5        (d)  41

23. Two atoms that are isotopes have the same

    (a)  number of protons

    (b)  electron configuration

    (c)  mass numbers

    (d)  atomic numbers

24. The isotope of an atom of atomic number 16 and mass number 32 might have

    (a)  32 protons              (c)  16 neutrons

    (b)  17 protons              (d)  17 neutrons

25. Without consulting the Periodic Table, decide which element would be in the same family as the element with atomic number 12. (The numbers in the choices are atomic numbers.)

    (a)  4          (b)  11        (c)  13        (d)  20

26. Without consulting the Periodic Table or referring to a table of the elements, choose the element(s) (by atomic number) that would most likely be a metal.

    (a)  8          (b)  19        (c)  13        (d)  15

27. An element is

    (a)  one of the kinds of matter

    (b)  one of the fundamental particles of matter

    (c)  consists of atoms identical in *every* respect

    (d)  consists of atoms in which electron configurations in the main levels are identical

28. An atom is

    (a)  one of the classes of matter

    (b)  one of the fundamental particles of matter

    (c)  a building block of an element

    (d)  an electrically neutral particle

29. A metal is likely

    (a)  to be in Group VA, VIA, or VIIA of the Periodic Table

    (b)  to have one, two, or three electrons in its outside shell

    (c)  to be a nonconductor of electricity

    (d)  to have examples among the transition elements

# ANSWERS

## ANSWERS TO SELF-TESTING QUESTIONS

### Completion

1.  matter
2.  solid, liquid, and gas
3.  elements, compounds, and mixtures
4.  elements and compounds
5.  exothermic
6.  physical
7.  chemical
8.  conservation of mass; 4 g
9.  (a) Matter consists of atoms.

    (b) Atoms are indestructible. (Even in chemical change they only rearrange.)

    (c) Atoms of the same element are identical, particularly in mass.

    (d) Atoms of different elements differ in mass.

    (d) In compounds, atoms occur in definite ratios or proportions.
10. whole
11. mass
12. subatomic particles

    electron, proton, neutron
13. protons
14. nucleus
15. neutron
16. electrons
17. neutral; nucleus; zero; element
18. compound
19. 108
20. atomic number
21. mass number
22. atomic numbers; masses (or atomic masses)
23. electron configuration
24. Niels Bohr
25. levels
26. energy
27. one

28.  second (or, of course, the first if there were room)
29.  emit
30.  two; eight; eighteen
31.  (a)  8        (b)  3        (c)  5        (d)  1        (e)  3        (f)  2
32.  (a)  nonmetal        (b)  nonmetal        (c)  metal
     (d)  nonmetal (borderline)        (e)  metal
33.  isotopes
34.  35
35.  80
36.  atom; element
37.  (a)  Hg        (f)  N        (k)  Na
     (b)  K        (g)  Cl        (l)  Pb
     (c)  P        (h)  Ca        (m)  Ne
     (d)  S        (i)  C        (n)  Fe
     (e)  O        (j)  Cu        (o)  Ag
38.  (a)  copper        (f)  platinum        (k)  phosphorus
     (b)  magnesium        (g)  aluminum        (l)  sodium
     (c)  lithium        (h)  lead        (m)  bromine
     (d)  potassium        (i)  zinc        (n)  fluorine
     (e)  manganese        (j)  barium        (o)  chlorine
39.  One isotope of hydrogen; the atomic number is one and its mass number is one.
40.  Periodic; periodic; numbers
41.  groups (or families)
42.  periods
43.  seven; two
44.  metals; one to three (sometimes four)
45.  two

## Multiple-Choice

| | | |
|---|---|---|
| 1. b | 11. a | 21. a |
| 2. a | 12. a, c | 22. d |
| 3. d | 13. a | 23. a, b, d |
| 4. c | 14. a, d | 24. d |
| 5. d | 15. c | 25. a, d |
| 6. b | 16. b | 26. b, c |
| 7. c | 17. a | 27. a, d |
| 8. c | 18. b | 28. b, c, d |
| 9. b | 19. b, c | 29. b, d |
| 10. d | 20. d | |

# CHAPTER 3

# THE NATURE OF MATTER: COMPOUNDS AND BONDS

## OBJECTIVES

No chapter has a greater number of basic terms and concepts than this chapter. After you have studied its content, worked the exercises in the text and in this Study Guide, and gained the ability to do well on the Self-Testing Questions, you should also be able to do the following.

1. Name two kinds of compounds and describe the kinds of particles that make up each type.

2. Describe the difference between

    (a) a molecule and an atom

    (b) a molecule and a polyatomic ion

    (c) an atom and an ion

(d)   a molecular compound and an ionic compound

(e)   an ionic bond and a covalent bond

(f)   a polar molecule and a nonpolar molecule

(g)   oxidation and reduction

(h)   an oxidizing agent and a reducing agent

3.   Given only the atomic numbers of any two elements (from 1 to 20), predict if the two can form a compound. If so, state whether the compound will be ionic or molecular, and give the correct formula.

4.   Give the names and formulas (including the electrical charges) of all the ions in Tables 3.1 and 3.4.

5.   Write formulas from the names of ionic compounds and names from their formulas, limiting ourselves to information in Tables 3.1 and 3.4.

6.   Give the covalences of the elements listed in Table 3.3 and use these numbers to write structures of molecular compounds if given only their molecular formulas.

7.   Use the octet rule to predict the likely oxidation number or covalence for any element of atomic number 1-20 from its electron configuration or its location in the Periodic Table.

8.   Define "oxidation" and "reduction" and give an illustration of each.

9.   Use the order of relative electronegativities to predict whether a covalent bond will be polar and in what direction if it is.

10.   Explain how molecules can sometimes stick to each other even though they are neutral.

11.   Define all the terms in the Glossary.

## GLOSSARY

**Bond, Chemical.**   A net force of attraction that holds atomic nuclei within compounds near each other.

**Chemical Bond.**   See **Bond**.

**Covalence.**   The number of electron-pair (covalent) bonds an atom can have when forming a molecule.

**Covalent Bond.**   The net force of attraction that arises when two (or sometimes more) atomic nuclei share a pair of electrons between them. In a single bond, one pair of electrons is shared; in a double bond, two pairs; in a triple bond, three.

**Dipole, Electrical.**   A pair of equal but opposite electrical charges separated by a small distance in a molecule.  The electrical charges are usually fractional and are commonly symbolized by δ+ andδ–.

**Double Bond.**   A chemical bond arising from the sharing of two pairs of electrons between two atomic nuclei and commonly represented by two parallel lines between the atomic symbols.

**Electrolyte.**   A substance that in solution conducts electricity.

**Electron Cloud.**   A region within a molecule having a relatively high density of negative charge as in electron-pair (covalent) bonds.

**Electron Density.**   The concentration of the negative charge carried by electrons residing in molecules and their bonds.

**Electron Sharing.**   The joint attraction of two atomic nuclei toward a pair of electrons situated between the nuclei and between which, therefore, a covalent bond exists.

**Electronegativity.**   The ability of an atom joined to another by a covalent bond to attract toward itself the electrons of that bond.

**Formula, Chemical.**   The shorthand symbol for a substance written by combining the abbreviations of its elements in the correct number and proportion to indicate the substance's elementary composition.

**Formula Unit.**   A small particle - an atom, a molecule, or a set of ions - that has the composition given by the chemical formula of the substance.

**Ion.**   An electrically charged particle having one or a few atomic nuclei and either one or two (seldom three) too many or too few electrons to render the particle electrically neutral.

**Ionic Bond.**   The force of attraction between oppositely charged ions in an ionic compound.

**Ionic Compound.**   One kind of compound; a substance made up of a more or less orderly combination of oppositely charged ions occurring in a definite ratio that ensures that all of the electrical charges on the assembled ions cancel out so that the substance is electrically neutral.

**Molecular Compound.**   A compound whose smallest representative particle is a molecule; sometimes called a covalent compound.

**Molecular Formula.**   A formula of a molecular compound that states the composition of an individual molecule without (in most cases) revealing its structure.

**Molecule.**   An electrically neutral (but frequently polar) particle made up of the nuclei and electrons of two or more elements held together by covalent (or coordinate covalent) bonds.  The molecule is the smallest representative sample of a molecular compound.

**Octet Rule.** The atoms of the reactive elements tend to undergo those chemical reactions that most directly give them electron configurations of the nearest noble gas (all but one of which involve outer octets).

**Outer Octet.** Eight electrons in a main energy level when it is the outside level of the atom.

**Oxidation.** An event accompanied by the loss of one or more electrons from an atom, a molecule, or an ion. (In organic chemistry, oxidation may be regarded as either the loss of hydrogen or the gain of oxygen by an organic molecule or ion.)

**Oxidation Number.** The charge carried by an ion or by a particular atom within a polyatomic ion.

**Oxidizing Agent.** Any substance that can cause another substance to be oxidized. See **Oxidation**.

**Polar Bond.** A covalent bond with a partial positive charge at one end and a partial negative charge at the other (more electronegative) end.

**Polar Molecule.** A molecule possessing a permanent electrical dipole, that is, having sites with net partial positive and negative charges.

**Polyatomic Ion.** Any ion made up of two or more atoms. Examples are the hydroxide ion, $OH^-$; the sulfate ion, $SO_4^{2-}$; and the carbonate ion, $CO_3^{2-}$.

**Redox Reaction.** Abbreviation of "reduction-oxidation reaction." A redox reaction is one in which electrons have left one reactant and gone to another.

**Reducing Agent.** Any substance that can cause another substance to be reduced. See **Reduction**.

**Reduction.** An event accompanied by the gain of one or more electrons by an atom, a molecule, or an ion. (In organic chemistry, reduction may be regarded as the gain of hydrogen or the loss of oxygen by an organic molecule or ion.)

**Single Bond.** A chemical bond arising from the sharing of one pair of electrons between two atomic nuclei and normally symbolized by a single line between the two atomic symbols; a covalent bond.

**Structural Formula.** The nucleus-to-nucleus sequence within an individual molecule.

**Structure.** The structural formula of the molecules of a compound.

**Subscripts.** Numbers placed to the right and a half space below the atomic symbols in a chemical formula.

**Triple Bond.** A chemical bond arising from the sharing of three pairs of electrons between two atomic nuclei and generally represented by three parallel lines between the two atomic symbols.

## PREDICTING OXIDATION NUMBERS OR COVALENCES OF ELEMENTS 1 THROUGH 20

The oxidation number or the covalence of an element, at least in the series of atomic numbers 1-20, can be predicted by using the atomic number, the octet rule, and the following steps.

Step 1.    Write out the electron configuration of the main levels using the rules developed in the previous chapter.

Step 2.    Note the number of electrons in the highest occupied level. (If that level is level one, then skip to part D, "Special Cases.")

    A.    The highest occupied level has one, two, or three electrons. Such elements can form positive ions with charges 1+, 2+, 3+, respectively. They do not (in the scope of our study) form covalent bonds. Typically, these elements are in Groups IA, IIA, and IIIA of the Periodic Table.

After atoms of these elements have lost one, two, or three electrons, respectively, their original outside levels have been stripped away and they have new outside levels. They now also have the electron configuration of one of the noble gas atoms. Partway measures will not work. An atom with three outside-level electrons either loses all or none; an atom with two outer electrons loses two, no more no less.

    B.    The highest occupied level has four, five, six, or seven electrons. These elements can form four, three, two, and one covalent bonds, respectively.

An atom with four outside-level electrons does not pick up four electrons (and thus have a charge of 4-) or lose four electrons (and have a charge of 4+). All it can do is *share* four more electrons with another atom or atoms to reach an octet. In order to share four more electrons, the atom must form four covalent bonds.

An atom with five outside-level electrons normally does not lose the five (thus becoming charged with 5+) or gains three more (and a charge of 3-). It normally picks up a *share* of three to get an octet. Gaining a share of three electrons means forming three covalent bonds.

An atom with six outside-level electrons can pick up a share in two more electrons to form its octet by forming two covalent bonds. (See also part C, next.)

An atom with seven outside-level electrons can pick up a share in one more electron for its octet and can thus form only one covalent bond. (See also part C, next.)

THE COVALENCE NUMBER IS THE NUMBER OF ELECTRONS AN ATOM MUST GET BY SHARING IN ORDER TO ACQUIRE AN OUTER OCTET.

C.    The number of outside-level electrons equals six or seven.  Not only can these form covalent bonds (as described in part B), they can also form ions with charges of 2– and 1–, respectively.  Such elements are in Groups VIA and VIIA of the Periodic Table.  Atoms in Group VIA have six outside-level electrons and can acquire two more outright, *if given the right partner*.  Elements in Group VIIA have seven outside-level electrons and can acquire one more outright, *if given the right partner*.  The *right partner* in both cases is a metal, an element that must either give up electrons or not react.

NONMETALS IN GROUPS VIA AND VIA TEND TO FORM NEGATIVE IONS WHEN THE PARTNER IS A METAL.  IF THE PARTNER IS A NONMETAL FROM GROUPS IVA, VA, VIA, OR VIIA,  THE NONMETALS IN GROUPS VIA AND VIIA TEND TO FORM COVALENT BONDS.

D.    Special Cases - helium and hydrogen.  Helium is unreactive toward all elements; its outside level is level one, which holds only two electrons.  When the outer level is level one, then two electrons in it is the condition of stability.

Hydrogen has one electron in level one, its only electron.  If it loses that electron, only its nucleus is left and that is nothing more than a proton.  We often symbolize it as $H^+$ and call it a proton.  We also use the terms "hydrogen ion" and "proton" interchangeably.

If given the right partner (i.e., a metal), hydrogen can acquire another electron, fill its outside level, and become heliumlike in configuration.  The resulting particle is called the "hydride ion" and has the symbol H:⁻ (the two dots are included to indicate the two electrons.

Given a nonmetal partner, hydrogen can acquire a share in one more electron to fill level one with two electrons.  Hydrogen then has a covalence number of 1.  In short, hydrogen can have a charge of zero (the atom), 1+, 1–, or it can form a covalent bond.

**Examples:**  Predict reasonable oxidation numbers and covalences for each of the following.  (By "reasonable," we mean in terms of the rules we have laid down.)

1.    An element of atomic number 11.

Step 1.    Its electron configuration is 2   8   1.

Step 2.    Having only one electron in its outside level, it can only form an ion having an oxidation number of 1+.  It can't form a covalent bond; therefore it has no covalence number.

2.    An element of atomic number 6.

Step 1.     Its electron configuration is 2   4.

Step 2.    Having four electrons in its outside level, it can't form a discrete ion;

thus, it has no oxidation number. It must share four more electrons to gain an octet; its covalence number is 4.

3.  An element of atomic number 9.

Step 1.    Its electron configuration is 2   7.

Step 2.    Having seven electrons it its outside level, it can accept one electron, thereby acquiring a net charge of 1– and becoming an ion; its oxidation number is –1. It can also get an outer octet by accepting a share of another electron, so its covalence is 1.

4.  An element of atomic number 4.

Step 1.    Its electron configuration is 2   2.

Step 2.    If the two outer electrons are lost, the new outer level is level one, which is stable when filled with two electrons (as with helium). The ion formed has a charge of 2+; the oxidation number is +2. According to our theory, an element with two electrons in its outside level cannot be involved in covalent bonds; thus, it has no covalence number.

5.  An element of atomic number 10.

Step 1.    Its electron configuration is 2   8.

Step 2.     Since it already has an outer octet, this element has neither an oxidation number nor a covalence.


## SELF-TESTING QUESTIONS


### Completion

1.  The kind of natural force that operates to hold the pieces of atoms together in compounds is called the _____ force.

2.  A small particle with one or a few atomic nuclei but bearing a net negative or positive charge is called _____.

3.  A compound consisting of an orderly aggregation of oppositely charged ions is classified as _____.

4.  The formula unit of a molecular compound is called _____.

5.  When the nuclei and electrons of atoms are reorganized to form particles of opposite electrical charge, the product is classified as _____.

6.  The net force of attraction that operates in an ionic compound to keep the ions from flying apart has the special name of _____.

7.  The reaction of lithium, Li, with chlorine, $Cl_2$, produces lithium chloride, LiCl. This reaction changes the lithium atom (Li) to the _____, which has the symbol: _____. It changes the chlorine atoms (in $Cl_2$) into _____, which have the symbol _____. In this oxidation-reduction reaction, the lithium atom is _____ and the chlorine atom is
    (reduced or oxidized)

    _____. The oxidizing agent is _____ and
    (name)

    its symbol is _____. The reducing agent is _____
    (name)

    and its symbol is _____.

8.  The formula unit of an ionic compound consists of a pair or a small cluster of

    _____.

9.  The ions $Al^{3+}$ and $SO_4^{2-}$ can aggregate in a ratio of _____ ions of $Al^{3+}$ to _____ ions of $SO_4^{2-}$.

10. They *must* aggregate in this ratio because chemical compounds are, in general, electrically _____.

11. Test how well you have learned the names and formulas for the ions in Tables 3.1 and 3.4 in the text by completing the following.  Give the formula of each ion; for example,

    sulfite ion      $\underline{SO_3^{2-}}$

    (You must include the electric charge; for instance, $SO_3$ by itself is the symbol of sulfur trioxide, an air pollutant.)

    (a)  chloride ion                          _____

    (b)  sodium ion                            _____

(c)  magnesium ion                       _____

(d)  hydroxide ion                        _____

(e)  sulfate ion                          _____

(f)  potassium ion                        _____

(g)  iodide ion                           _____

(h)  calcium ion                          _____

(i)  sulfide ion                          _____

(j)  nitrate ion                          _____

(k)  acetate ion                          _____

(l)  permanganate ion                     _____

(m)  hydronium ion                        _____

(n)  nitrite ion                          _____

(o)  ammonium ion                         _____

(p)  dihydrogen phosphate ion             _____

12. Using your knowledge of the names and formulas of ions, write formulas for each of the following ionic compounds.

(a)  sodium iodide                        _____

(b)  sodium sulfate                       _____

(c)  sodium phosphate                     _____

(d)  sodium nitrate                       _____

(e)  magnesium sulfate                    _____

(f)  calcium sulfate                      _____

(g)  potassium sulfate                    _____

(h)  aluminum sulfate                     _____

(i)  ammonium nitrate          _____

(j)  ammonium sulfate          _____

(k)  calcium phosphate         _____

(l)  sodium monohydrogen phosphate   _____

(m)  sodium acetate            _____

(n)  silver nitrate            _____

(o)  potassium permanganate    _____

(p)  iron(III) oxide           _____

13.  Write the name of each of the following compounds.

(a)  NaBr          _____

(b)  $Na_2SO_4$          _____

(c)  $NaNO_3$          _____

(d)  $Na_2CO_3$          _____

(e)  $NaMnO_4$          _____

(f)  $Na_3PO_4$          _____

(g)  NaOH          _____

(h)  $KH_2PO_4$          _____

(i)  KCl          _____

(j)  $KHCO_3$          _____

(k)  $KNO_2$          _____

(l)  $MgCl_2$          _____

(m)  $MgHPO_4$          _____

(n)  $Ca(NO_3)_2$          _____

(o)  $CaSO_4$ _____

(p)  $(NH_4)_2SO_4$ _____

(q)  $NH_4Cl$ _____

(r)  $KC_2H_3O_2$ _____

(s)  $FeCl_3$ _____

(t)  $HgCl_2$ _____

(u)  $AgCl$ _____

(v)  $FeCl_2$ _____

(w)  $HgCl$ _____

(x)  $K_2CO_3$ _____

(y)  $CaCO_3$ _____

(z)  $MgCO_3$ _____

(aa) $NaHCO_3$ _____

(bb) $LiHCO_3$ _____

(cc) $Ca(HCO_3)_2$ _____

(dd) $Al(SO_4)_3$ _____

14. The particle that results when two or more atoms have their nuclei and electrons reorganized so that some electrons are shared between the nuclei is called

_____.

15. A small particle having two or more nuclei that is capable of isolated existence and that is electrically neutral is called _____.

16. A small particle that is the smallest representative sample of a molecular compound is called _____.

17. The net force of attraction that operates in molecules to keep their nuclei from flying apart is called the _____.

18.  If two electrically neutral molecules can still attract each other, they must be

_____.

19.  When two atoms are held together by a triple bond, as in nitrogen (:N≡N:), there is a region of high electron _____ between the two.  We could also say that the electron _____ is particularly dense between the two atoms.

20.  Using their atomic symbols, arrange these elements in the order of their relative electronegativities:  carbon, oxygen, nitrogen, and hydrogen.

| _____ | < | _____ | < | _____ | < | _____ |
|---|---|---|---|---|---|---|
| least | | | | | | most |
| electronegative | | | | | | electronegative |

21.  If element X has a relative electronegativity higher than that of Y, then in the molecule X-Y, the δ+ will be close to _____
(X or Y)
and the δ– will be close to _____.
(X or Y)

22.  The covalence of oxygen is two, not three or four, because the outside level of an oxygen atom has _____ electrons, just _____ electrons short of an outer octet.

23.  Complete the following table according to the pattern provided by the example. Write the electron configuration (main levels only) of the atom whose atomic number is given.  Deduce if the atom can form an ion; if so, write the formula of the ion using just the general symbol X for the atom in each case.  Deduce next if the atom can participate in covalent bond formation; if so, write the most reasonable covalence number it will have on the basis on the octet theory. (Many elements have multiple covalences, but we will not study this here.)

| | Atomic Number | Electron Configuration | Symbol of Ion (if any) | Covalence Number (if any) |
|---|---|---|---|---|
| Example | 16 | 2  8  6 | $X^{2-}$ | 2 |
| (a) | 9 | _____ | _____ | _____ |
| (b) | 20 | _____ | _____ | _____ |
| (c) | 5 | _____ | _____ | _____ |
| (d) | 17 | _____ | _____ | _____ |
| (e) | 8 | _____ | _____ | _____ |
| (f) | 13 | _____ | _____ | _____ |

24.  Since selenium, Se, is in the same family of the Periodic Table as oxygen, one reasonable covalence of selenium must be _____, because members of the same family have the identical number of electrons in their _____.

25.  Using your knowledge of covalence numbers (review Table 3.3 in the text), write molecular formulas of the compounds of each of the following elements combined with *hydrogen*; for example,

oxygen    $\underline{\quad H_2O \quad}$

(a)  sulfur    _____

(b)  chlorine    _____

(c)  nitrogen    _____

(d)  iodine    _____

(e)  arsenic, As (arsenic is in the same family as nitrogen)    _____

(f)  phosphorus, P (phosphorus is in the same family as nitrogen)    _____

(g)  silicon, Si (silicon is in the same family as carbon)    _____

**Multiple-Choice**

1.  If an atom, X, can accept two electrons from another atom in the process of changing into a relatively stable ion, then an atom of X

    (a)  has two protons in its nuclei

    (b)  is a metal

    (c)  will be in Group IIA in the Periodic Table

    (d)  is also capable of forming a molecule of the formula $H_2X$

2.  Atoms of elements with atomic numbers 20 and 17 are known to combine. If X is the symbol of element 20 and Y is the symbol of element 17, the correct formula for the most stable compound between these two elements is

    (a)  $X_2Y$          (b)  $XY_2$          (c)  $YX_2$          (d)  $XY$

3.  Pure substances classified as molecular are characterized as being

    (a)  mixtures

    (b)  made of oppositely charged ions

    (c)  aggregations of molecules

    (d)  metallic

4.  The type of bonding found in sodium bromide is

    (a)  ionic          (b)  covalent          (c)  coordinate          (d)  nonpolar

5.  A particle having eleven protons and ten electrons bears a charge of

    (a)  1–          (b)  0          (c)  1+          (d)  11+

6.  If elements X and Y are in the same family and X forms the compound $Na_2X$, then one of the compounds of Y is

    (a)  $K_2Y$          (b)  $CaY$          (c)  $Na_2Y$          (d)  $Al_2Y_3$

7.  If X just precedes Y in a horizontal row of the Periodic Table (i.e., the atomic number of X is one less than that of Y), and if X forms the compound HX, then a compound of Y is

    (a)  HY

    (b)  $H_2Y$

    (c)  $H_3Y$

    (d)  Y is a noble gas; therefore it will form no compound.

8.  Each formula unit of $Mg(H_2PO_4)_2$ contains a total of how many atomic nuclei?

    (a)  sixteen          (b)  fifteen          (c)  eight          (d)  fourteen

9. In the reaction of zinc, Zn, with silver nitrate, $AgNO_3$, that produces silver, Ag, and zinc nitrate, $Zn(NO_3)_2$,

   (a) Zn is the reducing agent

   (b) Ag is the oxidizing agent

   (c) Zn is oxidized

   (d) $Ag^+$ is reduced

10. In the reaction of problem 9,

   (a) Zn is an oxidizing agent

   (b) Zn is a reducing agent

   (c) $Ag^+$ is an oxidizing agent

   (d) $Ag^+$ is a reducing agent

11. The substance calcium chloride, $CaCl_2$, is an aggregation of which particles?

   (a) calcium atoms and chlorine molecules $(Cl_2)$

   (b) calcium atoms and chlorine atoms

   (c) calcium ions and chlorine molecules

   (d) calcium ions and chloride ions

12. On the basis of electronegativities, the bonds in the molecule $NH_3$ would be in which state of polarity?

   | | | | |
   |---|---|---|---|
   | $\delta+$ $\delta-$ | $+$ $-$ | $\delta-$ $\delta+$ | $-$ $+$ |
   | (a) N–H | (b) N–H | (c) N–H | (d) N–H |

13. If element A has a relative electronegativity of 3.7 and that of B is 2.7, and the molecule whose formula is $A_2B$ is nonpolar, then the structure of $A_2B$ is most likely

   (a) A–A–B    (b) A–A \\ B    (c) B / A  A \\    (d) A–B–A

14. A substance, Z, consists of molecules. Z is a gas at room temperature and remains a gas even at $-140°C$. The molecules of Z, therefore, are most likely

   (a) very polar

   (b) made of oppositely charged ions

   (c) relatively nonpolar

   (d) negative ions

# ANSWERS

## ANSWERS TO SELF-TESTING QUESTIONS

### Completion

1. electrical
2. an ion
3. an ionic compound
4. a molecule
5. an ionic compound
6. the ionic bond
7. lithium ion, $Li^+$; chloride ions, $Cl^-$; oxidized, reduced; chlorine, $Cl_2$; lithium, $Li$
8. oppositely charged ions
9. two, three
10. neutral
11. (a) $Cl^-$      (g) $I^-$      (l) $MnO_4^-$
    (b) $Na^+$      (h) $Ca^{2+}$      (m) $H_3O^+$
    (c) $Mg^{2+}$      (i) $S^{2-}$      (n) $NO_2^-$
    (d) $OH^-$      (j) $NO_3^-$      (o) $NH_4^+$
    (e) $SO_4^{2-}$      (k) $C_2H_3O_2^-$      (p) $H_2PO_4^-$
    (f) $K^+$
12. (a) $NaI$      (g) $K_2SO_4$      (l) $Na_2HPO_4$
    (b) $Na_2SO_4$      (h) $Al_2(SO_4)_3$      (m) $NaC_2H_3O_2$
    (c) $Na_3PO_4$      (i) $NH_4NO_3$      (n) $AgNO_3$
    (d) $NaNO_3$      (j) $(NH_4)_2SO_4^-$      (o) $KMnO_4$
    (e) $MgSO_4$      (k) $Ca_3(PO_4)_2$      (p) $Fe_2O_3$
    (f) $CaSO_4$
13. (a) sodium bromide
    (b) sodium sulfate
    (c) sodium nitrate
    (d) sodium carbonate
    (e) sodium permanganate
    (f) sodium phosphate
    (g) sodium hydroxide
    (h) potassium dihydrogen phosphate

(i)     potassium chloride

(j)     potassium bicarbonate

(k)     potassium nitrite

(l)     magnesium chloride

(m)    magnesium monohydrogen phosphate

(n)     calcium nitrate

(o)     calcium sulfate

(p)     ammonium sulfate

(q)     ammonium chloride

(r)     potassium acetate

(s)     iron(III) chloride (ferric chloride)

(t)     mercury(II) chloride (mercuric chloride)

(u)     silver chloride

(v)     iron(II) chloride (ferrous chloride)

(w)    mercury(I) chloride (mercurous chloride)

(x)     potassium carbonate

(y)     calcium carbonate

(z)     magnesium carbonate

(aa)   sodium bicarbonate

(bb)   lithium bicarbonate

(cc)   calcium bicarbonate

(dd)   aluminum sulfate

14.   a molecule

15.   a molecule

16.   a molecule

17.   covalent bond

18.   polar

19.   density; cloud

20.   H < C < N < O

21.   Y, X (since X is more electronegative than Y)

22.   six, two

23.

| Atomic Number | Electron Configuration | Symbol of Ion | Covalence Number |
|---|---|---|---|
| (a) 9 | 2 7 | X$^-$ | 1 |
| (b) 20 | 2 8 8 2 | X$^{2+}$ | none |
| (c) 5 | 2 3 | X$^{3+}$ | none |
| (d) 17 | 2 8 7 | X$^-$ | 1 |
| (e) 8 | 2 6 | X$^{2-}$ | 2 |
| (f) 13 | 2 8 3 | X$^{3+}$ | none |

24. 2, outside levels

25. (a) $H_2S$
    (b) HCl
    (c) $NH_3$ ($H_3N$ is all right, too)
    (d) HI
    (e) $AsH_3$
    (f) $PH_3$
    (g) $SiH_4$

## Multiple-Choice

1. d (X has six electrons in its outer level and therefore can also form two covalent bonds.)
2. b
3. c
4. a
5. c [(11+) + (10−) = 1+]
6. a, b, c, and d (X must have an oxidation number of 2−; therefore, the oxidation number of Y must be 2−, also.)
7. d
8. b
9. a, c, and d
10. b and c
11. d
12. c
13. d (Choices a and b could not exist because they show element A with two covalences. If the compound were choice c, the substance would be polar. Only in choice d, the linear molecule, will the bond polarities cancel each other out and make the molecule nonpolar. In choice d, the center of density of the positive charge and the center of density of the negative charge coincide.
14. c

# CHAPTER 4

# CHEMICAL REACTIONS: EQUATIONS AND MASS RELATIONSHIPS

## OBJECTIVES

After you have studied this chapter, worked the exercises both in the text and in this *Study Guide*, learned the basic definitions, and correctly answered the Self-Testing Questions, you should be able to do the following.

1.  *Recognize when a chemical equation is balanced.*

2.  *Write a chemical equation (for relatively simple reactions) and balance it.*

3.  *Restate the information of a balance chemical equation in words.*

4.  *Give the value of Avogadro's number and the significance of this number in chemistry.*

5.  *Calculate the number of atoms in a given mass of an element, or calculate the mass of a given number of atoms.*

6.  *Calculate formula weights (molecular weights) from formulas and atomic weights.*

7. *Give the relationship between a mole of a chemical and Avogadro's number.*

8. *Describe what the coefficients in a balanced chemical equation disclose in terms of mole relationships of the chemicals involved.*

9. *Given a balanced equation and given a specific number of moles of one of its chemicals, calculate the numbers of moles of all other chemicals involved in the reaction.*

10. *Calculate the moles of a substance in a given number of grams.*

11. *Calculate the grams of a substance in a given number of moles.*

12. *Given a balanced equation and given a specific number of grams of one of the chemicals involved, calculate the numbers of grams of all of the other chemicals involved.*

13. *Use the appropriate technical terms to describe the components of a solution and its relative concentration.*

14. *Use the value of a solution's molar concentration (molarity) to calculate the moles and the grams of the solute in a given volume.*

15. *Calculate how many grams of a solute must be taken to prepare a given volume of a solution at a stated molar concentration.*

16. *Calculate how many milliliters of a solution of a specified molar concentration have to be taken to obtain a certain number of moles or number of grams of the solute.*

17. *Use data on volumes and molar concentrations to calculate other quantities (volumes, molar concentrations, moles, or grams) that are involved in a given balanced chemical equation.*

## GLOSSARY

**Avogadro's Number.**  $6.023 \times 10^{23}$. The number of formula units in one mole.

**Coefficient.**  A small whole number placed in front of the formula of a substance in a chemical equation to indicate the relative amounts of the substances (by formula units) that interact or form.

**Concentration.**  The quantity of one component in a unit amount of a mixture.

**Equation, Balanced.**  An equation in which the number of each kind of atom found among the reactants equals the number present among the products.

**Equation, Chemical.**  The shorthand representation of a chemical reaction using the correct formulas for the reactants and products and the smallest possible whole numbers as coefficients, placed in front of the formulas, to indicate the proportions of the chemicals involved.

**Formula Weight.**   The sum of the atomic weights of the elements present in the formula of a substance.  A molecular weight.

**Molar Concentration (M).**   Describing a solution's concentration in units of moles of solute per liters of solution.  1 M means a one-molar concentration - one mole per liter.

**Molarity.**  See **Molar Concentration.**

**Molar Mass.**   The formula weight in grams; the grams per mole of a substance.

**Mole.**   The mass of a chemical equal to its formula weight in grams and that has Avogadro's number of its formula units.

**Molecular Weight.**  See **Formula Weight**.

**Precipitate.**   A solid that separates from a solution.

**Precipitation.**   The formation of a precipitate.

**Products.**   Substances that form in reactions.

**Reactants.**   Substances that react in reactions.

**Solubility.**   The extent to which a given substance will dissolve in a fixed volume or weight of a solvent at a given temperature.

**Solute.**   That component of a solution that is understood to be dissolved or dispersed in a continuous solvent.

**Solution.**   A reasonably stable, uniformly distributed mixture of two or more substances at the smallest particle level (ions, atoms, or molecules).

**Solution, Aqueous.**   A solution having water as the solvent.

**Solution, Concentrated.**   A solution with a large ratio of solute to solvent.

**Solution, Dilute.**   A solution with a small ratio of solute to solvent.

**Solution, Saturated.**   A solution in which the solvent cannot dissolve any more of the specified solute at a particular temperature.

**Solution, Unsaturated.**   A solution in which more of the solute could be dissolved at a particular temperature.

**Solution, Supersaturated.**   A solution (unstable) that holds more solute than it ordinarily can hold at a given temperature.

**Solvent.**   That component of a solution present as a continuous phase; the medium in which the solute is dissolved or dispersed. (Commonly, the term is applied to liquids.)

## DRILL EXERCISES

### I.    EXERCISES IN BALANCING CHEMICAL EQUATIONS

Convert the following sentences into balanced chemical equations.  Always remember that once you've set down the correct formula for a substance you are not allowed to alter the formula to balance the equation.  All you can alter are the coefficients.  Also remember that many nonmetals occur as diatomic molecules, for example, $O_2$, $H_2$, $N_2$, $Cl_2$, etc.

1.    Magnesium reacts with oxygen to form magnesium oxide.

2.    Calcium reacts with oxygen to form calcium oxide.

3.    Sulfur reacts with oxygen to form sulfur dioxide ($SO_2$).

4.    Sulfur reacts with oxygen to form sulfur trioxide ($SO_3$).

5.    Hydrogen reacts with chlorine to form hydrogen chloride.

6.    Hydrogen reacts with oxygen to form water.

7.    Carbon reacts with oxygen to form carbon dioxide ($CO_2$).

### II.    EXERCISES IN BALANCING CHEMICAL EQUATIONS

Balance these equations by writing the correct coefficients on the blank lines.

1.    ____$NaOH$ + ____$H_3PO_4$ $\longrightarrow$ ____$Na_3PO_4$ + ____$H_2O$

2.    ____$AgNO_3$ + ____$AlCl_3$ $\longrightarrow$ ____$AgCl$ + ____$Al(NO_3)_3$

3.    ____$Pb(NO_3)_2$ + ____$NaCl$ $\longrightarrow$ ____$PbCl_2$ + ____$NaNO_3$

4.    ____$Ni(NO_3)_2$ + ____$K_2S$ $\longrightarrow$ ____$NiS$ + ____$KNO_3$

5.    ____$Na_2CO_3$ + ____$Cu(NO_3)_2$ $\longrightarrow$ ____$NaNO_3$ + ____$CuCO_3$

### III.    EXERCISES IN CALCULATING FORMULA WEIGHTS

Calculate the formula weights of the following substances.  Round atomic weights to their first decimal places before using them in these calculations, and round the answers to the first decimal place.  Then write the two conversion factors made possible for each substance by its formula weight.  The answers will be used in later drill exercises.

1.    $NH_3$               _____    Conversion factors:

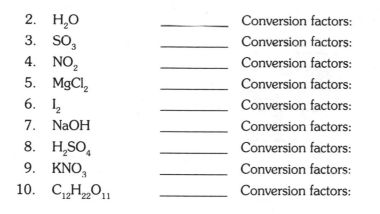

2.  $H_2O$ _____ Conversion factors:

3.  $SO_3$ _____ Conversion factors:

4.  $NO_2$ _____ Conversion factors:

5.  $MgCl_2$ _____ Conversion factors:

6.  $I_2$ _____ Conversion factors:

7.  NaOH _____ Conversion factors:

8.  $H_2SO_4$ _____ Conversion factors:

9.  $KNO_3$ _____ Conversion factors:

10. $C_{12}H_{22}O_{11}$ _____ Conversion factors:

## IV.    EXERCISES IN CALCULATING GRAMS FROM MOLES

Calculate the number of grams in the following samples. The conversion factors were prepared in the previous exercise. Show the full setup for the solution and make the correct cancel lines. (Notice that the mole quantities that are specified limit each answer to three significant figures.)

1.  3.00 mol of $NH_3$ _____ g $NH_3$

2.  9.00 mol of $H_2O$ _____ g $H_2O$

3.  0.200 mol of $SO_3$ _____ g $SO_3$

4.  0.0100 mol of $NO_2$ _____ g $NO_2$

5.  6.00 mol of $MgCl_2$ _____ g $MgCl_2$

6.  0.300 mol of $I_2$ _____ g $I_2$

7.  0.0500 mol of NaOH _____ g NaOH

8.  0.300 mol of $H_2SO_4$ _____ g $H_2SO_4$

9.  1.50 mol of $KNO_3$ _____ g $KNO_3$

10. 0.100 mol of $c_{12}H_{22}O_{11}$ _____ g $C_{12}H_{22}O_{11}$

## V.    EXERCISES IN CALCULATING MOLES FROM GRAMS

Calculate the number of moles in the following samples. The conversion factors were prepared in Drill Exercise III. Show the full setup for the solution and make the correct cancel lines. Include the right unit with the answer, and leave only the correct number of significant figures.

1.  34.0 g of $NH_3$    _____

2.  54.0 g of $H_2O$    _____

3.  400 g of $SO_3$    _____

4.  4.60 g of $NO_2$    _____

5.  0.950 g of $MgCl_2$    _____

6.  12.7 g of $I_2$    _____

7.  8.00 g of NaOH    _____

8.  5.65 g of $H_2SO_4$    _____

9.  32.0 g of $KNO_3$    _____

10.  28.4 g of $C_{12}H_{22}O_{11}$    _____

## VI.    EXERCISES IN MASS-RELATION PROBLEMS INVOLVING BALANCED EQUATIONS

We can summarize the basic steps for working these problems as follows.

Step 1.    Be sure you are working with a balanced equation.

Step 2.    Calculate the formula weights and set them off to one side for reference.

Step 3.    Beneath the formulas of the balanced equation, draw two rows of blank lines, the upper row representing the "mole level" and the lower the "gram level."  For example:

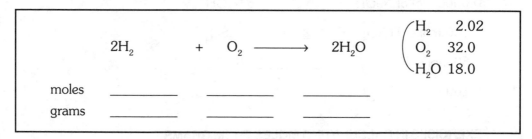

Step 4.    Write in the "given" on the proper line.  For example, let the question be as follows.  If you start with 64.0 grams of oxygen, how much hydrogen can be consumed and how much water will be produced (both in grams)?  If the

given is in a mass, convert it at once to moles and write in the answer on the proper line.

To convert the given into moles:

$$64.0 \; \cancel{g \; O_2} \times \frac{1}{32.0} \; \frac{\boxed{\text{mole } O_2}}{\cancel{g \; O_2}} = 2.00 \text{ moles } O_2$$

Here is how the setup should now look:

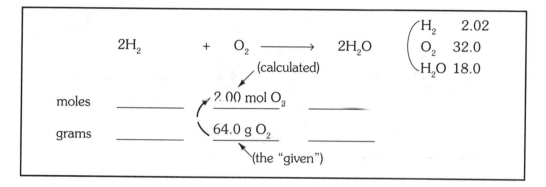

This step illustrates the most important rule in mass relation problems; SOLVE THE PROBLEM AT THE MOLE LEVEL; THEN GO BACK TO THE GRAM LEVEL AS NEEDED.

Step 5.    Use the coefficients of the equation to construct conversion factors that are used in calculating the numbers of moles of anything else requested. This step fills in the blanks of the mole-level of the answer. For example, to find out how many moles of hydrogen are needed for a reaction with 2.00 mol of oxygen:

$$2.00 \text{ mol } O_2 \times \frac{2 \text{ mol } H_2}{1 \text{ mol } O_2} = 4.00 \text{ mol } H_2$$

↑

( conversion factor that uses the coefficients of $O_2$ and $H_2$ )

To find the number of moles of water made by the reaction of 2.00 mol of $O_2$:

$$2.00 \text{ mol } O_2 \times \frac{2 \text{ mol } H_2O}{1 \text{ mol } O_2} = 4.00 \text{ mol } H_2O$$

↑

conversion factor that uses
the coefficients of $O_2$ and $H_2O$

The setup should now look like this:

| | $2H_2$ | + | $O_2$ | ⟶ | $2H_2O$ | $\begin{pmatrix} H_2 & 2.02 \\ O_2 & 32.0 \\ H_2O & 18.0 \end{pmatrix}$ |
|---|---|---|---|---|---|---|
| moles | 4.00 mol $H_2$ | | 2.00 mol $O_2$ | | 4.00 mol $H_2O$ | (all calculated) |
| grams | _____ | | 64.0 g $O_2$ | | _____ | |

(the given)

Step 6.   Convert moles into grams for the final answers to the original question.

For hydrogen:   $4.00 \text{ mol } H_2 \times \dfrac{2.01 \text{ g } H_2}{1 \text{ mol } H_2} = 8.04 \text{ g } H_2$

For water:   $4.00 \text{ mol } H_2O \times \dfrac{18.0 \text{ g } H_2O}{1 \text{ mol } H_2O} = 72.0 \text{ g } H_2O$

When completed, the setup should look like this:

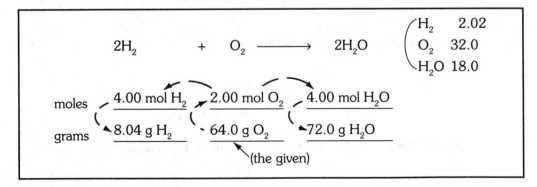

As a final check, add up the masses of the chemicals used as reactants. Then add up the masses of the chemicals made as products. These two sums must be identical. Remember: matter is neither created nor destroyed by a chemical reaction.

$$8.04 \text{ g} + 64.0 \text{ g} = 77.0 \text{ g (when correctly rounded)}$$

Work the following problems using the steps that we have just covered. Check your answers at the end of the chapter.

## PROBLEM A

The combustion of methane is shown by the following equation:

$$CH_4 \quad + \quad 2O_2 \quad \longrightarrow \quad CO_2 \quad + \quad 2H_2O$$

methane    oxygen            carbon dioxide    water

If 48.0 grams methane are burned in this way, how much oxygen is needed and how much carbon dioxide and water are produced? State your answer both in moles and in grams.

## PROBLEM B

The combustion of ethane in shown by the following equation:

$$2C_2H_6 + 7O_2 \longrightarrow 4CO_2 + 6H_2O$$

ethane

If 15.0 grams of ethane are burned this way, how much oxygen is needed and how much carbon dioxide and water are produced? State your answer both in moles and in grams.

## VII.    EXERCISES IN MOLAR CONCENTRATIONS

Complete the following table by doing the necessary calculations.  (The needed formula weights were calculated for Exercise III.)

| Solution No. | Molarity | Volume of Solution | Moles of Solute | Grams of Solute |
|---|---|---|---|---|
| 1. | 2.00 M $NH_3$ | 1.00 L | _____ | _____ |
| 2. | 0.440 M NaOH | 500 mL | _____ | _____ |
| 3. | _____ | 1.00 L | _____ | 10.0 g $H_2SO_4$ |
| 4. | 0.100 M $KNO_3$ | 250 mL | _____ | _____ |
| 5. | _____ | 750 mL | _____ | 25.0 g $MgCl_2$ |

## SELF-TESTING QUESTIONS

### Completion

1.    An amount of carbon containing Avogadro's number of carbon atoms has a mass of _____ grams.

2.    The formula weight of carbon dioxide ($CO_2$) is 44.0.  Therefore, one mole of carbon dioxide has a mass of _____ grams, and 0.500 mole has a mass of _____.

3.    A sample of water containing $6.02 \times 10^{23}$ molecules has a mass of _____ (include the unit); this amount of water is one standard reacting unit of water or, to use the scientific term, one _____.

4.    Sodium chloride has the formula NaCl, and it is an ionic compound.  Its formula weight is 58.5.  One formula unit of sodium chloride consists of one _____, whose chemical symbol is _____, and one _____, whose chemical symbol is _____.

5.  A sample of sodium chloride containing $6.02 \times 10^{23}$ of these formula units has a mass of _____ (include the unit); this amount is one standard reacting unit of sodium chloride, or one _____.

6.  The smallest representative sample of a covalent substance such as water is called a _____; a sample of a covalent compound that contains $6.02 \times 10^{23}$ of these tiny particles makes up one _____ of that substance.

7.  If you have a relatively large sample of any pure substance - an element or a compound - that consists of $6.02 \times 10^{23}$ particles of its own particular smallest representative sample, then you would have one _____ of it.

8.  Compounds X, Y, and Z have the following formula weights; X = 50.0; Y = 100; and Z = 150. We'll assume they are all covalent compounds and therefore consist of molecules.

    (a)  Suppose you had 50.0 g of X, 100 g of Y, and 150 g of Z in separate containers. What would each of these samples have in common?

    (b)  Suppose you had 100 g of Y and 100 g of X in separate containers. Which container would have the greatest number of molecules? _____

    _____

    (c)  Suppose that X and Y react to produce Z and that you elect to use 50.0 g of X in a particular experiment. How much of Y (in grams) would you have to use (assuming you want a "clean" reaction, one in which nothing of X or Y was left over)? _____

    The equation of the reaction is $X + Y \longrightarrow Z$

    (d)  An individual molecule of Z has a mass that is _____ times as much as the mass of a molecule of X; one billion molecules of Z would have a mass that is _____ times as much as one billion molecules of X.

    (e)  If you wanted to prepare 300 g of Z, how many moles of X and Y would you need? _____ moles X    _____ moles Y

How many grams?  _____ g X,  _____ g Y

9.  How many molecules of water are in 1.00 g of water? _____

10. How many moles of water are in 1.00 g of water? _____

11. Calculate (to 3 significant figures) the molar concentration of water in pure water. _____

12. Describe how you could prepare 500 mL of an aqueous solution of sodium carbonate at a concentration of 0.150 M.

_____

_____

13. If you needed 0.250 mol of NaCl for an experiment and it is available only as an aqueous solution with a concentration of 0.400 M, how many milliliters of this solution would you have to take?  _____

14. Calcium carbonate is essentially insoluble in water, and it can be made by the following reaction in which it precipitates from the solution as it forms.

$$CaCl_2(aq) + K_2CO_3(aq) \longrightarrow CaCO_3(s) + 2KCl(aq)$$

Suppose that reactants are available as 0.500 M $CaCl_2$ and 0.750 M $K_2CO_3$, calculate the milliliters of each solution that should be mixed to prepare 12.0 g of $CaCO_3$.

_____

## Multiple-Choice

1.  In the reaction, $CaCl_2 + 2AgNO_3 \longrightarrow Ca(NO_3)_2 + 2AgCl$
    if 1 mole of $AgNO_3$ is to be consumed, then
    (a)  1 mole of AgCl will form
    (b)  1/2 mole of $CaCl_2$ will be needed
    (c)  1 mole of $Ca(NO_3)_2$ will also be produced
    (d)  1/2 mole of calcium will also be produced

2. Suppose that the formula weight of X is 30 and the formula weight of Y is 60. Which of the following statements are correct?

   (a) X weighs 30 g and Y weighs 60 g.

   (b) 30 g of X will have the same number of molecules as 60 g of Y.

   (c) 60 g of X will have the same number of molecules as 120 g of Y.

   (d) Molecules of Y are twice as heavy as molecules of X.

   (e) Molecules of Y weigh five times as much as atoms of carbon–12.

3. How many molecules of water in 2.00 mol of water?

   (a) $6.02 \times 10^{46}$     (c) 2.00

   (b) $12.04 \times 10^{23}$     (d) $2.00 \times 10^{23}$

4. If a 40-g sample of substance X is known to contain the same number of molecules as a 120-g sample of substance Y, then the formula weight of X must be related to the formula weight of Y in which way(s)? The formula weight of X is

   (a) equal to the formula weight of Y

   (b) one-third the formula weight of Y

   (c) three times the formula weight of Y

   (d) 4.8 times the formula weight of Y

5. The formula units of X are three times as heavy as the atoms of carbon–12. The formula weight of X is

   (a) 36          (b) 3          (c) 4          (d) 12

6. The number of moles of NaOH in 40.0 g of NaOH is

   (a) 40.0          (b) $6.02 \times 10^{23}$ (c) 80.0          (d) 1.00

7. The number of grams of $NaHCO_3$ in 2.00 mol of $NaHCO_3$ is

   (a) 2.00          (b) 42.0          (c) 84.0          (d) 168

8. If 40.0 g of NaOH reacted with $H_2SO_4$ by the following equation: $2NaOH + H_2SO_4 \longrightarrow Na_2SO_4 + 2H_2O$ the number of grams of $H_2SO_4$ that would be used is

   (a) 49.1 g          (b) 98.1 g          (c) 196.2 g          (d) 80.0 g

9. To prepare 500 mL of 0.125 M $CaCl_2$ requires that you take how many grams of $CaCl_2$?

   (a) 0.0625 g  (b) 6.94 g          (c) 13.9 g          (d) 1.78 g

10. The number of moles of NaOH in 250 mL of 1.38 M NaOH is

    (a) 0.345 mol (b) 13.8 mol  (c) 5.52 mol  (d) 6.61 mol

11.  The number of milliliters of 0.124 M $H_2SO_4$ that hold 0.00244 mol of $H_2SO_4$ is

(a)   24.4 mL    (b)   12.4 mL    (c)   19.7 mL    (d)   16.8 mL

12.  How many milliliters of 0.112 M HCl are required to react with the NaOH in 34.2 mL of 0.142 M NaOH?  The equation is:

$$NaOH + HCl \longrightarrow NaCl + H_2O$$

(a)   43.4 mL    (b)   21.7 mL    (c)   27.0 mL    (d)   34.2 mL

13.  $H_2SO_4$ reacts with $NaHCO_3$ as follows:

$$H_2SO_4 + 2NaHCO_3 \longrightarrow Na_2SO_4 + 2CO_2 + 2H_2O$$

If 18.2 mL of a solution of $H_2SO_4$ reacts exactly with 24.8 mL of 0.148 M $NaHCO_3$ according to this equation, the molar concentration of the $H_2SO_4$ solution is

(a)   0.202 M    (b)   0.101 M    (c)   0.0501 M   (d)   0.151 M

# ANSWERS

## ANSWERS TO DRILL EXERCISE

I.   **Exercises in Balancing Chemical Equations**

1.   $2Mg + O_2 \longrightarrow 2MgO$
2.   $2Ca + O_2 \longrightarrow 2CaO$
3.   $S + O_2 \longrightarrow SO_2$
4.   $2S + 3O_2 \longrightarrow 2SO_3$
5.   $H_2 + Cl_2 \longrightarrow 2HCl$
6.   $2H_2 + O_2 \longrightarrow 2H_2O$
7.   $C + O_2 \longrightarrow CO_2$

II.  **Exercises in Balancing Chemical Equations**

1.   $3\,NaOH + H_3PO_4 \longrightarrow Na_3PO_4 + 3\,H_2O$
2.   $3\,AgNO_3 + AlCl_3 \longrightarrow 3\,AgCl + Al(NO_3)_3$
3.   $Pb(NO_3)_2 + 2\,NaCl \longrightarrow PbCl_2 + 2\,NaNO_3$
4.   $Ni(NO_3)_2 + K_2S \longrightarrow NiS + 2\,KNO_3$
5.   $Na_2CO_3 + Cu(NO_3)_2 \longrightarrow 2\,NaNO_3 + CuCO_3$

## III.  Exercises in Calculating Formula Weights

1.  $\dfrac{17.0 \text{ grams NH}_3}{1 \text{ mole NH}_3}$   or   $\dfrac{1 \text{ mole NH}_3}{17.0 \text{ grams NH}_3}$

2.  $\dfrac{18.0 \text{ grams H}_2\text{O}}{1 \text{ mole H}_2\text{O}}$   or   $\dfrac{1 \text{ mole H}_2\text{O}}{18.0 \text{ grams H}_2\text{O}}$

3.  $\dfrac{80.1 \text{ grams SO}_3}{1 \text{ mole SO}_3}$   or   $\dfrac{1 \text{ mole SO}_3}{80.1 \text{ grams SO}_3}$

4.  $\dfrac{46.0 \text{ grams NO}_2}{1 \text{ mole NO}_2}$   or   $\dfrac{1 \text{ mole NO}_2}{46.0 \text{ grams NO}_2}$

5.  $\dfrac{95.3 \text{ grams MgCl}_2}{1 \text{ mole MgCl}_2}$   or   $\dfrac{1 \text{ mole MgCl}_2}{95.3 \text{ grams MgCl}_2}$

6.  $\dfrac{254 \text{ grams I}_2}{1 \text{ mol I}_2}$   or   $\dfrac{1 \text{ mole I}_2}{254 \text{ grams I}_2}$

7.  $\dfrac{40.0 \text{ grams NaOH}}{1 \text{ mole NaOH}}$   or   $\dfrac{1 \text{ mole NaOH}}{40.0 \text{ grams NaOH}}$

8.  $\dfrac{98.1 \text{ grams H}_2\text{SO}_4}{1 \text{ mole H}_2\text{SO}_4}$   or   $\dfrac{1 \text{ mole H}_2\text{SO}_4}{98.1 \text{ grams H}_2\text{SO}_4}$

9.  $\dfrac{101.1 \text{ grams KNO}_3}{1 \text{ mole KNO}_3}$   or   $\dfrac{1 \text{ mole KNO}_3}{101.1 \text{ grams KNO}_3}$

10. $\dfrac{342.2 \text{ grams C}_{12}\text{H}_{22}\text{O}_{11}}{1 \text{ mole C}_{12}\text{H}_{22}\text{O}_{11}}$   or   $\dfrac{1 \text{ mole C}_{12}\text{H}_{22}\text{O}_{11}}{342.2 \text{ grams C}_{12}\text{H}_{22}\text{O}_{11}}$

## IV.  Exercises in Calculating Grams from Moles

1.  51.0 g $NH_3$ $\left(3.00 \text{ \sout{mol NH}}_3 \times \dfrac{17.0 \text{ g NH}_3}{1 \text{ \sout{mol NH}}_3}\right)$

2.  162 g $H_2O$

3.  16.0 g $SO_3$                7.  2.00 g NaOH

4.  0.460 g $NO_2$               8.  29.4 g $H_2SO_4$

5.  572 g $MgCl_2$               9.  152 g $KNO_3$

6.  76.2 g $I_2$                 10. 34.2 g $C_{12}H_{22}O_{11}$

## V.  Exercises in Calculating Moles from Grams

1.  2.00 mol $NH_3$ $\left(34.0 \text{ \sout{g NH}}_3 \times \dfrac{1 \text{ mol NH}_3}{17.0 \text{ \sout{g NH}}_3}\right)$

2.  3.00 mol $H_2O$              7.  0.200 mol NaOH

3.  4.99 mol $SO_3$              8.  0.0576 mol $H_2SO_4$

4.  0.100 mol $NO_2$             9.  0.317 mol $KNO_3$

5.  0.00997 mol $MgCl_2$         10. 0.0830 mol $C_{12}H_{22}O_{11}$

6.  0.0500 mol $I_2$

## VI.    Exercises in Mass-Relation Problems Involving Balanced Equations

**Problem A.**   When completed the setup should look like this.

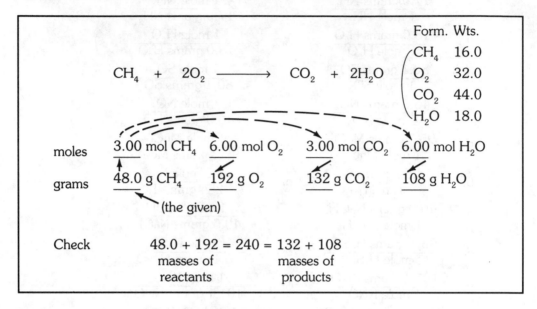

Form. Wts.

CH$_4$   16.0
O$_2$    32.0
CO$_2$   44.0
H$_2$O   18.0

CH$_4$ + 2O$_2$ $\longrightarrow$ CO$_2$ + 2H$_2$O

moles    3.00 mol CH$_4$   6.00 mol O$_2$    3.00 mol CO$_2$   6.00 mol H$_2$O

grams    48.0 g CH$_4$    192 g O$_2$    132 g CO$_2$    108 g H$_2$O

(the given)

Check         48.0 + 192 = 240 = 132 + 108
              masses of          masses of
              reactants          products

**Problem B.**   When completed the setup should look like this.

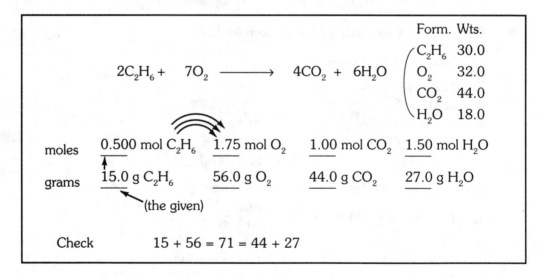

Form. Wts.

C$_2$H$_6$   30.0
O$_2$    32.0
CO$_2$   44.0
H$_2$O   18.0

2C$_2$H$_6$ + 7O$_2$ $\longrightarrow$ 4CO$_2$ + 6H$_2$O

moles    0.500 mol C$_2$H$_6$   1.75 mol O$_2$    1.00 mol CO$_2$   1.50 mol H$_2$O

grams    15.0 g C$_2$H$_6$    56.0 g O$_2$    44.0 g CO$_2$    27.0 g H$_2$O

(the given)

Check         15 + 56 = 71 = 44 + 27

**VII.  Exercises in Molar Concentrations.**
  1.   2.00 mol, 34.0 g
  2.   0.220 mol, 8.80 g
  3.   0.102 M, 0.102 mol
  4.   0.0250 mol, 2.53 g
  5.   0.349 M, 0.262 mol

## ANSWERS TO SELF-TESTING QUESTIONS

### Completion
  1.   12.0
  2.   44.0 g; 22.0 g
  3.   18.0 g; mole
  4.   sodium ion, $Na^+$, chloride ion, $Cl^-$
  5.   58.5 g; mole
  6.   molecule; mole
  7.   mole
  8.   (a)   Each would consist of 1.00 mole of that substance (or each would contain Avogadro's number - $6.02 \times 10^{23}$ - of molecules.)
       (b)   X (Since each molecule of X is lighter by half than each molecule of Y, you get twice as many X molecules as Y in the same mass of samples.)
       (c)   100 g
       (d)   three; three
       (e)   2.00, 2,00; 100, 200
  9.   $3.34 \times 10^{22}$ molecules
  10.   0.0556 mol
  11.   55.6 mol/L
  12.   Dissolve 7.95 g of sodium carbonate in water and make the final volume 500 mL.
  13.   625 mL
  14.   240 mL of 0.500 M $CaCl_2$ and 160 mL of 0.750 M $K_2CO_3$

**Multiple-Choice**

1. a and b
2. b, c, d and e
3. b
4. b
5. a
6. d
7. d
8. a
9. b
10. a
11. c
12. a
13. b

# CHAPTER 5

# KINETIC THEORY AND CHEMICAL REACTIONS

## OBJECTIVES

After you have studied this chapter, worked the problems and exercises both in the text and in this *Study Guide*, learned the basic definitions, and handled the Self-Testing Questions successfully, you should be able to do the following.

*Concerning States of Matter and Changes of State*

1. Give the relation between the millimeter of mercury as a unit of pressure and the pressure of 1 atm.

2. Describe how the pressure and volume of a gas are related (at constant volume), in both words and equation.

3. Describe how the temperature and volume of a gas are related (at constant pressure), in both words and equation.

4. Describe the relation between the partial pressures of the gases comprising a mixture and the total pressure, in both words and equation.

5. Describe the relation between the pressure and the temperature of a gas (at constant volume), in both words and equation.

6. Describe the relation between the volume of a gas and its number of moles (under constant temperature and pressure).

7.  Give the equation for the general gas law, defining each term.

8.  Describe an ideal gas (the postulates of the kinetic theory).

9.  In terms of the kinetic theory explain how each gas law arises - Boyle's, Charles', and Gay-Lussac's, as well as Dalton's law of partial pressure.

10. Describe how the kinetic theory explains gas temperature.

11. Explain how the heat of vaporization of water helps to make it especially useful as the body fluid.

12. Name the ways by which the body loses heat (becomes hypothermic) and explain how each works.

Concerning Rates of Reactions

13. Use a progress-of-reaction drawing to define "energy of activation" and "energy of reaction."

14. Use the same drawing to explain how a reaction is exothermic or endothermic.

15. Name the chief factors that affect the rate of a chemical reaction.

16. Discuss how these factors work in terms of a kinetic theory.

17. Define and explain a catalyst and its effect on energy of activation.

18. Define each of the terms in the Glossary.


## GLOSSARY

**Atmosphere.**   See **Standard Atmosphere**.

**Boiling Point, Normal.**   The temperature at which a substance boils when the outside pressure is exactly 760 mm Hg.

**Boyle's Law.**   See **Pressure-Volume Law**.

**Catalysis.**   The phenomenon of the increase in the rate of a chemical reaction that is brought about by the presence of a usually small amount of some chemical - the catalyst - that can be recovered chemically unchanged at the end of the reaction.

**Catalyst.**   A substance that in relatively small amounts can accelerate a chemical reaction without being permanently changed in a chemical sense itself.

**Charles' Law.**   See **Temperature-Volume Law**.

**Conduction.**   In the physics of heat, the transfer of heat by the transfer of kinetic energy from an atom, a molecule, or an ion to its neighbors.

**Convection.**   The transfer of heat by the circulation of a warmer fluid stream throughout the remainder of the fluid.

**Dalton's Law.**   See **Law of Partial Pressures**.

**Energy of Activation.**   The minimum amount of energy that must be received by particles (ions, molecules, or atoms) engaged in a potential chemical reaction before they can actually undergo that reaction.

**Enzyme.**   A catalyst in living things.

**Evaporation.**   The change of a liquid into its vapor form.

**Gas Constant (R).**   $R = 6.23 \times 10^4 \dfrac{\text{mm-Hg mL}}{\text{mol K}}$

The ratio of PV to nT, where P = pressure in mm Hg, V = volume in mL, n = number of moles of ideal gas, and T = Kelvin temperature.

**Gay-Lussac's Law.**   See **Pressure-Volume Law**.

**General Gas Law.**   $\dfrac{P_1 V_1}{T_1} = \dfrac{P_2 V_2}{T_2}$   where P = pressure, V = volume, and T = Kelvin temperature.

**Heat of Fusion.**   The number of calories that 1 g of a substance will absorb in going from the solid state to the liquid state without changing its temperature. (These calories are released in the reverse change from liquid to solid.)

**Heat of Reaction.**   The net energy difference between the reactants and the products of a reaction.

**Heat of Vaporization.**   The number of calories that 1 g of a substance will absorb in going from the liquid state to the gaseous state without changing its temperature. (These calories are released in the reverse change from gas to liquid.)

**Hyperthermia.**   A body condition in which the body temperature is higher than normal.

**Hypothermia.**   Lower than normal temperature of the body caused by the rapid loss of heat during exposure to cold air or water which, if not reversed, will end in death; in medicine, the deliberate lowering of a patient's temperature prior to certain types of surgery.

**Ideal Gas.**   A gas that obeys the gas laws exactly. (Real gases approach the ideal gas in behavior as they are allowed to be at lower and lower pressures and at temperatures well above their boiling points.)

**Kinetic Theory of Gases.**   A series of postulates about gases on which the laws of the behavior of gases are based. The postulates are; that gases consist of a large number of very small particles in constant random motion; that the collisions of the particles are perfectly elastic; that between collisions, the particles move in straight lines; and that the motions and collisions of the particles obey all the laws of physics.

**Kinetics.**   The science of the study of reaction rates.

**Law of Partial Pressures.**   The total pressure exerted by a mixture of gases is the sum of the individual partial pressures.

**Melting Point.**   The temperature at which dynamic equilibrium exists between the solid and the liquid states of a substance; the temperature at which a solid changes, without chemical change, into the liquid form.

**Metabolism.**   The sum total of all the continuing chemical activity occurring in the body.

**Millimeter of Mercury (mm Hg).**   A unit of pressure equal to 1/760 atm. There are 760 mm Hg in 1 atm of pressure.

**Molar Volume, Standard.**   The volume occupied by 1 mole of a gas at standard conditions of temperature and pressure (0 °C and 1 atm of pressure).  For an ideal gas this volume is 22.414 liters.

**Partial Pressure.**   The pressure contributed by an individual gas in a mixture of gases.

**Pressure.**   Force per unit area.

**Pressure-Temperature Law.**   The pressure of a confined gas is directly proportional to its Kelvin temperature when the volume is constant.

**Pressure-Volume Law.**   The volume of a fixed quantity of gas kept at constant temperature is inversely proportional to its pressure.

**Radiation.**   In the physics of light and heat, the process of emitting light or heat rays; the emitted heat or light.  (In atomic and nuclear physics, an emission of some ray such as an alpha ray, a beta ray, or a gamma ray.)

**Rate of Reaction.**   The number of successful (product-forming) events per second per unit of volume in a given reaction.

**Standard Atmosphere (atm).**   The pressure that will support a column of mercury 760 mm high when the mercury is at a temperature of 0 °C and is at sea level.

**Standard Conditions of Temperature and Pressure.**   STP. 0 °C (or 273 K) and 1 atm (or 760 mm Hg).

**Temperature-Volume Law.**   The volume of a fixed quantity of gas kept at constant pressure is directly proportional to its Kelvin temperature.

**Universal Gas Law.**   $PV = nRT$, where P = pressure, V = volume, n = moles, R = the gas constant, and T = the Kelvin temperature.

**Vapor.**   The gaseous form of a liquid.

**Vapor Pressure.**   The equilibrium pressure exerted by the vapor over a liquid at a given temperature.

**Volume-Mole Law.**   The volume of a gas at constant temperature and pressure is directly proportional to the number of moles of the gas.

# HOW TO STUDY THE GAS LAWS

The properties of gases are described by various laws.  What is important here are the properties of gases, not the names of the laws.  Even after you forget which law is Boyle's and which is Charles' or Gay-Lussac's, you should remember the basic properties of gases.

Start by learning the list of variables used in the gas laws.  These variables are the properties of a sample of gas that can be measured and that can be varied or adjusted by the experimenter.  They are:

Pressure     P          Temperature     T  (in kelvins)
Volume       V          Quantity        n  (in moles)

Often the subscripts 1 and 2 are used to designate the two different states of the gas.  Usually, a fixed quantity of a gas is involved.  This means that n, the number of moles of the gas, is a constant.  That leaves just three variables: P, V, and T.  Always remember that T must be in kelvins.  (kelvins = degrees Celsius + 273).

If a question involves changes in all three variables - P, V, and T - then the general gas law is generally used:

$$\frac{P_1 V_1}{T_1} = \frac{P_2 V_2}{T_2}$$

We are also concerned with Dalton's Law and Avogadro's Law.  Be especially certain to understand and learn Dalton's Law of Partial Pressure.

The kinetic theory of gases is particularly helpful in its explanation of the gas laws; it gives us something of a "molecule's-eye view" of gases.  It is also a beautiful illustration of the interplay between experimental work and theoretical concepts.  The theory tells us that the pressure of a gas is proportional to the mean kinetic energy of the gas particles.  Likewise, it tells us that the temperature of a gas is proportional to the mean kinetic energy of the gas particles.  If we raise the temperature of a gas, we increase its mean kinetic energy; if we raise the temperature of a *confined* gas, we also raise its pressure.

The kinetic theory also gives us a "molecule's-eye view" of the motion or vibration of atoms and molecules in liquids and solids.  Concentrate on learning the vocabulary associated with liquids and solids.

The kinetic theory also helps us understand how the rate of a chemical reaction is influenced by the concentration of the reactants, the temperature, and the presence or absence of a catalyst.  Be sure to learn what is meant by the energy of activation and what, in general, a catalyst does to it.  We'll need these concepts to understand how enzymes work.

## SELF-TESTING QUESTIONS

### Completion

1.  The atmosphere at sea level presses down with a force of _____lb/in.$^2$ and will hold up a column of mercury that is _____ mm in height.

2.  Another name for 1 mm Hg (with virtually no error) is 1 _____.

3.  Robert Boyle discovered the relation between the pressure and the _____ of a fixed amount of gas when its _____ is constant.

4.  Using symbols of $P_1$, $P_2$, $V_1$, and $V_2$, Boyle's law may be written in mathematical form as: _____.

5.  In a general way, Boyle's law tells us that if greater pressure is placed on a gas, its volume will _____.

6.  Alternatively, if the pressure on a gas is reduced, its volume will

    _____.

7.  If the pressure on a gas is reduced by one-half, its volume will _____.

8.  If a mixture of gases is made up of three individual gases, A, B, and C, and the total pressure on the gas mixture is 760 mm Hg, what is the partial pressure of C if the partial pressures of A and B are each 350 mm Hg?

    _____

9.  Comparing inhaled air with exhaled air, the partial pressure of _____ is less in exhaled air partly because it has been consumed during chemical reactions in the body.

10. At STP, 1 mole of oxygen occupies _____ liters, 1 mole of hydrogen occupies _____ liters, and 1 mole of nitrogen occupies _____ liters.

11. Hydrogen and oxygen combine to form water:

    $2H_2 + O_2 \longrightarrow 2H_2O$.  For a complete reaction, 1 liter of oxygen at STP would require _____ liters of hydrogen also at STP because, according to Avogadro's law, "when measured at the same T and P, equal volumes of gases contain equal numbers of _____.

12.  The basic postulates of the kinetic theory of gases are:

(a) _____

(b) _____

(c) _____

13.  If somehow the average kinetic energy of all the particles in a sample of helium gas contained in a tight flask is made to increase, then both the _____ and the _____ of the helium will increase.

14.  When molecules of a liquid escape from it more rapidly than they return to it, the liquid will eventually _____.

15.  If a liquid is cooled enough so that its molecules have constant neighbor molecules, the liquid has changed into a _____.

16.  The three ways by which an object, such as a bar of metal, can lose some of its heat to its surroundings are _____, _____, and _____.

17.  Which of these would be ruled out if the metal were in a vacuum (a complete absence of air) but still in contact with a colder object? _____

18.  Which of these means of heat loss could still take place even if the bar of metal were in a vacuum and suspended in such a way as to be out of contact with any other object (except for a very thin string holding it suspended that could absorb no heat)? _____

19.  Each year, on the average, the planet earth loses just as much heat to outer space as it receives from the sun. (Otherwise the earth would either heat up and melt or cool down and everything would freeze.) Which of the three means of losing heat is involved? _____

20. In winter, the uncovered human head loses heat principally by _____ and _____ when little physical activity is in progress.

21. When a substance changes from its liquid state to its vapor state, we call this, in general, a _____ change; The specific name for this change is

    _____.

22. To change a specified quantity of a substance from its liquid to its vapor state requires a certain amount of heat; the amount of heat required per gram is called the _____ of the liquid.

23. If 1 g of water at 100 °C is changed to its vapor at 100 °C, _____ cal are absorbed by the water and are present in the water vapor.

24. What happens to the heat in water vapor when the vapor changes back into a liquid at the same temperature? _____

    _____

25. Besides radiation, conduction, and convection, the human body loses heat energy by _____.

26. The set of chemical changes that occur in the body is called the body's

    _____.

27. If insensible perspiration is insufficient to help cool the body, then

    _____ occurs.

28. The minimum collision energy of two reacting particles above which all collisions of the proper orientation will produce products is called the

    _____.

29. If a reaction has a high energy of activation, it probably will proceed

    _____.
       (slowly or rapidly)

30. We could speed up a slow reaction by one or more of these operations.

    (a) Increase the _____ of the reactants.

    (b) Raise the _____ of the reacting mixture.

    (c) Introduce a special _____.

31. Which of the three factors in question 30 best explains why substances burn more vigorously in pure oxygen than in air? _____

32. Which of the three factors in question 30 has the greatest influence on how rapidly reactions proceed in living things? _____

33. Increasing the concentration of the reactants increases the rates of most reactions because higher concentrations mean higher frequencies of _____ between the reactants.

34. Raising the temperature of a reacting mixture increases the rate of the reaction primarily by increasing the _____ of the collisions.

35. Introducing a catalyst into a reacting mixture increases the rate of the reaction largely by lowering the reaction's _____.

36. The special catalysts found in living things are called _____.

37. The combustion of coal in air is an example of an energy-releasing or _____ event.

38. A reaction normally requiring a high temperature in order to take place is an energy-consuming or _____ event.

**Multiple-Choice**

1. The two gases important in respiration are
    (a) nitrogen              (c) oxygen
    (b) carbon dioxide        (d) water vapor

2.  Pressure is the same as
    (a)  force                          (c)  force per unit distance
    (b)  force per unit volume          (d)  force per unit area

3.  One atmosphere of pressure may be expressed as
    (a)  1 atm     (b) 760 mm Hg     (c)  14.7 mm Hg     (d)  20 lb/in.$^2$

4.  If 1 liter of a gas at 20°C is made to change its pressure from 1 atm to 2 atm, the new volume will be
    (a)  2 liters     (b)  1 liter     (c)  0.5 liter     (d)  0.1 liter

5.  If 1 liter of gas at 50°C is made to change its temperature to 100 °C under conditions where it is allowed to freely expand in order to maintain a constant pressure, the new volume of the gas will be
    (a)  2 liters     (b)  0.5 liter     (c)  4 liters     (d)  1.15 liters

6.  If a gas mixture consists of 1 mole of helium and 1 mole of argon at STP, the partial pressure of helium will be
    (a)  1 atm     (b)  0.5 atm     (c)  2 atm     (d)  1.15 atm

7.  A sample of an unknown gas occupies 22.4 liters at STP and the sample weighs 32 g.  The formula weight of the gas is, therefore,
    (a)  22.4     (b)  32     (c)  11.2     (d)  16

8.  The postulate of the kinetic theory stating that the molecules of a gas move in straight lines, not curved lines, between collisions is another way of postulating that between the molecules
    (a)  no forces of attraction exist
    (b)  no forces of repulsion exist
    (c)  balanced forces of attraction and repulsion exist
    (d)  collisions are perfectly elastic

9.  If the molecules of a gas confined within a sealed container are somehow made to have a lower average velocity but still remain in the gaseous state,
    (a)  the density of the gas will decrease
    (b)  the pressure of the gas will decrease
    (c)  the temperature of the gas will decrease
    (d)  the volume of the gas will decrease

10. To understand how smoke will spread throughout an entire room, which of these concepts or theories is most useful?
    (a)  octet theory                   (c)  kinetic theory
    (b)  Bohr theory                    (d)  mole concept

11.    The temperature of a substance is a measure of

   (a)    the average kinetic energy of its formula units

   (b)    the number of its formula units present

   (c)    the gram-formula weight

   (d)    the chemical energy of the formula units

12.    When liquid water at 0°C changes to ice at 0°C, the water will give up to the surroundings its

   (a)    specific heat capacity          (c)    heat of fusion

   (b)    heat of vaporization           (d)    specific gravity

13.    When liquid water at 0°C changes to ice at 0 °C, the water gives up to the surroundings

   (a)    80 cal/g    (b)    80 kcal/g    (c)    540 cal/g    (d)    540 kcal/g

14.    When ethyl chloride is sprayed on the skin to cool it, the heat drawn from the skin goes mostly to the ethyl chloride to

   (a)    warm it                         (c)    sublime it

   (b)    evaporate it                    (d)    condense it

15.    Ice is a better coolant than water (even when the water is at 0°C) because ice has a relatively large

   (a)    specific heat                   (c)    heat of vaporization

   (b)    heat capacity                   (d)    heat of fusion

16.    When the body allows the evaporation of water to carry away body heat, it takes advantage of water's relatively large

   (a)    specific heat                   (c)    heat of vaporization

   (b)    heat capacity                   (d)    heat of fusion

17.    A breeze has a cooling effect because it aids in the loss of body heat by

   (a)    convection                      (c)    radiation

   (b)    conduction                      (d)    condensation

18.    Hot steam at 100 °C is more dangerous to exposed skin than hot water at 100 °C because the steam contains the water's

   (a)    heat of vaporization            (c)    kinetic energy

   (b)    specific heat                   (d)    sensible heat

19. A chemical reaction that is extremely rapid at room temperature
    (a) is probably highly endothermic
    (b) probably has a very low energy of activation
    (c) very likely has an unusually high energy of activation
    (d) may have the benefit of a catalyst

## ANSWERS

### ANSWERS TO SELF-TESTING QUESTIONS

**Completion**
1. 14.7; 760
2. torr
3. volume, temperature
4. $P_1V_1=P_2V_2$
5. decrease
6. increase
7. double
8. 60 mm Hg (350 + 350 + 60 = 760; Dalton's law)
9. oxygen
10. 22.414, 22.414, 22.414
11. 2, moles (or molecules)
12. (a) Gases consist of large numbers of small particles in random motion.
    (b) These particles are very hard and are perfectly elastic.
    (c) These particles move in straight lines between collisions.
13. pressure, temperature
14. evaporate
15. solid
16. radiation, conduction, and convection
17. convection
18. radiation
19. radiation
20. radiation and convection
21. physical; evaporation

22.  heat of vaporization
23.  540 (rounded from 539.5)
24.  The heat is released to the surrounding air.
25.  evaporation (or perspiration)
26.  metabolism
27.  sensible perspiration
28.  energy of activation
29.  slowly
30.  (a)  concentrations
     (b)  temperature
     (c)  catalyst
31.  a (increased concentration)
32.  c (special catalysts)
33.  collisions
34.  energies (or average kinetic energies)
35.  energy of activation
36.  enzymes
37.  exothermic
38.  endothermic

## Multiple-Choice

 1.  b and c
 2.  d
 3.  a and b
 4.  c
 5.  d
 6.  b
 7.  b
 8.  a and b
 9.  b and c
10.  c
11.  a
12.  c
13.  a
14.  b (See Note 1)
15.  d (See Note 2)
16.  c

17.   a
18.   a (See Note 3)
19.   b and d

**Notes**

1.   Some warming (choice a) also occurs, but usually heats of vaporization are generally much greater than specific heats.  Only the specific heat of ethyl chloride is involved in warming; the heat of vaporization is involved in evaporation.

2.   Ice can cool an object in contact with it simply by drawing heat from the object in order to melt, and ice's heat of fusion is much larger than liquid water's heat capacity.

3.   Both forms, of course, are very dangerous.  However, cooler skin in contact with the steam would make some of the steam condense, releasing the heat of vaporization which is relatively large.  This extra heat could then cause more injury.

# CHAPTER 6

# WATER, SOLUTIONS, AND COLLOIDS

## OBJECTIVES

The topics in this chapter that are most important for an understanding of the molecular basis of life are:

1.  The hydrogen bond

    Because it occurs in all carbohydrates, proteins, and genes.

2.  How water can dissolve some substances well and others poorly.

    Because water is the basic fluid of cells and blood.

3.  LeChâtelier's principle and equilibria

    Because living systems must maintain some kind of equilibrium in the face of forces that unbalance equilibria.

| | | |
|---|---|---|
| 4. | Dialysis | Because that is how many substances migrate into and out of cells. |
| 5. | Ways to express percentage concentrations | Because these are sometimes used to describe fluids found in living systems as well as fluids administered to patients. |

After you have studied this chapter and worked the exercises in it, you should be able to do the following.

1. *Explain how the water molecule gets its bond angle and its polarity.*

2. *Explain how the high polarity of water helps it to function as a solvent.*

3. *Draw an illustration of the hydrogen bond and define it, citing specific evidence for its existence in the liquid state.*

4. *Explain how surface tension arises.*

5. *Describe the effect of a surfactant and cite two examples in which surfactants are vital to our lives.*

6. *Compare and contrast true solutions, colloidal dispersions, and suspensions with respect to particle sizes and four physical properties.*

7. *Explain why hydrates are called compounds, not mixtures, and give formulas for plaster of paris and gypsum.*

8. *State LeChâtelier's principle and use it to explain how temperature affects solubility.*

9. *Describe how the solubility of a gas is related to its partial pressure.*

10. *Do the calculations required for preparing and using solutions of known concentrations (a) from pure solutes and solvents and (b) from more concentrated solutions.*

11. *Give the circumstances necessary for osmosis or dialysis.*

12. *Describe how dialysis is necessary for life.*

13. *Predict the direction in which water will flow in osmosis or dialysis, given the concentrations of solutes expressed as molarities or osmolarities.*

14. *Define each of the terms in the Glossary.*

## GLOSSARY

**Active Transport.** The use of energy-consuming reactions in cell membranes to cause ions and molecules to migrate from one side of the membrane to the other.

**Anhydrous.**   Without water; in chemistry, referring to a substance without water of hydration.

**Brownian Movement.**   The random, chaotic movements of particles in a colloidal dispersion that can be seen with a microscope.

**Colligative Property.**   A property of a solution that depends only on the concentrations of the solute and the solvent and not upon their chemical identities.   Examples are changes in boiling point, melting point, and osmotic pressure that accompany changes in concentration.

**Colloidal Dispersion.**   A reasonably stable, uniform distribution in some dispersing medium of colloidal particles, those that are one size larger than an atom, an ion, or a small molecule and that have at least one dimension between one and 1000 nanometers.

**Colloidal Osmotic Pressure.**   The contribution made to the osmotic pressure of a solution by substances colloidally dispersed in it.   See **Osmotic Pressure**.

**Crenation.**   The shrinkage of red blood cells when they are in contact with a hypertonic solution.

**Deliquescent.**   Descriptive of a substance whereby it can gradually dissolve and change into a concentrated solution by drawing moisture to itself from humid air.

**Desiccant.**   A substance that will reduce the concentration of water vapor in an enclosed space by combining with the water to form a hydrate.

**Dialysis.**   The passage of water and particles in solution, but not particles of colloidal size, through a dialyzing membrane.

**Emulsion.**   A colloidal dispersion of tiny microdroplets of one liquid in another liquid.

**Equilibrium.**   The situation in which two opposing events are occurring at the same rates, resulting in no further net change.

**Gas Tension.**   The partial pressure of a gas over its solution in some liquid when the system is in equilibrium.

**Gel.**   A colloidal dispersion of a solid in a liquid that has adopted a semisolid form.

**Hemolysis.**   The bursting of a red blood cell and the liberation of its hemoglobin as the result of the cell being in contact with a hypotonic solution or as the result of some other cause.

**Henry's Law.**   See **Pressure-Solubility Law**.

**Homogeneous Mixture.**   A mixture in which a small sample taken in one place has the same composition and properties as a sample of the same size taken from another place.

**Hydrate.**   A chemical compound in which intact molecules of water are trapped within a crystalline lattice in a definite ratio to the other components.

**Hydration.**  The association of molecules of water with dissolved ions or polar molecules which helps these ions or molecules to remain in solution.

**Hydrogen Bond.**  The force of attraction between a $\delta$+ on a hydrogen bound by a covalent bond to oxygen, nitrogen, or fluorine and a $\delta$– on a nearby oxygen, nitrogen, or fluorine.

**Hygroscopic.**  Descriptive of a substance that can reduce the concentration of water vapor in air by forming a hydrate.

**Hypertonic.**  The quality of having an osmotic pressure greater than some reference; having a concentration higher than the reference when measured according to the ability to undergo osmosis.

**Hypotonic.**  The quality of having an osmotic pressure less than some reference; having a concentration lower than the reference when measured according to the ability to undergo osmosis.

**Isotonic.**  Having an osmotic pressure identical to that of some reference with respect to the ability to undergo osmosis.

**Le Châtelier's Principle.**  If a system is in stable equilibrium and something happens to change the conditions, the system will alter in such a way as to restore equilibrium.

**Osmolarity (Osm).**  The molar concentration of all osmotically active solute particles in a solution.

**Osmosis.**  The passage of water only, without any of the solute, from a less concentrated solution (or pure water) to a more concentrated solution when the two solutions are separated by a semipermeable membrane.

**Osmotic Membrane.**  A semipermeable membrane that permits only osmosis, not dialysis.

**Osmotic Pressure.**  The pressure that would have to be applied to a solution to prevent osmosis if that solution were separated from pure water by an osmotic membrane.

**Percent Concentration (%).**  A measure of concentration.

   **Vol/vol percent:**  the number of volumes of solute in 100 volumes of solution (often used for mixtures in the gaseous state and sometimes used when a liquid is dissolved in a liquid.)

   **Wt/wt percent:**  the number of grams of solute in 100 grams of the solution.

   **Wt/vol percent:**  the number of grams of solute in 100 mL of the solution (very commonly used when a liquid or a solid is dissolved in a liquid).

   **Milligram percent:**  the number of milligrams of the solute in 100 mL of the solution.

**Physiological Saline Solution.** A solution of sodium chloride with an osmotic pressure equal to that of the blood.

**Pressure-Solubility Law.** (Henry's Law)  The solubility of a gas at a given temperature is directly proportional to partial pressure of the gas.

**Reagent.** Any mixture of chemicals, usually a solution, that is used to carry out a chemical test.

**Semipermeable.** Descriptive of a membrane that permits only certain kinds of molecules to pass through but not others.

**Solution.** A homogeneous mixture of two or more substances that are at the smallest levels of their states of subdivision - at the ion, atom, or molecule level.

**Surface Active Agent.** See **Surfactant**.

**Surface Tension.** The quality of a liquid's surface by which the surface acts as a thin, invisible, elastic membrane.

**Surfactant.** A surface-active agent, such as a detergent, that reduces water's surface tension.

**Suspension.** An homogeneous mixture in which the particles of at least one component have average diameters over 1000 nm.

**Tyndall Effect.** The scattering of light by colloidal size particles in a colloidal dispersion.

**Water of Hydration.** The water molecules held in a hydrate in some definite molar ratio to the rest of the substance.

## SELF-TESTING QUESTIONS

### Completion

1. Write the structure of the water molecule. _____

2. In terms of the net electrical charge for the whole molecule, the water molecule is _____.

3. However, the water molecule is _____, because the center of density of all its _____ charges (contributed by its atomic nuclei) does not coincide with the center of density of all its _____ charges (contributed by its _____).

4.   Write the structure of the water molecule again and place the symbols δ+ and δ– by the appropriate parts of the structure to show the locations of sites of relative electron deficiency and electron richness.

_____

5.   Write the structure of two water molecules, orienting them in a manner we would expect because they can attract each other.

6.   This force of attraction in water is called the _____. It extends from a site of δ+ associated with the _____ end of one molecule to a site of δ–
          (H or O)

     associated with the _____ end of the other molecule.
                (H or O)

7.   Because of water's _____, the surface of water in contact with air behaves as thin, elastic membrane.

8.   The surface tension of water would be enough to collapse the lungs by collapsing the _____ in the lungs if the water in the lungs did not contain some dissolved _____.

9.   Bile contains a _____ agent that aids in the digestion of

     _____.

10.  _____ are common surfactants used in cleaning.

11.  Potassium chloride is similar in structure to sodium chloride. At 20 °C the solubility of KCl in water is 24.7 g/100 ml solution.

     (a)  What particles make up the formula unit of KCl? (Give both their names and their symbols). _____, whose symbol is _____, and _____, whose symbol is _____.

     (b)  What particles separate from the crystal of KCl as it goes into solution in water? (Formulas only) _____ and _____

     (c)  What happens to them as they enter the solution (something involving the water molecules)? _____

(d)   In the crystal of KCl, the potassium ions were generally surrounded by what electron-rich particles? (Name) _____

(e)   By going into solution, have the potassium ions given up being surrounded by an electron-rich envelope? _____

What electron-rich substance surrounds each $K^+$ ion in the solution?

_____

(f)   As for the chloride ions in KCl, they are surrounded by what relatively electron-poor particles? (Give name) _____

(Formula) _____

(g)   When the chloride ions go into solution in water, what is it that is relatively electron-poor that crowds around them? (Be specific; it is not just the particle's name we seek but also a particular part of it.)

_____

12.   Le Châtelier's principle enables us to predict that the following equilibrium will shift to the _____ if it is cooled.
         left or right

$$\text{heat} + CuSO_4(s) \rightleftharpoons Cu^{2+} (aq) + SO_4^{2-} (aq)$$

The symbol (s) stands for the solid substance and (aq) for the dissolved particles.

13.   Le Châtelier's principle helps us to predict that the gas tension of carbon dioxide over water will _____ if the following equilibrium is
                              increase or decrease
heated.

$$CO_2 (aq) + \text{heat} \rightleftharpoons CO_2(g)$$

The symbol (g) stands for the gaseous state, and (aq) means the aqueous solution.

14.   At 25 °C, about 20.0 mg of $N_2$ dissolve in water when the partial pressure of $N_2$ is 1.0 atm. How many milligrams of $N_2$ will dissolve if the partial pressure is increased to 2.0 atm? _____

15. Suppose you have a solution of KCl at 20 °C with a concentration of 34.7 g/100 g (which is also the solubility of KCL in water at 20 °C).

(a) What qualitative expression describes this solution most accurately?

_____

(b) What is the percent of concentration (w/w)? _____

(c) What would you see if you added some crystals of KCl to this solution?

_____

(d) What would you see (eventually) if you cooled this solution to, say, 5 °C?

_____

If you did not see this, then the cooler solution is best described as

_____

(e) If you separated this solution from pure water by a semipermeable membrane, _____ would occur and the volume of this solution would _____ and its
            (increase or decrease)
concentration would _____.
                    (increase or decrease)

(f) If we had a 30% (w/w) solution of KCl in water, we might describe it in two ways: _____ and _____

(g) A 0.3% (w/w) solution of KCl in water might be described qualitatively as

_____

16. How many grams of glucose are in 250 g of a solution with a concentration of 0.800% (w/w) glucose? _____

17. How many milliliters of 0.150% (w/v) NaCl contain 1.25 g of NaCl?

_____

18. If 80.0 mL of a 0.500 M solution are diluted with water to a final volume of 400 mL, the final concentration will be _____

19. How many milliliters of 0.400% (w/v) sodium hydroxide solution must be diluted to a final volume of 500 mL to give a final concentration of 0.250% (w/v)?

_____

20.  If water contains a dispersion of particles larger than simple ions or molecules but not heavy enough to be forced by gravity to settle or otherwise coagulate, this mixture is called a _____

21.  If you direct a light beam through this mixture (question 20), you will observe the _____ effect.

22.  A colloidal dispersion of one liquid in another is called an

_____.

23.  The motions of colloidal dispersed particles in a fluid are called the

_____.

24.  Three colligative properties of solutions containing nonvolatile solutes are

(a) _____

(b) _____

(c) _____

25.  A membrane that lets water molecules and small ions or molecules pass through it in either direction is called a _____ membrane.

26.  The concentration of all osmotically active particles in a solution is called the _____ of the solution.

27.  If pure water were fed intravenously, a patient's red blood cells would tend to _____; technically, this is called _____.
      (shrink or burst)

28.  A compound with the formula $Na_3PO_4 \cdot 12H_2O$ is an example of a

_____.

29.  If this compound (in question 28) were heated intensely, it would lose its water of _____ and be changed into its _____ form.

30.  Calcium chloride, $CaCl_2$, will take water vapor from humid air and change to $CaCl_2 \cdot 6H_2O$. What would this product be called?

_____

31.   Because anhydrous calcium chloride will do this, it is called a

_____.

## Multiple-Choice

1.   The curved arrow points to what kind of bond?

   (a)   covalent

   (b)   ionic

   (c)   hydrogen bond

   (d)   nonpolar

   $$H \quad H \cdots O \overset{\displaystyle O}{\underset{\displaystyle H}{\diagdown}}$$

2.   Hydrogen bonds between molecules will be the strongest for which substance?

   (a)   $NH_3$      (b)   $CH_4$      (c)   $H_2O$      (d)   H–H

3.   If water tends to spread out and form a very thin film on a surface, then the molecules in that surface are very likely

   (a)   nonpolar   (b)   polar        (c)   dense        (d)   mobile

4.   Substance A is a molecular compound that dissolves in gasoline but not in water. The molecules of A are very likely

   (a)   polar          (b)   nonpolar   (c)   solvated    (d)   crenated

5.   When water molecules in liquid water are attracted to and crowd around solute ions (or molecules), the phenomenon is called (one choice)

   (a)   hydrolysis                      (c)   hydration

   (b)   solvation                        (d)   hydrogenation

6.   The property of the surface of water that can be likened to its having a thin, elastic invisible membrane is called

   (a)   surface action                (c)   surface pressure

   (b)   surface polarity              (d)   surface tension

7.   When the partial pressure of a gas is reduced by half, its solubility is

   (a)   reduced by half              (c)   doubled

   (b)   unchanged                      (d)   none of the above

8.   One property a colloidal dispersion has that a solution does not have is

   (a)   homogeneity                    (c)   the Tyndall effect

   (b)   filterability                      (d)   osmotic pressure

9.   The dispersion of oil droplets in water is called

   (a)   a smoke   (b)   an emulsion   (c)   a sol        (d)   a gel

10. In a dilute solution of sugar in water, water is the

    (a) solute   (b) solvent   (c) emulsifying agent   (d) desiccant

11. An aqueous solution whose concentration effectively matches the concentration of blood is said to be

    (a) isotopic   (b) isobaric   (c) isotonic   (d) isoelectronic

12. The osmolarity of 0.500 M $Mg(NO_3)_2$ is

    (a) 0.500 Osm   (b) 1.00 Osm (c) 1.50 Osm (d) 4.50 Osm

13. A red cell surrounded by pure water would undergo

    (a) crenation (b) hemolysis (c) no change (d) dehydration

14. Physiological saline solution has a sodium chloride concentration of about

    (a) 0.9% (w/w)   (b) 0.9 mg-% (c) 0.9 Osm   (d) 0.9 M

15. If a bag made of a dialyzing material and containing a mixture of water, dissolved salt, and colloidally dispersed protein were placed in a beaker of pure water, the water would eventually contain

    (a) dissolved salt          (c) both salt and protein

    (b) protein                 (d) no additional material

16. The weak force of attraction acting between water molecules is called

    (a) a hydrogen to oxygen covalent bond

    (b) a hydrogen bond

    (c) an ionic bond

    (d) a hydration bond

17. To make 300 mL of a 4.00% solution of sugar we need to weigh out

    (a) 4.00 g of sugar

    (b) 40.0 g of sugar

    (c) 12.0 g of sugar

    (d) we need to know the formula weight of sugar before making an answer

18. In the substance whose formula is $Na_2CO_3 \cdot 10H_2O$, for every five formula units of $Na_2CO_3$ there are how many water molecules?

    (a) 50          (b) 2          (c) 10          (d) 100

19. To prepare 250 mL of 0.0400% (w/w) glucose $(C_6H_{12}O_6)$ we would have to obtain a sample of glucose with a mass of how many grams?

    (a) 0.180 g glucose          (c) 1.80 g glucose

    (b) 0.100 g glucose          (d) 10.0 g glucose

20. If we obtained 125 mL of 0.0450% (w/v) NaCl, the sample would contain how much solute?

(a) 0.0563 g NaCl          (c) 56.3 g NaCl

(b) 0.00231 mol NaCl       (d) 0.0112 g NaCl

21. Which is the most dilute solution of NaCl?

(a) 0.100% (w/w)           (c) 0.100% (v/v)

(b) 0.100% (w/v)           (d) 0.100 mg-%

22. If 10.5 mL of 0.150 M HCl were diluted to a final volume of 100 mL, the concentration of the resulting solution would be

(a) 1.43 M HCl             (c) 0.00150 M HCl

(b) 0.0158 M HCl           (d) 10.5 M HCl

# ANSWERS

## ANSWERS TO SELF-TESTING QUESTIONS

### Completion

1. 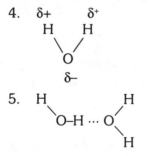 (Special Topic 3.2 explained the bond angle in water. If you wrote H–O–H as the answer, this is acceptable at this place.

2. electrically neutral

3. polar, positive, negative, electrons

4. $\delta+$      $\delta+$
   H      H
    \    /
     O
    $\delta-$

5. H            H
    \          /
     O–H ··· O
              \
               H

6. hydrogen bond; H, O

7. surface tension

8. alveoli, surfactant (or surface-active agent)

9.  surface-active, fats and oils (or lipids) as well as other foods coated with fats and oils

10. Detergents (soaps)

11. (a)  the potassium ion, $K^+$, the chloride ion, $Cl^-$

    (b)  $K^+$ and $Cl^-$

    (c)  They become hydrated.

    (d)  chloride ions

    (e)  No, they have not; water molecules

    (f)  potassium ions, $K^+$

    (g)  the hydrogen ends of water molecules

12. left

13. increase

14. 40.0 mg

15. (a)  saturated

    (b)  34.7% (w/w)

    (c)  The crystals would sink to the bottom and not dissolve.

    (d)  Crystals of KCl would appear; supersaturated

    (e)  osmosis, increase, decrease

    (f)  unsaturated and concentrated

    (g)  dilute (also, unsaturated)

16. 2.00 g of glucose

17. 833 mL

18. 0.100 M

19. 313 mL

20. colloidal dispersion

21. Tyndall

22. emulsion

23. Brownian movement

24. (a)  lowering of the freezing point of the solvent

    (b)  raising of the boiling point of the solvent

    (c)  osmotic pressure of the solution

25. semipermeable (or dialyzing)

26. osmolarity

27.  burst; hemolysis
28.  hydrate
29.  hydration, anhydrous
30.  a hydrate (calcium chloride hexahydrate)
31.  desiccant

**Multiple-Choice**

| | | | | | |
|---|---|---|---|---|---|
| 1. | a | 9. | b | 16. | b |
| 2. | c | 10. | b | 17. | c |
| 3. | b | 11. | c | 18. | a |
| 4. | b | 12. | c | 19. | b |
| 5. | c | 13. | b | 20. | a |
| 6. | d | 14. | a | 21. | d |
| 7. | a | 15. | a | 22. | b |
| 8. | c | | | | |

# CHAPTER 7

# ACIDS, BASES, AND SALTS

## OBJECTIVES

In our study of the molecular basis of life, we need to be familiar with the major topics of this chapter for several reasons.

| | |
|---|---|
| Acids and bases | Because they are involved in most reactions of metabolism. |
| Ionic equations | Because from now on we shall use them almost exclusively to describe reactions. |
| The reactions of ions | Because we need to know how to tell which ions can remain unaffected in the presence of others and which cannot, since ions occur in all body fluids. |

Before you attempt any problems or exercises, you should commit to memory the names and formulas for the five strong acids plus phosphoric acid listed in Table 7.1 and the strong bases listed in Table 7.2. These are short lists, and we make the important but simplified assumption that any acids or bases not on the lists of the strong acids or bases will be weak. Also commit to memory the solubility rules for salts (Section 7.5). Then you will have automatically learned what the insoluble salts are - any salts not mentioned in the solubility rules. Next learn the properties of acids and bases in water and the reactions of strong acids with active metals and their hydroxides, carbonates, and bicarbonates. Practice writing both full and net ionic equations.

The summary of the circumstances in which the various ions react with one another given in Section 7.5 is very important. If only one goal were to be stated for this chapter, it would be to learn the chief chemical properties of ions and ion-producing substances - acids, bases, and salts.

After studying this chapter and working the exercises in it, you should be able to do the following.

1.   *Give the definitions of acids and bases according to Brønsted.*

2.   *Define and give examples of strong and weak electrolytes.*

3.   *Write the names and formulas of five strong acids and two strong, highly soluble bases and explain why they are classified as "strong".*

4.   *Given the formula of an acid or base, tell whether it is strong or weak.*

5.   *Write the structures of the hydronium ion, the hydroxide ion, the ammonium ion, the carbonate ion, and the bicarbonate ion as well as the structures of water, ammonia, and carbonic acid.*

6.   *Write the equation for the dynamic equilibrium involved in the self-ionization of water.*

7.   *Write equations, full and net ionic, for the reactions of strong aqueous acids with active metals and their hydroxides, bicarbonates, and carbonates.*

8.   *Describe what it means when a metal is high in the activity series.*

9.   *Write net ionic equations involving proton transfers between (a) the ammonia molecule and the hydronium ion and (b) the ammonium ion and the hydroxide ion.*

10.  *Use the relative strength of an acid to describe the relative strength of the acid's anion as a base.*

11.  *Recognize from a formula if a compound is a salt and if it is, write equations for making that salt in four different ways using a strong acid.*

12.  *Use the solubility rules to predict double decompositions.*

13.  *Summarize the three circumstances under which ions usually react, giving the equations of two specific examples of each.*

14.  *Calculate the equivalent weight of an ion.*

15.  *Define (and, where substances are involved, illustrate) all of the terms listed in the Glossary.*

## GLOSSARY

**Acid.**  Any substance that can donate $H^+$ (a proton). ("Acids are proton-donors.")

**Acid-Base Neutralization.**  The reaction of an acid with a base to produce a solution that is neither acidic nor basic, but neutral.

**Acidic Solution.**  A solution in which $[H^+]>[OH^-]$.

**Activity Series.**  A list of elements (or other substances) in order of the ease with which they release an electron under standard, specified conditions.

**Alkali.**  A strongly basic substance; generally sodium or potassium hydroxide.

**Anion.**  A negatively charged ion.

**Anode.**  An electrode to which negatively charged ions will be attracted because it bears a deficiency of electrons.

**Base.**  Any substance that will accept a proton ($H^+$). ("Bases are proton-acceptors.")

**Basic Solution.**  A solution in which $[H^+]<[OH^-]$.

**Brønsted Acids and Bases.**  See **Acid**. See **Base**.

**Cathode.**  An electrode to which positively charged ions will be attracted because it bears an excess of electrons.

**Cation.**  A positively charged ion.

**Diprotic.**  Having two protons per molecule that are capable of neutralizing a base, as in a diprotic acid such as sulfuric acid ($H_2SO_4$).

**Double Decomposition Reactions.**  A reaction in which a compound is made by the exchange of "partner" ions between two salts.

**Electrical Balance.**  The condition of a net ionic equation wherein the algebraic sum of the electrical charges shown on the reactants equals the algebraic sum of those shown on the products. See **Net Ionic Equation**.

**Electrode.**  A metal object, usually a wire, suspended in an electrical conducting medium through which electricity passes to or from an external circuit.

**Electrolysis.**  A procedure in which an electrical current is passed through a solution containing ions, or through a molten salt, for the purpose of bringing about a chemical change.

**Electrolyte.** Any substance whose solution in water will conduct an electric current; in some contexts, the solution itself.

**Equivalent.** The molar mass of an ion divided by its electrical charge.

**Equivalent Weight.** The mass in grams of one equivalent of a chemical species.

**Hydronium Ion.** $H_3O^+$.

**Hydroxide Ion.** $OH^-$.

**Ionic Equation.** A chemical equation that explicitly shows all of the particles - ions, atoms, or molecules - that are involved in a reaction even if some are only spectator particles.

**Ionization.** A reaction, usually involving solvent molecules, whereby molecules are broken apart into ions.

**Material Balance.** The condition of an equation wherein all of the atoms present among the formulas of the reactants appear again in identical numbers among the formulas of the products.

**Milliequivalent.** 1/1000th of an equivalent.

**Molecular Equation.** An equation that shows the complete formulas of all of the substances present in a mixture undergoing a reaction.

**Monoprotic.** Having one proton per molecule that is capable of neutralizing a base. (For example, the monoprotic acid, H–Cl.)

**Net Ionic Equation.** A chemical equation in which all spectator particles have been omitted leaving only the particles (which can be ions or molecules) that actually participate. See **Electrical Balance** and **Material Balance**.

**Neutral Solution.** A solution in which the molar concentration of hydronium ions exactly equals the molar concentration of hydroxide ions.

**Nonelectrolyte.** Any substance that cannot furnish ions when dissolved in water or when melted.

**Salt.** Any crystalline substance consisting of a reasonably orderly aggregation of oppositely charged ions (other than $H^+$ or $OH^-$) occurring in a definite ratio.

**Simple Salt.** A salt made up of just one kind of cation and one kind of anion.

**Strong.** When applied to acids, bases, or electrolytes, strong describes a high percentage of ionization, or dissociation, in water. A strong Brønsted acid is any particle that has a strong tendency to release a proton; a strong Brønsted base is any particle that will bind an accepted proton strongly.

**Triprotic.** Having three protons per molecule that are capable of neutralizing a base. (For example, $H_3PO_4$.)

**Weak.**   When applied to acids, bases, or electrolytes, weak describes a low percentage of ionization, or dissociation, in water.  A weak Brønsted acid is one that has a weak tendency to release a proton; a weak Brønsted base is one that cannot accept and hold a proton very well.

## SELF-TESTING QUESTIONS

### Completion

1.   The self-ionization of water can be written by what equilibrium equation?

_____

2.   The three broad classes of substances that can liberate ions in water are

_____, _____, and _____.

3.   Negative ions are called _____ and positive ions are called

_____.

4.   When water is the solvent and its molecules react with molecules of hydrogen chloride to form hydronium ions and chloride ions, water is acting as a Brønsted _____ and hydrogen chloride is acting as a Brønsted _____.

5.   The hydronium ion has the formula _____; the hydrogen ion has the formula _____.

6.   After dissolving 100 g of the covalent compound HX in a liter of water, an average of 10% of the HX molecules were always in their intact molecular form and an average of 90% were split up into the ions $X^-$ and $H^+$ (actually, $H_3O^+$). HX may be classified as a _____ _____.
                                      (strong or weak)                    (acid, base, or salt)

What kind of an electrolyte is it? _____

7.  Give the chemical formulas for:

    (a)  sulfuric acid          _____

    (b)  carbonic acid          _____

    (c)  nitric acid            _____

    (d)  phosphoric acid        _____

    (e)  hydrochloric acid      _____

    (f)  ammonia                _____

8.  Write the names and formulas of the strong acids and bases.

### STRONG ACIDS

| NAME | FORMULA |
| --- | --- |
| _____ | _____ |
| _____ | _____ |
| _____ | _____ |
| _____ | _____ |
| _____ | _____ |

### STRONG BASES

| NAME | FORMULA |
| --- | --- |
| _____ | _____ |
| _____ | _____ |
| _____ | _____ |
| _____ | _____ |
| _____ | _____ |

9.  If HCl is a strong Brønsted acid, this must mean that the ion $Cl^-$ is a

    _____ binder of a proton and that $Cl^-$ is therefore a
    (good or poor)

    _____ Brønsted base.
    (weak or strong)

10. We can generalize from this.  The particle obtainable when any <u>strong</u> acid gives

    up $H^+$ is a relatively _____ Brønsted base.  The base obtainable from any

    <u>weak</u> acid is a relatively _____ Brønsted base.

11. When the water molecule loses $H^+$, the particle remaining is named the

    _____ and has the formula _____.

    It is a _____ Brønsted base, and water is a
    (strong or weak)

    _____ Brønsted acid.

12. Suppose we contrive to make a neutral particle, B, take a proton to form BH.
    What will be the net electrical charge on BH, if any?

    _____

13. If B holds the proton very weakly, would we classify $HB^+$ as a weak or a strong

    acid? _____

14. Which is the stronger Brønsted base, $PO_4^{3-}$ or $H_2PO_4^-$ _____;

    $NH_3$ or $NH_4^+$ _____?

15. When metals are arranged in an order corresponding to their relative tendencies

    to become ionic, we call this series the _____.

16. The most reactive metals are in Group _____ of the Periodic Table.

17. Write the net ionic equation for the reaction (if any) of

    (a)  calcium with nitric acid

    _____

(b)    potassium hydroxide with hydrochloric acid

_____

(c)    sodium chloride with sulfuric acid

_____

(d)    sodium bicarbonate with hydrochloric acid

_____

18.    Test your knowledge of the solubility rules for the salts given in the text by predicting whether each of the following would be soluble in water.  (Use a plus sign if soluble, a minus sign if insoluble.)

(a)    sodium carbonate        _____

(b)    calcium bromide        _____

(c)    potassium nitrate        _____

(d)    lead nitrate        _____

(e)    magnesium nitrate        _____

(f)    nickel sulfide        _____

(g)    copper(II) arsenate        _____

(h)    ammonium sulfate        _____

(i)    lead chloride        _____

(j)    sodium sulfide        _____

19.    Test your knowledge of the solubility rules, of the typical reactions of acids and bases, and of which acids and bases are strong by predicting the chemical reactions (if any) that would occur if you mixed together aqueous solutions of the following pairs.  Write net ionic equations for reactions you predict.  (If there is no reaction, write "none.")

(a)    $LiCl$ and $AgNO_3$        _____

(b)    $NaNO_3$ and $CaCl_2$        _____

(c)    $KOH$ and $H_2SO_4$        _____

(d)  $Pb(NO_3)_2$ and NaCl   _____

(e)  $K_2S$ and $CuSO_4$   _____

(f)  $Na_2CO_3$ and HBr   _____

20.  The equivalent weight of $Al^{3+}$ is _____. (The atomic weight of Al is 27.0.)

21.  A solution contained the following ions: $Na^+$, $Ca^{2+}$, $SO_4^{2-}$, and $Cl^-$. The concentrations were 25.0 meq $Na^+$/L, 12.0 meq $Ca^{2+}$/L, and 25.0 meq $SO_4^{2-}$/L. What was the concentration of $Cl^-$ in meq/L? _____

### Multiple-Choice

1.  When nitric acid reacts with sodium bicarbonate, which of the following is not formed?

(a)  $NaNO_3$    (b)  $CO_2$      (c)  Na      (d)  $H_2O$

2.  Which equation represents the neutralization of hydrochloric acid?

(a)  $HCl + AgNO_3 \longrightarrow AgCl + HNO_3$

(b)  $HCl + NaBr \longrightarrow NaCl + HBr$

(c)  $2HCl + Ca(NO_3)_2 \longrightarrow CaCl_2 + 2HNO_3$

(d)  $HCl + NaHCO_3 \longrightarrow NaCl + H_2O + CO_2$

3.  The net ionic equation for $Mg + 2HBr \longrightarrow MgBr_2 + H_2$ (assuming that $MgBr_2$ is a water-soluble salt) is

(a)  $Mg^{2+} + 2H^+ \longrightarrow Mg + H_2$

(b)  $Mg + 2H^+ \longrightarrow Mg^{2+} + H_2$

(c)  $Mg + 2HBr \longrightarrow MgBR_2 + 2H^+$

(d)  $Mg + 2HBr \longrightarrow Mg^{2+} + Br_2 + H_2$

4.  After a solution of potassium hydroxide has been carefully neutralized by hydrochloric acid, the solution contains (besides water)

(a)  $K^+$ and $Cl^-$                (c)  $K^+$ and $OH^-$

(b)  K and Cl                (d)  $K^-$, $H^+$, and $Cl_2$

5.  The covalent substance HA was soluble enough in water to make a very concentrated solution; however, whether concentrated or dilute, the percentage of ionization of HA into $A^-$ and $H^+$ (actually $H_3O^+$) was less than 5%. Another

covalent substance, HB, was virtually insoluble in water, but what did dissolve ionized by about 95% into B⁻ and H⁺. Which would be classified as the stronger acid?

(a)   the concentrated solution of HA

(b)   the dilute solution of HA

(c)   simply HA

(d)   HB

6.   When $PO_4^{3-}$ acts once as a Brønsted base, it changes to

(a)   $HPO_4^{2-}$      (b)   $H_2PO_4^-$      (c)   $H_3PO_4$      (d)   $H_4PO_4^+$

7.   The strongest base among the following is

(a)   $NO_3^-$      (b)   $HSO_4^-$      (c)   $CO_3^{2-}$      (d)   $I^-$

8.   Which of the following substances could be added to aqueous hydrobromic acid to reduce the concentration of hydronium ions?

(a)   silver nitrate                          (c)   ammonium chloride

(b)   ammonia                               (d)   potassium nitrate

9.   An example of a weak acid is

(a)   HBr         (b)   $HNO_3$         (c)   $H_2SO_4$         (d)   $HNO_2$

10.   If you mixed a solution of sodium bicarbonate with a solution of sulfuric acid, which particles would interact to form new products?

(a)   $SO_4^{3-}$ and $Na^+$                    (c)   $Na^+$ and $H_3O^+$

(b)   $HCO_3^-$ and $SO_4^{2-}$                 (d)   $H_3O^+$ and $HCO_3^-$

11.   If you mixed a solution of hydrochloric acid with ammonia water, which particles would be involved in the primary chemical change?

(a)   $NH_3$ and $Cl^-$                     (c)   $NH_3$ and $H_2O$

(b)   $NH_3$ and $H_3O^+$                 (d)   $NH_3$ and HCl (molecules)

12.   The hydronium ion would readily transfer a proton to which of the following?

(a)   $PO_4^{3-}$       (b)   $CO_3^{2-}$       (c)   $OH^-$       (d)   $NO_3^-$

13.   The reaction of hydrochloric acid with sodium hydroxide is called

(a)   hydrolysis                             (c)   neutralization

(b)   crenation                              (d)   proton transfer

14.   Reactions involving acids, bases, or salts can be predicted to take place if there is the possibility of forming

(a)   an insoluble precipitate             (c)   a gas

(b)   a strong acid                         (d)   an un-ionized (but soluble) substance

15. The atomic weight of oxygen is 16.0. Therefore the equivalent weight of $O^{2-}$ is

    (a)  16.0 g/eq                (c)  2.0 g/eq

    (b)  32.0 g/eq                (d)  8.0 g/eq

# ANSWERS

## ANSWERS TO SELF-TESTING QUESTIONS

**Completion**

1. $2H_2O(l) \rightleftharpoons H_3O^+(aq) + OH^-(aq)$
2. acids, bases, salts (any order)
3. anions; cations
4. base, acid
5. $H_3O^+$; $H^+$
6. strong acid; strong electrolyte
7. (a) $H_2SO_4$     (b) $H_2CO_3$     (c) $HNO_3$
    (d) $H_3PO_4$     (e) HCl       (f) $NH_3$
8.

| Strong Acids | Strong Bases |
|---|---|
| hydrochloric acid HCl | sodium hydroxide NaOH |
| hydrobromic acid HBr | potassium hydroxide KOH |
| hydriodic acid HI | magnesium hydroxide $Mg(OH)_2$ |
| nitric acid $HNO_3$ | calcium hydroxide $Ca(OH)_2$ |
| sulfuric acid $H_2SO_4$ | |

9. poor, weak
10. weak; strong
11. hydroxide ion, $OH^-$; strong; weak
12. +1 (as in $BH^+$)
13. strong
14. $PO_4^{3-}$; $NH_3$
15. activity series
16. IA
17. (a) $Ca + 2H^+ \longrightarrow Ca^{2+} + H_2$ (or $Ca + 2H_3O^+ \longrightarrow Ca^{2+} + H_2 + 2H_2O$)
    (b) $OH^- + H^+ \longrightarrow H_2O$ (or $OH^- + H_3O^+ \longrightarrow 2H_2O$)
    (c) none
    (d) $HCO_3^- + H^+ \longrightarrow H_2O + CO_2$ (or $HCO_3^- + H_3O^+ \longrightarrow 2H_2O + CO_2$)

18. (a) +      (b) +      (c) +      (d) +      (e) +
   (f) –      (g) –      (h) +      (i) –      (j) +

19. (a) $Cl^- + Ag^+ \longrightarrow AgCl$
    (b) none
    (c) $OH^- + H^+ \longrightarrow H_2O$
    (d) $Pb^{2+} + 2Cl^- \longrightarrow PbCl_2$
    (e) $S^{2-} + Cu^{2+} \longrightarrow CuS$
    (f) $CO_3^{2-} + 2H^+ \longrightarrow H_2O + CO_2$

20. 9.0 g/eq

21. 12 meq/L

## Multiple Choice

1. c
2. d (In the three other choices the concentration of hydronium ions does not change. Therefore, regardless of the fate of the chloride ion the acid has not been neutralized.)
3. b
4. a
5. d
6. a
7. c
8. b
9. d
10. d
11. b
12. a, b, and c
13. c and d
14. a, c, and d
15. d

# CHAPTER 8

# ACIDITY: DETECTION, CONTROL, MEASUREMENT

## OBJECTIVES

Broadly speaking, the most important topics in this chapter are:

| | |
|---|---|
| The pH concept | Because we use pH to describe the relative acidities of body fluids. |
| Substances that affect the pH of a solution of ions other than $H_3O^+$ and $OH^-$ | Because anything that can affect the pH of a solution must be taken seriously when dealing with life at the molecular level. |
| Dissolved substances that can stabilize or buffer the pH of a solution | Because the carbonate buffer in the blood is vital to health. |

| How ventilation responds to the $CO_2$ level in the blood | Because expelling or retaining $CO_2$ helps to regulate the pH of the blood. |

After you have studied this chapter and have worked the Practice and Review Exercises, you should be able to do the following.

1. *Write the equation of $K_w$ and use it to calculate a value of $[H^+]$ from a value of $[OH^-]$ or vice versa.*

2. *Relate $[H^+]$ to pH.*

3. *State the range of pH values that corresponds to an acidic solution and the range that corresponds to a basic solution.*

4. *Describe the purpose of an indicator.*

5. *Give the color of phenolphthalein and litmus in an acid and in a base.*

6. *Use the facts of which cations and anions hydrolyze to predict whether an aqueous solution of a given salt will test acidic, basic, or neutral.*

7. *State what is meant by the expression, "this aqueous solution is buffered."*

8. *Write the net ionic equations that show how a carbonate buffer works; do the same for a phosphate buffer.*

9. *Explain how the instability of carbonic acid helps it function in preventing a major upset in the acid-base balance of blood and write the equations for all of the relevant equilibria.*

10. *Explain how the process of ventilation works together with the carbonate buffer to regulate the pH of the blood.*

11. *Explain why involuntary hypoventilation leads to acidosis.*

12. *Explain why involuntary hyperventilation leads to alkalosis.*

13. *Explain the difference between "neutralizing capacity" and "pH."*

14. *Describe what takes place when a titration is carried out.*

15. *Describe the difference between "end point" and "equivalence point."*

16. *Calculate how much acid (or base) should be weighed out to prepare a specific volume of a solution that is to have a certain molarity.*

17. *Calculate molarities of "unknown" solutions from titration data and the equation for the reaction.*

18. *Define all of the terms listed in the Glossary.*

# GLOSSARY

**Acid-Base Indicator.**   See **Indicator**.

**Acidic Solution.**   One in which pH < 7.00 at room temperature.

**Acidosis.**   An increasing acidity of the blood.

**Alkalosis.**   An increasing basicity of the blood.

**Basic Solution.**   One in which pH > 7.00 at room temperature.

**Buffer.**   A combination of solutes in an aqueous solution that holds the pH relatively constant even when small amounts of acids or bases are added.

**Carbonate Buffer.**   A mixture or solution that includes bicarbonate ions ($HCO_3^-$) - which can neutralize hydrogen ions - and carbonic acid ($H_2CO_3$) - which can neutralize hydroxide ions.

**End Point.**   In a titration, the stage of the operation when you stop the titration.

**Equivalence Point.**   In an acid base titration, the stage in the operation at which the moles of hydrogen ions from the acid equals the moles of base.

**Hydrolysis of Ions.**   Any reaction of a solute ion with water in which a metallic ion (other than $H^+$) or a nonmetallic ion (other than $OH^-$) changes $[H^+]$ relative to $[OH^-]$ in an aqueous solution.

**Hyperventilation.**   Breathing (ventilating) considerably faster and deeper than is usual.

**Hypoventilation.**   Breathing (ventilating) less deeply and at a much slower rate than is normal; shallow breathing.

**Indicator.**   A dye that has one color in an acidic medium and a different color in a basic medium.  Litmus and phenolphthalein are examples.

**Ion-Product Constant of Water ($K_w$).**   $K_w = [H^+][OH^-]$ where the brackets, [ ], refer to molar concentrations.  (At 25 °C, $K_w = 1.00 \times 10^{-14}$.)

**$K_w$.**   See **Ion-Product Constant of Water**.

**Neutralizing Capacity.**   The capacity of a solution or a substance to neutralize an acid or a base; expressed as a molar concentration.

**Neutral Solution.**   A solution in which pH = 7.00 at room temperature.

**pH.**   pH is the negative power to which the base 10 must be raised to express the concentration of hydrogen ions in an aqueous solution in moles per liter.  The defining equation is: $[H^+] = 1 \times 10^{-pH}$.

**Phosphate Buffer.**   A mixture or solution containing dihydrogen phosphate ions ($H_2PO_4^-$) - which can neutralize hydroxide ions - and monohydrogen phosphate ions ($HPO_4^{2-}$) - which can neutralize hydrogen ions.

**Respiration.**   The entire physical and chemical apparatus for bringing oxygen into the living system, using it, and removing carbon dioxide.

**Standard Solution.**   Any solution in which the concentration of the solute is accurately known.

**Titration.**   The measured addition of a standard solution to an "unknown"; a laboratory procedure for determining the concentration of an acid or a base (or some other solute in other situations) in a solution.

**Ventilation.**   The circulation of air into and out of the lungs.


## SELF-TESTING QUESTIONS


### Completion

1.  The pH of a solution for which $[H^+] = 1.00 \times 10^{-5}$ is _____.

2.  A solution with a pH of 6.50 is _____.
    (acidic, basic, neutral)

3.  The concentration of hydroxide ions in pure water at room temperature is (in moles per liter) _____.

4.  If the pH of a solution is 7.42, then the concentration of hydrogen ions (in moles per liter) is between $1.0 \times 10^{-7}$ and $1.0 \times$ _____.

5.  If $[OH^-] = 4.0 \times 10^{-9}$, the solution is _____.
    (acidic or basic)

6.  If $K_2CO_3$ were added to water:

    (a)   Which ions would it liberate in the solution? (Formulas) _____
          and _____

    (b)   Could either ion react with water to produce $OH^-$? _____
          If yes, which one? (Formula) _____

    (c)   Could either ion react with water to produce $H^+$? _____
          If yes, which one? (Formula) _____

(d)   In what way will $K_2CO_3$ change the pH - leave it the same, lower it, or raise it? _____

7.   Will the addition of $Na_3PO_4$ to water make the solution more basic, more acidic, or will it not affect the pH? _____

8.   Write the ionic equation for the buffering of an acid by the carbonate buffer.

_____

9.   Write the ionic equation for the buffering of a base by the carbonate buffer.

_____

10.   If too much acid invades a solution buffered by the carbonate buffer, the buffer will be destroyed as ____  ____ breaks down to _____ (which leaves the solution) and water.

11.   At a considerable sacrifice in accuracy, parts (a) through (d) ignore the fact that the following equilibrium and many others exist in the solution described in part

$$PO_4^{-3} + H_2PO_4^- \rightleftharpoons 2HPO_4^{2-}$$

(a)   If you had a solution containing both $NaH_2PO_4$ and $Na_3PO_4$ as solutes, would the pH of that solution be less than or greater than 7? _____

(b)   Whatever the pH is, would the solution be buffered? _____

(c)   If so, what ionic equation represents the reaction that will occur if some acid is added? _____

(d)   If so, what ionic equation represents the reaction that will occur if some alkali is added? _____

12.   The general process whereby air is moved into and out of the body is called _____. When its rate increases we say that _____ occurs.

13.   Involuntary hyperventilation can occur when someone experiences _____ or _____ and it results in the excessive loss of _____ from the blood and causes a blood pH condition
                (formula)

called _____.

14. Involuntary hypoventilation in people suffering from _____ means the
retention of _____ by the blood and causes a blood pH condition
         (formula)
called _____.

15. To measure the value of the pH of a solution by any technique involves determining the value of the molar concentration of a _____.
                                                    (symbol)

To measure the neutralizing capacity of a solution we have to use a technique
called _____.

16. An aqueous solution of a weak acid has a molar concentration of acid that is
considerably _____ than the molar concentration of
              (smaller or larger)
hydronium ion.

17. A trace of dye called in general an _____ is used in an acid-base titration
to tell the _____ of the operation. The names of two common dyes used for
this purpose are _____ and _____.

18. If you titrate acetic acid with sodium hydroxide, what salt is present at the
equivalence point? (Name) _____

19. Would the pH of this solution (question 18) be less than 7, greater than 7, or
exactly equal to 7? _____

20. What is the molarity of $H_2SO_4$(aq) if 42.0 mL of it exactly neutralize 46.0 mL of
a 0.100 M NaOH(aq)? _____

**Multiple-Choice**

1. The ion-product constant for water is given by the equation

   (a) $K_w = \dfrac{[H^+][OH^-]}{[H_2O]}$
   
   (c) $K_w = [H^+][OH^-]$

   (b) $K_w = \dfrac{[H_3O^+][OH^-]}{[H_2O]^2}$
   
   (d) $K_w = [\log pH][\log pOH]$

2. If ventilation changes in such a way that $CO_2$ is retained in the blood,
   (a) the blood's pH decreases
   (b) involuntary hyperventilation occurs
   (c) alkalosis develops
   (d) hydrogen ion is excessively neutralized

3. Which of the following make an acidic solution when added to pure water?
   (a) $H_2SO_4$      (c) $NaCl$      (e) $AlCl_3$
   (b) $NH_3$      (d) $Na_2CO_3$      (f) $Na_2HPO_4$

4. Which of the choices in question 3 make a basic solution when added to pure water?

5. Dissolving $NaOH$ in water produces a solution that
   (a) turns blue litmus red
   (b) turns phenolphthalein red
   (c) has a tart taste
   (d) turns red litmus blue

6. The pH of a solution is 6.00. Hence,
   (a) it is more acidic than a solution whose pH is 4.00
   (b) its concentration of hydrogen ions is $1.00 \times 10^{-8}$ moles per liter
   (c) its concentration of hydrogen ions is $1.00 \times 10^{-6}$ moles per liter
   (d) its concentration of hydrogen ions is $1.00 \times 10^6$ moles per liter

7. If the pH of an aqueous solution cannot be changed by adding small amounts of a strong acid or a strong base, the solution contains
   (a) an indicator
   (b) a buffer
   (c) a protective colloid
   (d) a strong acid and a strong base already

8.  The pH of a solution with a concentration of 0.001 M nitric acid is closest to

    (a)  $10^{-3}$        (b)  $10^{-11}$        (c)  3        (d)  11

9.  If a 0.10 M solution of the monoprotic acid, HX, is known to have a pH of 4.50, we know that the acid HX is

    (a)  strong        (b)  weak        (c)  neutral        (d)  slightly soluble

10. The end point in a titration of any acid by any base is reached when (pick the one best choice)

    (a)  the pH of the solution is 7.00 (at room temperature)

    (b)  the equivalence point is reached

    (c)  the indicator is halfway through its characteristic change in color

    (d)  an equal volume of acid has been mixed with an equal volume of base

11. When the system's respiration and ventilation is working and acidosis develops, then

    (a)  the pH of the blood is tending to decrease

    (b)  the $[OH^-]$ of blood is increasing

    (c)  hyperthermia also occurs

    (d)  hyperventilation also occurs

## ANSWERS

### ANSWERS TO SELF-TESTING QUESTIONS

**Completion**

1.  5.00
2.  acidic
3.  $1.00 \times 10^{-7}$ mol/L
4.  $10^{-8}$
5.  acidic
6.  (a)  $K^+$, $CO_3^{2-}$
    (b)  Yes; $CO_3^{2-}$
    (c)  No
    (d)  Raise it. (If some of the $CO_3^{2-}$ ions manage to take $H^+$ ions from molecules of water, then there will be a slight excess of $OH^-$ ions; the solution will have become slightly basic and, therefore, the pH will have gone up.)
7.  more basic

8.  $HCO_3^- + H^+ \longrightarrow H_2CO_3$
9.  $H_2CO_3 + OH^- \longrightarrow HCO_3^- + H_2O$
10. carbonic acid ($H_2CO_3$); carbon dioxide ($CO_2$)
11. (a)  greater than 7
    (b)  Yes
    (c)  The most likely event for neutralizing $H^+$ is $H^+ + PO_4^{3-} \longrightarrow HPO_4^{2-}$
    (d)  The most likely event for neutralizing $OH^-$ is
         $$OH^- + H_2PO_4^- \longrightarrow HPO_4^{2-} + H_2O$$
12. ventilation; hyperventilation
13. hysterics or overbreathing at high altitude; $CO_2$; alkalosis
14. emphysema; $CO_2$; acidosis
15. $H_3O^+$ (writing $H^+$ is all right, too); titration
16. larger
17. indicator; end point; litmus and phenolphthalein
18. sodium acetate
19. greater than 7 (Sodium acetate hydrolyzes in water.)
20. 0.0548 $M$

**Multiple-Choice**

1.  c
2.  a
3.  a and e
4.  b, d, and f
5.  b and d
6.  c
7.  b
8.  c  (Since $HNO_3$ is a strong acid we may assume that it is 100% ionized and that 0.001 M = [$H^+$]. Hence, pH = 3 because $0.001 = 10^{-3}$.)
9.  b  (If it were strong, its pH would be much closer to 1 since $0.10 = 1.0 \times 10^{-1}$.)
10. c
11. a and d

# CHAPTER 9

# INTRODUCTION TO ORGANIC CHEMISTRY

## OBJECTIVES

After you have studied this chapter and worked the Exercises and Review Questions in it, you should be able to do the following. Objectives 6 through 9 are particularly important as they concern the symbols we shall use in later work with organic compounds.

1. *State what the vital force theory was and how it affected research in its day.*

2. *Describe Wöhler's experiment and explain how it helped to overthrow the vital force theory.*

3. *List the main differences between organic and inorganic compounds.*

4. *Give the ways in which carbon is a unique element.*

5. *State how a molecular formula and a structural formula are alike and how they are different.*

6. *Explain how each possible conformation of a carbon chain does not represent a different compound.*

7. *Give an example (the names and structures) of two compounds that are related as isomers.*

8. *Write condensed structures from full structures.*

9. *Examine a pair of structures and tell if they are identical, related as isomers, or different.*

10. *Describe the molecular geometry of a carbon atom bonded to four other atoms.*

11. *Give definitions for each of the terms listed in the Glossary.*

## GLOSSARY

**Branched Chain.**  A sequence of carbon atoms in which additional carbon atoms are attached at points other than the ends.

**Condensed Structure.**  See **Structure**.

**Free Rotation.**  The lack of a barrier to the rotation of two groups with respect to each other when they are joined by a single bond in an open-chain molecule.

**Functional Group.**  An atom or group of atoms in an organic molecule that is responsible for a set of chemical reactions characteristic of all molecules having such a group.

**Inorganic Compound.**  Any compound that is not an organic compound.

**Isomer.**  One of a set of compounds having identical molecular formulas but different structures.

**Isomerism.**  The existence of two or more compounds with identical molecular formulas but different structures.

**Nonfunctional Group.**  The part of an organic molecule that remains unchanged in structure as a chemical reaction occurs at a functional group.  See **Functional Group**.

**Ring.**  A feature of molecular structure in which several atoms are covalently connected to form a cyclic structure.

**Straight Chain.**  A continuous sequence of covalently bound carbon atoms from which no additional carbon atoms are attached at interior locations of the sequence.

**Structure.** In chemistry, a synonym for structural formula; the nucleus-to-nucleus sequence within an individual molecule.

**Organic Compounds.** Compounds of carbon other than those related to carbonic acid and its salts, or to the oxides of carbon, or to the cyanides.

## DRILL EXERCISES

### I.    EXERCISES ON CONDENSED STRUCTURES

The rules for converting from a full structure to a condensed structure are:

1.    $H-\overset{\displaystyle H}{\underset{\displaystyle H}{\overset{|}{\underset{|}{C}}}}-$    becomes $CH_3$ (sometimes $H_3C$)

2.    $H-\overset{|}{\underset{|}{C}}-H$  or    $-\overset{\displaystyle H}{\underset{\displaystyle H}{\overset{|}{\underset{|}{C}}}}-$    becomes $CH_2$ (sometimes $H_2C$)

3.    $H-\overset{|}{\underset{|}{C}}-$  or  $-\overset{\displaystyle H}{\overset{|}{\underset{|}{C}}}-$  becomes $CH$

4.    All bonds of hydrogen to carbon, oxygen, nitrogen, or sulfur can be "understood."

5.    Any single bond between heavy atoms (C–C, C–O, C–N, C–S, for example) can be shown or it can be understood, provided that it would normally be drawn horizontally and, therefore, is part of the main chain.

6.    Any single bond between heavy atoms that hold substituents onto the main chain must be shown. (In other words, we'll always show a substituent group attached to a main chain by writing the group either above the chain or below it, and by connecting the group to the main chain by a line representing the bond.)

7.    All double and triple bonds must be shown. (A modification of this rule will appear in a later chapter when we study carbonyl compounds.)

Practice converting between full and condensed structures with these additional exercises.

**SET A.** Condense each of the following structures.

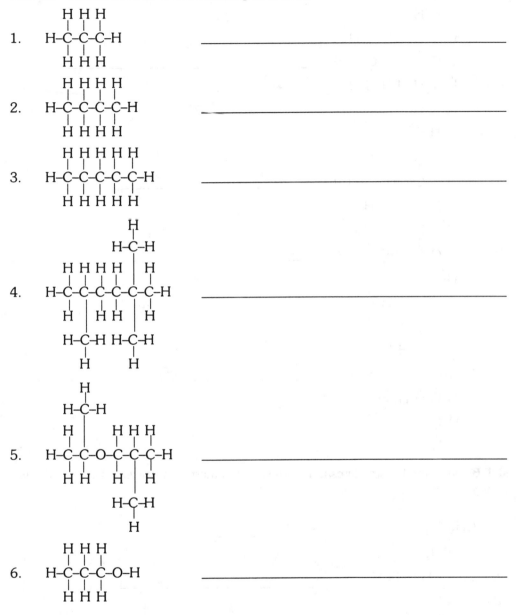

1. _____

2. _____

3. _____

4. _____

5. _____

6. _____

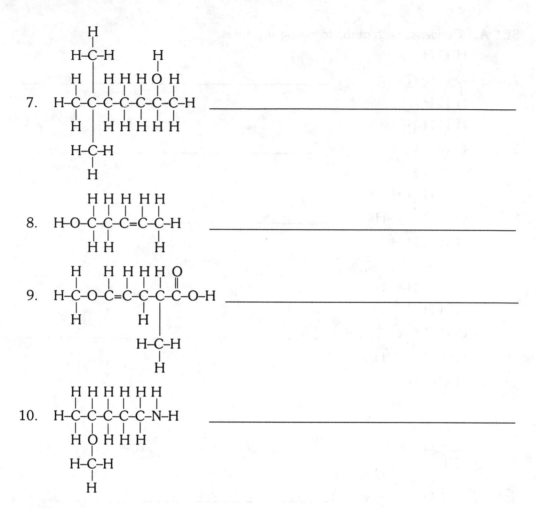

7.

8.

9.

10.

**SET B**. Write the full structures that correspond to these condensed structures.  Show all single bonds.

1.   $CH_3CH_3$ _____

2.   $CH_3CCH_2CH_2CH_3$ with $CH_3$ above and $CH_3$ below   _____

3.   $CH_3CH_2CH_2CH_2CCH_2CH_2CHCH_2CH_2CH_2CH_3$

with substituents:

$\overset{\displaystyle CH_3}{|}$ and $\overset{\displaystyle CH_2CH_3}{|}$ on the main chain,

and $CH_3CCH_2CH_3$

$\overset{\displaystyle |}{CH_3}$

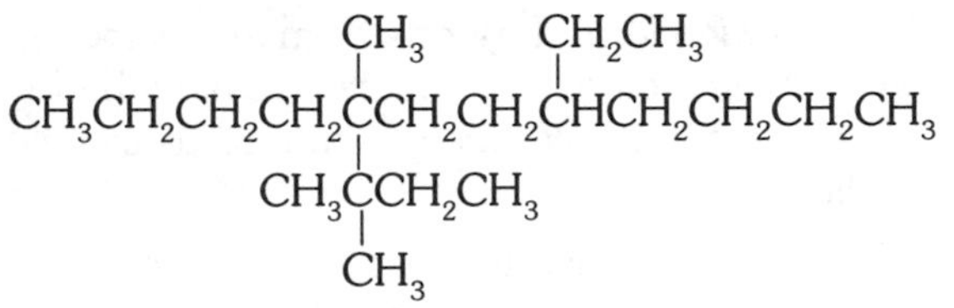

_____

4.   $CH_3CH=CH-CH_3$

_____

5.   $HOCH_2CH_3$        _____

6. 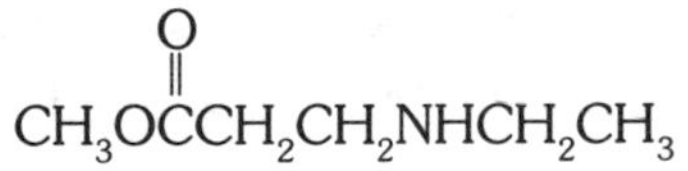   $CH_3O\overset{\displaystyle O}{\overset{\|}{C}}CH_2CH_2NHCH_2CH_3$

_____

7. 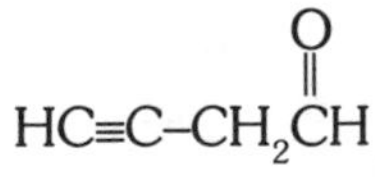   $HC{\equiv}C-CH_2\overset{\displaystyle O}{\overset{\|}{C}}H$        _____

8.   $NH_2CH_2CH_2NH_2$

_____

## II.    EXERCISES ON DEVISING STRUCTURES FROM MOLECULAR FORMULAS

Review Exercise 9.12 in Chapter 9 in the text ask you to devise a structural formula from each molecular formula given.  An answer cannot be correct unless each and every atom has exactly the correct number of bonds going from it - four from carbon, three from nitrogen, two from oxygen (or sulfur), one from any halogen, and one from hydrogen.  If the question in the text gives you trouble, see how the following semi-systematic approach works.  The number of rules may seem a bit awesome, but you will quickly advance beyond needing them.

Rule 1.    Identify those atoms in the given molecular formula that must have two, three, or four bonds - all of the atoms other than a halogen or hydrogen. (We'll work with these to establish a "skeleton" for the structure upon which the hydrogen and halogen atoms will be hung.  Think of each covalence of every atom (4 for carbon, etc.) as an unused arm.  A correct structure is one in which all the arms are in some way linked.  There can be "no unjoined arms" could be another way of stating it.)

Rule 2.    The number of unused valences ("free arms") left over on the skeleton - *all being single bonds* - must equal the number of hydrogens or halogens to be appended.

Rule 3.    Do not put a halogen on anything but carbon - not on oxygen, not on sulfur, and not on nitrogen.  (Although such possibilities do exist, we'll never encounter them.)

Rule 4.    In writing a structure, assemble the carbon atoms along a horizontal line as much as possible and pin the substituents to the resulting chain of carbon atoms.  (This rule has nothing to do with rules of valence, only with looks.)

When atoms have more than one bond, options exist as to how they may occur in structures.  All of the bonds may be single, or some may be single and some double or even triple.  Here are all the possible options for carbon, nitrogen, oxygen, the halogens, and hydrogen.

| | | | |
|---|---|---|---|
| For carbon | $-\overset{\textstyle|}{\underset{\textstyle|}{C}}-$ | $\diagdown C=$ | $-C\equiv$ |
| For nitrogen | $-\overset{\textstyle|}{N}-$ | $-N=$ | $N\equiv$ |
| For oxygen | $-O-$ | $O=$ | |
| For halogen | $-X$ (X = F, Cl, Br, or I) | | |
| For hydrogen | $-H$ | | |

Now let's work a few sample exercises.

**Sample Exercise 1:**  Write a structural formula for $CH_4O$.

Step 1.    The multivalent atoms are C and O.  (Rule 1)

Step 2.    Assemble the options:

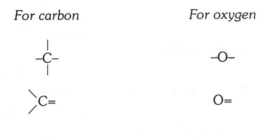

Step 3.    Try the possible combinations of those in the first column with those in the second.

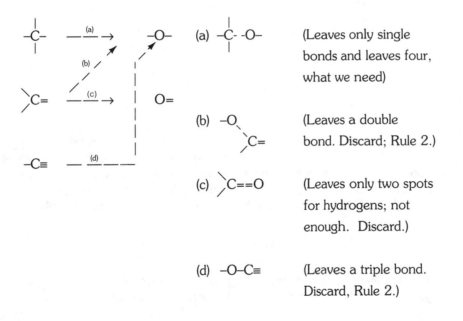

(a) –C̣- -O–   (Leaves only single bonds and leaves four, what we need)

(b) –O̦ C=   (Leaves a double bond. Discard; Rule 2.)

(c) C==O   (Leaves only two spots for hydrogens; not enough.  Discard.)

(d) –O–C≡   (Leaves a triple bond. Discard, Rule 2.)

Step 4.    Identify the skeleton, or skeletons, (rules 1 and 2) and append the hydrogens.

In this sample exercise only one skeleton meets the rules, the one generated by combination (a).

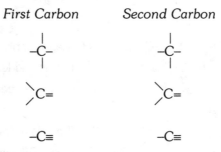

(the answer)

We may condense the answer to $CH_3OH$.

Step 5.    Run a mental check; does the carbon have 4 bonds; the oxygen, 2; and the hydrogen 1 each?  They do; therefore, the structure obeys the rules of covalence.

**Sample Exercise 2:**   Convert $C_2H_4$ to a structural formula.

Step 1.    The multivalent atoms are just the two Cs.

Step 2.    Assemble the options:

| First Carbon | Second Carbon |
|:---:|:---:|
| $-\overset{\mid}{\underset{\mid}{C}}-$ | $-\overset{\mid}{\underset{\mid}{C}}-$ |
| $\overset{\backslash}{\underset{/}{C}}=$ | $\overset{\backslash}{\underset{/}{C}}=$ |
| $-C\equiv$ | $-C\equiv$ |

Step 3.    Try combinations from each column.  Remember that the remaining unused bonds must equal four in number, for the four hydrogens, and that these remaining bonds must all be single bonds.  Mentally discarding all combinations that leave double or triple bonds "open," we have these possibilities:

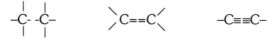

The first leaves room for six hydrogens, too many; the last for two hydrogens, too few.  The middle will work; room is left for four hydrogens

Step 4.    Put the pieces together:   $\overset{\backslash}{\underset{/}{C}}{=}{=}\overset{/}{\underset{\backslash}{C}}$  + 4H–  $\longrightarrow$  $\overset{H}{\underset{H}{}}\overset{\backslash}{\underset{/}{C}}{=}\overset{/}{\underset{\backslash}{C}}\overset{H}{\underset{H}{}}$

The condensed structure is: $CH_2{=}CH_2$

Step 5.    Mentally check for obedience to the rules of covalence. Each carbon has four bonds; each hydrogen has one.

**Sample Exercise 3:**   Convert $C_2H_4O$ into a structural formula that obeys the rules of covalence.

Step 1.    The multivalent atoms are two carbons and one oxygen.

Step 2.    Assemble the options:

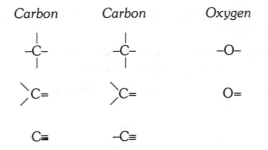

Step 3.    Try the possible combinations, remembering that the skeleton must have no double or triple bond left "hanging" and must have, via single bonds only, room for four hydrogens.  Here are the possibilities that have only single bonds remaining. (Beneath each is a number equalling the spaces for hydrogens.)

$$-\overset{|}{\underset{|}{C}}\text{- -}\overset{|}{\underset{|}{C}}\text{- -O-} \qquad -\overset{|}{\underset{|}{C}}\text{- -O- -}\overset{|}{\underset{|}{C}}\text{-} \qquad \overset{|}{\underset{/}{\diagdown}}C\text{==}\overset{|}{C}\text{- -O-} \qquad -\overset{|}{\underset{|}{C}}\text{- -C==O}$$

$$\quad\ (6) \qquad\qquad\ (6) \qquad\qquad\ (4) \qquad\qquad (4)$$

Any others you may have drawn, that avoid leaving open double or triple bonds, will be duplicates of one or more of these. (For instance,

$$-O\text{- -}\overset{|}{\underset{|}{C}}\text{- -}\overset{|}{\underset{|}{C}}\text{-} \text{ is the same as } -\overset{|}{\underset{|}{C}}\text{- -}\overset{|}{\underset{|}{C}}\text{- -O-;} \quad O\text{==}\overset{|}{C}\text{- -}\overset{|}{\underset{|}{C}}\text{- is the same as}$$

$$-\overset{|}{\underset{|}{C}}\text{- -}\overset{|}{C}\text{==O.)} \text{ The last two combinations have four spaces for hydrogens.}$$

We must consider them both.

Step 4.    Put the pieces together:

$$\overset{|}{\underset{/}{\diagdown}}C\text{==}\overset{|}{C}\text{- -O- } + \ 4H\text{-} \ \longrightarrow \ \overset{H}{\underset{\underset{H}{|}}{\overset{\diagdown}{C}}}\text{=}\overset{\overset{H}{|}}{C}\text{-O-H} \text{ or } CH_2\text{=CHOH}$$

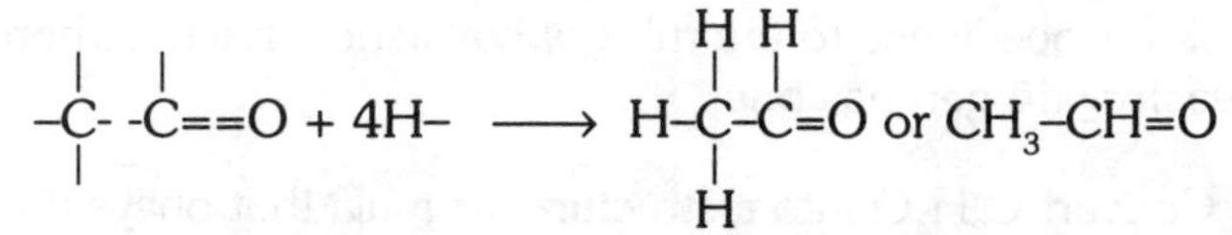

$$\underset{|}{\overset{|}{-C}} - \overset{|}{\underset{|}{C}}==O + 4H- \longrightarrow \overset{H\ H}{\underset{H}{H-\overset{|}{C}-\overset{|}{C}}=O} \text{ or } CH_3-CH=O$$

Usually, you will see carbon-oxygen double bonds aligned vertically, and the last structure will usually be seen as

$$CH_3-\overset{\overset{O}{\|}}{C}-H.$$ (Remember that not all of the single bonds in the main chain need to be "understood.")

Step 5.     Check the answer: four bonds from the carbons; two from the oxygens. Either structure is correct; the two are isomers.

Write structural formulas for the following molecular formulas.

1.     $H_2O_2$     _____

2.     $N_2H_4$     _____

3.     $CH_4S$     _____

4.     $C_2H_5Br$     _____

5.     $C_2H_3Cl$     _____


## SELF-TESTING QUESTIONS

### Completion

1.     If compound X boils at 176 °C, it is almost certainly in the family of

_____ compounds.  We would assign it to
(ionic or molecular)

the *organic* class if, when analyzed, it was found to contain _____.

2.    The structure

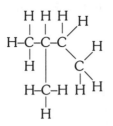

could be most neatly condensed to _____.

3.    The structure

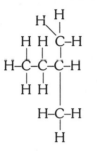

could be most neatly condensed to _____.

4.    Are the compounds whose formulas are given in questions 2 and 3 *isomers* or
are they *identical*? _____

5.    Write full structural formulas for each of the following:


_____    _____    _____    _____

(a)  $CH_5N$         (b)  $CH_2Br_2$      (c)  $C_3H_8$       (d)  $C_3H_4$

6.    Write the condensed structure of at least one isomer of

$CH_3CH_2CH_2OH$ ($C_3H_8O$).


_____

**Multiple-Choice**

1.    The compounds in which of the following pairs are related as isomers?

(a)    $CH_2{=}CH{-}CH_2{-}OH$  and  $\overset{\overset{\displaystyle HO}{\displaystyle |}}{CH_2}{-}CH{=}CH_2$

(b)    $H_2$  and  $D_2$

(c)    $CH_3{-}CH_2{-}NH{-}CH_3$  and  $CH_3{-}CH_2{-}CH_2{-}NH_2$

(d)    $CH_3{-}C{\equiv}C{-}CH_2NH_2$  and  $CH_3{-}CH_2{-}CH_2{-}C{\equiv}N$

2.    According to all the rules, which compound(s) should *not* be possible?

(a)    $AlPO_4$

(b)    $CH_3CH{=}C{=}CH_2$

(c)    $H{-}O{-}\overset{\overset{\displaystyle O}{\displaystyle \|}}{C}{-}O{-}H$

(d)    $CH{=}NH$

3.    A structural formula for $C_5H_{12}$ would be

(a)    $CH_3(CH_2)_3CH_3$

(b)    $CH_3CH_2CH_2CH_2CH_3$

(c)    $CH_3\overset{\overset{\displaystyle }{\displaystyle |}}{\underset{\underset{\displaystyle CH_3}{\displaystyle |}}{C}}HCH_2CH_3$

(d)    $CH_3\overset{\overset{\displaystyle CH_3}{\displaystyle |}}{\underset{\underset{\displaystyle CH_3}{\displaystyle |}}{C}}CH_3$

4.    Which (if any) of the choices in question 3 are identical compounds?

5.    Which (if any) of the choices in question 3 are isomers?

# ANSWERS

## ANSWERS TO DRILL EXERCISES

### I.    Exercises on Condensed Structures

<u>Set A</u>

1.    $CH_3CH_2CH_3$ or $CH_3{-}CH_2{-}CH_3$ (Single bonds along the main chain may be shown or they may be understood.  You will see both practices often.)

2.    $CH_3CH_2CH_2CH_3$ or $CH_3{-}CH_2{-}CH_2{-}CH_3$

3.    $CH_3CH_2CH_2CH_2CH_3$ or $CH_3{-}CH_2{-}CH_2{-}CH_2{-}CH_3$

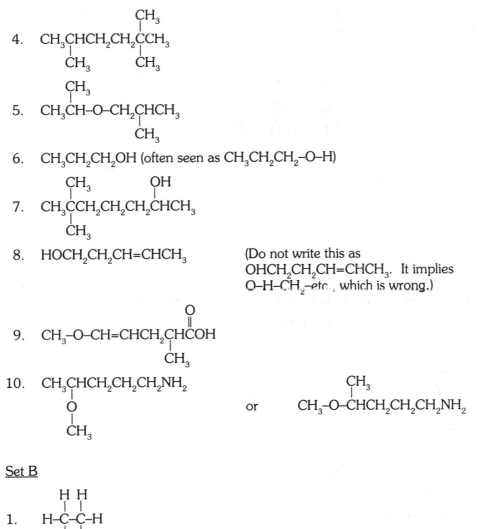

4.  $CH_3CHCH_2CH_2CCH_3$ with $CH_3$ groups on substituents

5.  $CH_3CH-O-CH_2CHCH_3$ with $CH_3$ substituents

6.  $CH_3CH_2CH_2OH$ (often seen as $CH_3CH_2CH_2-O-H$)

7.  $CH_3CCH_2CH_2CH_2CHCH_3$ with $CH_3$ and $OH$ substituents

8.  $HOCH_2CH_2CH=CHCH_3$    (Do not write this as
    $OHCH_2CH_2CH=CHCH_3$.  It implies
    $O-H-CH_2-etc.$, which is wrong.)

9.  $CH_3-O-CH=CHCH_2CHCOH$ with $O$ double bond and $CH_3$ substituent

10.  $CH_3CHCH_2CH_2CH_2NH_2$ with $O-CH_3$ substituent    or    $CH_3-O-CHCH_2CH_2CH_2NH_2$ with $CH_3$ substituent

Set B

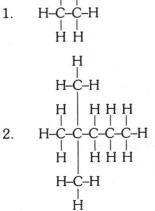

1.

2.

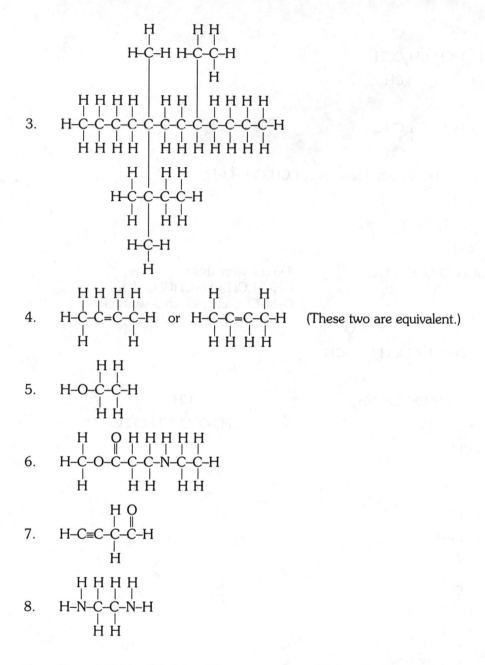

4. (These two are equivalent.)

## II.    Exercises on Devising Structures from Molecular Formulas

1.    H–O–O–H

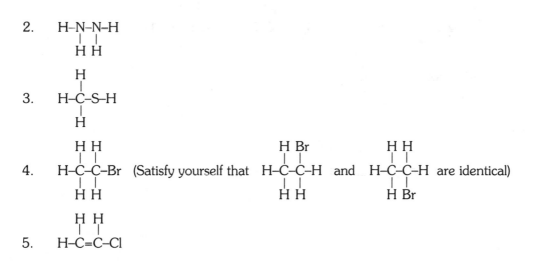

2.  H–N–N–H
       | |
       H H

       H
       |
3.  H–C–S–H
       |
       H

4.  H H
    | |
    H–C–C–Br   (Satisfy yourself that   H–C–C–H   and   H–C–C–H  are identical)
    | |
    H H

5.  H H
    | |
    H–C=C–Cl

## ANSWERS TO SELF-TESTING QUESTIONS

### Completion

1.  molecular; carbon

2.  $CH_3CHCH_2CH_3$
         |
        $CH_3$

3.         $CH_3$
           |
    $CH_3CH_2CHCH_3$

    (Your answers to 2 and 3 may be correct but still not look exactly like those given here.  All that matters is that they show the correct nucleus-to-nucleus sequence while obeying the accepted conventions for condensing.  One of these conventions is that as much of the structure as possible is written out on one line. Should you have questions about your answers, consult your instructor or one of the assistants.)

4.  identical

5.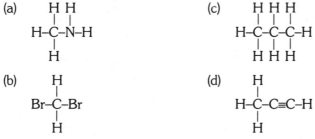

6.    There are two isomers:  $CH_3–O–CH_2CH_3$  and  $CH_3\underset{\underset{\displaystyle OH}{|}}{CH}CH_3$

**Multiple-Choice**

1.    c and d
2.    d
3.    a, b, c, and d
4.    a and b
5.    a (or b), c, and d are isomers.

# CHAPTER 10

# HYDROCARBONS

## OBJECTIVES

After you have studied this chapter, and worked the exercises in it, you should be able to do the following.

1.  Name the principal families of hydrocarbons, give an example of a member of each, and point out the structural features that put a compound into one of these families.

2.  List the sorts of physical properties we expect of a compound when its molecules are mostly or entirely hydrocarbonlike.

3.  Tell what it means for a compound to be saturated or unsaturated, aliphatic, or aromatic.

4.  Give names (and IUPAC) and structures for the $C_1$ to $C_{10}$ straight-chain alkanes. (Write condensed structures.)

5.  Give the structures and common names of the isomeric butanes.

6.   Give the structures and names of all of the alkyl groups having from one to four carbons.

7.   Write an IUPAC name for a given alkane or alkene.

8.   Give names and structures for cycloalkanes having rings from three to six carbons, and draw condensed symbols for their rings.

9.   State what happens when an alkane is mixed, at room temperature, with water, acids, or alkalies.

10.  Write the balanced equation for the complete combustion of any given hydrocarbon.

11.  Give structures and names for the $C_2$ to $C_4$ alkenes.

12.  Write structures that illustrate two alkenes related as geometric isomers.

13.  Describe what is true about a carbon-carbon double bond that makes geometric isomerism possible.

14.  Write equations that illustrate what happens when a simple alkene undergoes each of the following types of reactions. (a) the addition of hydrogen; (b) the addition of water; (c) polymerization.

15.  Write structures of products to be expected when the following reagents are mixed with any given hydrocarbon. (a) $H_2$ (with a catalyst, heat, and pressure); (b) $H_2O$ (with an acid catalyst). (In other words, be able to work problems like those in problems 10.41 and 10.42 in the text.)

16.  State Markovnikov's rule.

17.  Write an equation that illustrates the <u>kind</u> of reaction benzene undergoes.

18.  Define the terms in the Glossary.


## GLOSSARY

**Addition Reaction.**   In organic chemistry, a reaction in which a reagent adds to a double or a triple bond.

**Alkane.**   Any hydrocarbon, whether cyclic or open-chain, with only single bonds. A *normal alkane* is an open-chain alkane without any branches on its carbon chains - the straight-chain isomer.

**Alkene.**   An organic compound whose molecules have a carbon-carbon double bond.

**Alkyl Group.**   A substituent group that can be thought of as an alkane minus one hydrogen atom.

**Aromatic Compound.**   Any organic compound that contains the benzene ring system.

**Aliphatic Compound.**   An inorganic compound whose molecules lack a benzene ring or similar feature.

**Alkyne.**   A hydrocarbon with a triple bond.

**Butyl Group.**   $CH_3CH_2CH_2CH_2-$

**sec–Butyl Group.**   $CH_3CH_2CH-$
                                         |
                                       $CH_3$

**t-Butyl Group.**   $(CH_3)_3C-$

**Ethyl Group.**   $CH_3CH_2-$

**Geometric Isomerism.**   Isomerism caused by differences in the geometries of the isomers - differences that cannot be erased because of the lack of free rotation about either a double bond or about atoms joined in a ring.  Cis-trans isomerism.

**Geometric Isomers.**   Isomers whose molecules have identical atomic organizations but different geometries; cis-trans isomers.

**Heterocyclic Compounds.**   Compounds with rings that include atoms other than carbon.

**Homolog.**   In organic chemistry, any member of a homologous series.

**Homologous Series.**   In organic chemistry, a series of compounds in the same family whose members differ from each other by successive changes in the number of $CH_2$ groups in their molecules.

**Hydrocarbon.**   Any compound consisting entirely of carbon and hydrogen.

**International Union of Pure and Applied Chemistry System Rules (IUPAC Rules).**   A systematic set of rules for naming organic compounds designed to give each compound one unique name and for which only one structure is possible; the "Geneva system."

**Isobutyl Group.**   $(CH_3)_2CHCH_2-$

**Isopropyl Group.**   $(CH_3)_2CH-$

**Like-Dissolves-Like Rule.**   Polar solvents dissolve polar or ionic solutes; nonpolar solvents dissolve nonpolar or weakly polar solutes.

**Markovnikov's Rule.**   When an unsymmetrical reactant such as H–OH is added to an alkene having unequal numbers of H atoms at the ends of the double bond, the carbon with the greater number of H atoms gets one more.

**Methyl Group.**   $CH_3-$

**Monomer.**   A substance whose molecules can polymerize and be changed into a polymer.

**Nomenclature.**   The system of names and the rules for devising the names of compounds.

**Phenyl Group.** ⬡— or $C_6H_5$–

**Polymer.** A molecule whose structure consists of a large number of identical repeating units contributed by a monomer.

**Polymerization.** A reaction by which a monomer changes to a polymer.

**Primary Carbon (1° carbon).** In an organic structure, any carbon that is joined directly to just one other carbon.

**Saturated Compound.** In chemistry, a compound whose molecules are joined together by single bonds only; when an atom is holding a maximum number of groups, we say that its valence, or combining ability, is saturated.

**Secondary Carbon (2° carbon).** In an organic structure, any carbon that is joined directly to just two other carbons.

**Substitution Reaction.** A reaction in which one atom or group in the molecules of a compound is replaced by another atom or group (e.g., the conversion of benzene into nitrobenzene).

**Tertiary Carbon (3° carbon).** In an organic structure, any carbon that is joined directly to three other carbons.

**Unsaturated Compound.** In chemistry, any compound whose molecules contain one or more double or triple bonds.

## KEY MOLECULAR "MAP SIGNS"

| Key Molecular "Map Signs" in Organic Molecules | The Associated Chemical and Physical Properties |
| --- | --- |
| If the molecule is largely (or, of course, entirely) hydrocarbonlike | Expect the substance to be relatively insoluble in water and relatively soluble in organic solvents. Expect the substance to be less dense than water. |
| Alkanelike portions of all molecules | In these portions expect no chemical changes involving water, acids, alkalies, or chemical oxidizing agents and reducing agents. |

| Key Molecular "Map Signs" in Organic Molecules | The Associated Chemical and Physical Properties |
| --- | --- |
| An alkene double bond | Adds H–OH; forms an alcohol; Markovnikov's rule applies. Adds H–H; becomes saturated. Adds its own kind; polymerizes. |
| Benzene ring | Not like an alkene; gives substitution reactions, not addition reactions. |

## DRILL EXERCISES

### I.    EXERCISES ON STRUCTURAL FEATURES

The following set of structures illustrates several features discussed in the text. Supply the information requested by writing the identifying number(s) of the structure(s).

A.    $CH_3CH_2CH_2CH_2CH_2CH_3$

B.    $\underset{\displaystyle CH_3}{\overset{\displaystyle CH_3}{CH_3CH{-}CHCH_3}}$

C.    $CH_3CH{=}CHCH_2CH_2CH_3$

D.    $CH_3CH_2CH_2OH$

E.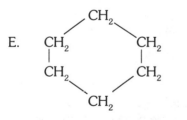

F.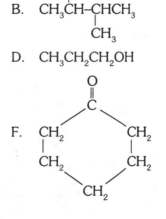

1.    Give the letters of

(a)   the saturated compounds          _____

(b)   the straight-chain compounds          _____

(c)   the branch-chain compounds          _____

(d)   any pairs of compounds related as isomers          _____

(e)   any compound with a 3° carbon          _____

2.   Write the IUPAC name of A   _____

of B   _____

of E   _____

3.   Write the structures of E and F using the geometric figure method illustrated in Figure 10.2 in the text.

_____          _____

E          F

4.   Write the structure and give the IUPAC name for the next lower homolog of A.          _____

5.   Write the full structure of $CH_3$

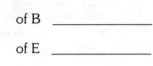

_____

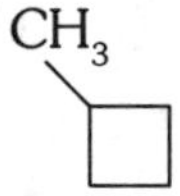

6.   Write the full structure of

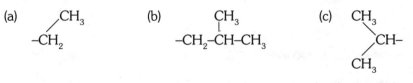

_____

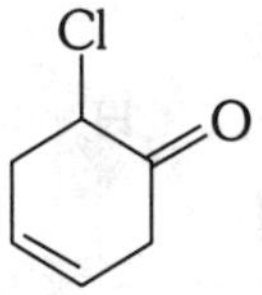

7.   To develop skill in the rapid recognition of alkyl groups regardless of how they are oriented in space, place the name of each group on the line beneath its structure.

(a)          $CH_3$
              /
        $-CH_2$

(b)          $CH_3$
              |
        $-CH_2-CH-CH_3$

(c)     $CH_3$
              \
               $CH-$
              /
        $CH_3$

_____   _____   _____

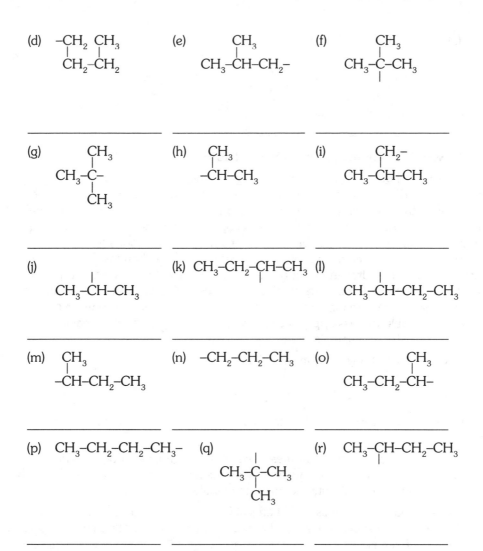

(d)   $-CH_2$ $CH_3$
      $CH_2-CH_2$

(e)        $CH_3$
      $CH_3-CH-CH_2-$

(f)        $CH_3$
      $CH_3-C-CH_3$

_____   _____   _____

(g)        $CH_3$
      $CH_3-C-$
           $CH_3$

(h)   $CH_3$
      $-CH-CH_3$

(i)        $CH_2-$
      $CH_3-CH-CH_3$

_____   _____   _____

(j)
      $CH_3-CH-CH_3$

(k) $CH_3-CH_2-CH-CH_3$

(l)
      $CH_3-CH-CH_2-CH_3$

_____   _____   _____

(m)   $CH_3$
      $-CH-CH_2-CH_3$

(n)   $-CH_2-CH_2-CH_3$

(o)        $CH_3$
      $CH_3-CH_2-CH-$

_____   _____   _____

(p)   $CH_3-CH_2-CH_2-CH_3-$

(q)
      $CH_3-C-CH_3$
           $CH_3$

(r)   $CH_3-CH-CH_2-CH_3$

_____   _____   _____

8.  To develop the skill of looking at a structure and visualizing it with its longest continuous chain on one horizontal line, write the IUPAC names of each of the following.

(a)        $CH_2CH_3$
      $CH_3CH_2$

(b)   $CH_3$
      $CH_2CH_3$

(c)   $CH_3CH_2$
      $CH_3CH_2$

_____   _____   _____

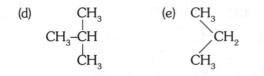

(d)

$CH_3-CH$ with $CH_3$ above and $CH_3$ below

(e)

$CH_3$ and $CH_3$ bonded to $CH_2$

## How to Work Problems in Organic Reactions

The overall goal of the chapters on organic chemistry in the text is to learn the chemical properties of functional groups. You know you've reached this goal when you can write the products made by the reaction of a given set of starting materials. In other words, it is not enough to be able to recognize the names and structural features of functional groups - although you can't do anything else without starting here. It's also not enough to memorize a sentence that summarizes a chemical fact, such as "alkenes react with hydrogen in the presence of a metal catalyst, heat, and pressure to give alkanes" - although memorizing such sentences is also absolutely essential to the goal. Such sentences are fundamental facts about the world in which we live. What you have to be able to do is *apply* such knowledge to a specific set of reactants. A typical problem, for example, is

"What forms, if anything, in the following reaction:

$$CH_3-CH=CH_2 + H_2 \xrightarrow[\text{heat, pressure}]{\text{Ni}} \quad ?"$$

Here's the strategy to follow in all of the reactions of organic compounds that we will study.

1.  Figure out the *family* or *families* to which the organic starting materials belong(s). You might even want to write the names of these families beneath the specific structures. (In our example, the family is "alkene.")

2.  Review your "memory list" of chemical facts about all alkenes. In your mental "storage," you should have the chemical-fact sentences that you will prepare in the next Exercise.

FOR EVERY FUNCTIONAL GROUP THAT WE STUDY YOU MUST PREPARE A MEMORIZED LIST OF CHEMICAL-FACT SENTENCES THAT SUMMARIZE THE CHEMICAL PROPERTIES THAT WE STUDY.

3.  If you go through the list and find no match to the stated problem, then assume that no reaction takes place and write "no reaction" as the answer.

4.  When you find the match between the specific problem and one of the listed chemical properties, stop and construct the structure of the answer. The many worked examples in the text develop the patterns for doing this, and study them thoroughly first.

## II.    SUMMARIZING THE CHEMICAL PROPERTIES OF ALKENES

When you have completed this exercise you will have a complete sentence summarizing each kind of chemical reaction of the carbon-carbon double bond that we have studied.  These statements are the chemical properties of this double bond that must be memorized, but they have to be learned in such a way that you can apply them to *specific* situations.  Hence, following this Exercise there are several drill problems to give you practice.  As you work each specific exercise among these drills, repeat to yourself the sentence statement that summarizes the property being illustrated.  This kind of repeated reinforcement will soon give you a surprisingly good working knowledge of these organic reactions, and you will be able to apply what you have learned to much more complicated situations with ease.

1.    The alkene double bond reacts with hydrogen (in the presence of a metal catalyst

and under pressure and heat) to give _____.

2.    The alkene double bond reacts with water (in the presence of an acid catalyst) to

give _____.

3.    Compounds with the double bond can polymerize to give _____.

4.    The complete combustion of any alkene gives _____ and _____.

Another way to organize chemical facts about a functional group such as the carbon-carbon double bond is by means of a 5 x 8" card.  An example is given below, but for the remaining functional groups it is vitally important that you prepare the cards yourself.  Part of the learning process is in this preparation, and having someone else do it for you robs you of that benefit.

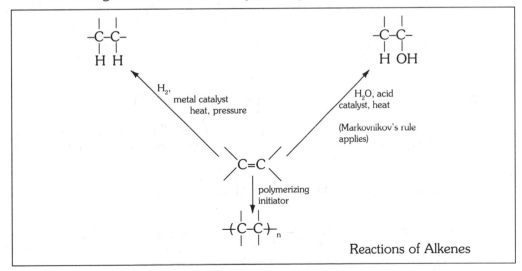

Reactions of Alkenes

Notice that the functional group is put in the center and its chemical properties are arranged about it. All of the arrows point outward. Later on, it will be useful to prepare cards summarizing key reactants, like $H_2O$ or $H_2$ on which you'll list all of the reactions studied involving these substances.

## III.    DRILL EXERCISE ON THE ADDITION OF HYDROGEN TO ALKENES

Write the structures of the products of the following reactions. If no reaction occurs, write "no reaction." (See Example 10.4 in the text.)

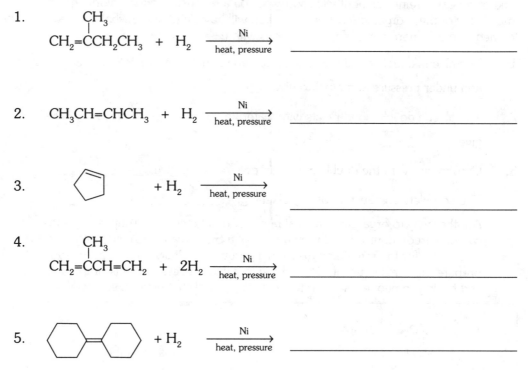

1.      $CH_2{=}\overset{\overset{\displaystyle CH_3}{|}}{C}CH_2CH_3$  +  $H_2$  $\xrightarrow[\text{heat, pressure}]{\text{Ni}}$  _____

2.    $CH_3CH{=}CHCH_3$  +  $H_2$  $\xrightarrow[\text{heat, pressure}]{\text{Ni}}$  _____

3.    $\bigcirc$  + $H_2$  $\xrightarrow[\text{heat, pressure}]{\text{Ni}}$ 
_____

4.      $CH_2{=}\overset{\overset{\displaystyle CH_3}{|}}{C}CH{=}CH_2$  +  $2H_2$  $\xrightarrow[\text{heat, pressure}]{\text{Ni}}$  _____

5.    $\bigcirc{=}\bigcirc$  + $H_2$  $\xrightarrow[\text{heat, pressure}]{\text{Ni}}$  _____

## IV.    DRILL EXERCISE ON THE ADDITION OF $H_2O$ TO ALKENES

Write the structures of the products of each reaction. If no reaction occurs, write "no reaction." Note: $H^+$ represents an acid catalyst. (See Example 10.5 in the text.)

1.    $CH_2{=}CHCH_2CH_3$  +  $H_2O$  $\xrightarrow{H^+}$ _____

2.    $CH_3CH{=}CHCH_3$  +  $H_2O$  $\xrightarrow{H^+}$ _____

3.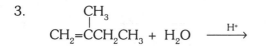

$$CH_2=\overset{\overset{\displaystyle CH_3}{\displaystyle |}}{C}CH_2CH_3 + H_2O \xrightarrow{\;H^+\;}$$ _____

4.  ⬠ $+ H_2O \xrightarrow{\;H^+\;}$ _____

5.  $CH_3CH_2$—⬡ $+ H_2O \xrightarrow{\;H^+\;}$ _____

Before going on to the Self-Testing Questions, be sure to work all the Practice and Review Exercises first.  As always, use the Self-Testing Questions as a "final exam" for the chapter.

## SELF-TESTING QUESTIONS

### Completion

1.  A substance whose molecules have the structure $CH_3CH_2CH_2CH_3$ is a member of the _____ family of hydrocarbons.  Its common name is _____.  Its IUPAC name is _____.

2.  A substance whose molecules have the structure $CH_3CH=CH_2$ is a member of the _____ family of hydrocarbons.  Its IUPAC name is _____.

3.  Complete the following equations by writing the structure of the organic product that would be expected to form in each case.  If no reaction is expected, write "none."

    (a)  $CH_3CH_2CH_2CH_3 + Na \longrightarrow$ _____

    (b)  $CH_3CH=CH_2 + H_2O \xrightarrow{\;H^+\;}$ _____

    (c)  $CH_2=CH_2 + H_2 \xrightarrow[\text{heat, pressure}]{\text{Ni}}$ _____

    (d)  $CH_3CH_2CH_3 + H_2O \xrightarrow{\;H^+\;}$ _____

(e)  $CH_3CH_3 + H_2SO_4 \longrightarrow$    _____

(f)  $C_6H_6$  +  $H_2O \longrightarrow$    _____
(benzene)

4.    Deduce the structure of the monomer from which this polymer may be made.

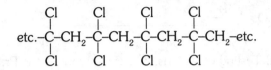

5.    Write condensed structures for each of the following:

(a)  ethylene    _____

(b)  pentane    _____

(c)  isobutane    _____

(d)  the isopropyl group    _____

(e)  the *trans* form of 2-butene    _____

6.    Write IUPAC names for each of the following:

(a)  $CH_3CH=CH_2$    _____

(b)  $CH_3CH_3$    _____

(c)    $CH_3$
      $CH_3\overset{|}{C}=CH_2$    _____

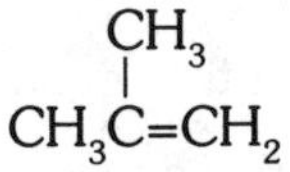

(d)    ⬠    _____

7.    Write IUPAC names for each of the following:

(a)          $CH_3$
      $CH_3CH_2\overset{|}{C}HCH_3$

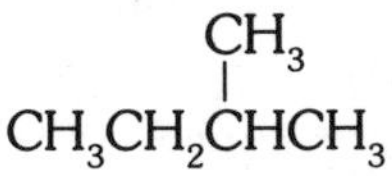

(b)   $CH_3CHCH_2CHCH_3$
        $\quad\ \ |\qquad\ \ |$
        $\quad\ \ CH_3\quad\ CH_3$

_____

(c)   $CH_3CHCH=CH_2$
        $\quad\ \ |$
        $\quad\ \ CH_3$

_____

(d)           $CH_3$
                $\ |$
     $Br–CH_2CHCH_3$

_____

## Multiple-Choice

1.  A family of organic compounds containing only carbon and hydrogen and having only single bonds are the

    (a)   alkenes    (b)   alkanes    (c)   alkynes    (d)   cycloalkanes

2.  A family of organic compounds whose molecules will add water (under an acid catalysis) and change into alcohols are the

    (a)   alkenes                         (c)   aromatic hydrocarbons

    (b)   alkanes                         (d)   cycloalkenes

3.  A chemist, handed a sample of an organic compound, was told that it was either $CH_2=CHCH_2CH_3$ or $CH_3CH_2CH_2CH_3$. He could decide which one it was by determining if the sample would react with

    (a)   sodium hydroxide

    (b)   hydrogen (with nickel present, heated under pressure)

    (c)   sodium chloride

    (d)   water (in the presence of an acid catalyst)

4.  An aromatic hydrocarbon can be expected to undergo

    (a)   substitution reactions         (c)   addition reactions

    (b)   reaction with water            (d)   no reactions

5.  The combustion of 1-butene will produce

    (a)   butylene and water             (c)   an alcohol

    (b)   carbon dioxide and water       (d)   cyclobutane

6.  An isomer of

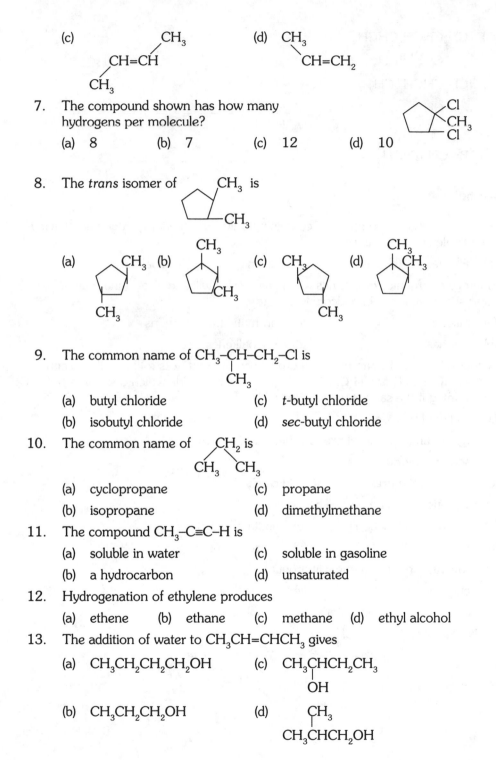

(c)
    $CH_3$
$CH=CH$
$CH_3$

(d)  $CH_3$
    $CH=CH_2$

7.  The compound shown has how many hydrogens per molecule?

    (a)  8          (b)  7          (c)  12          (d)  10

8.  The *trans* isomer of    $CH_3$   is
    $CH_3$

    (a)  $CH_3$  (b)   $CH_3$   (c)  $CH_3$   (d)   $CH_3$
         $CH_3$         $CH_3$        $CH_3$         $CH_3$

9.  The common name of $CH_3$–CH–$CH_2$–Cl is
    $CH_3$

    (a)  butyl chloride              (c)  *t*-butyl chloride
    (b)  isobutyl chloride           (d)  *sec*-butyl chloride

10. The common name of    $CH_2$ is
    $CH_3$   $CH_3$

    (a)  cyclopropane                (c)  propane
    (b)  isopropane                  (d)  dimethylmethane

11. The compound $CH_3$–C≡C–H is

    (a)  soluble in water            (c)  soluble in gasoline
    (b)  a hydrocarbon               (d)  unsaturated

12. Hydrogenation of ethylene produces

    (a)  ethene    (b)  ethane    (c)  methane    (d)  ethyl alcohol

13. The addition of water to $CH_3CH=CHCH_3$ gives

    (a)  $CH_3CH_2CH_2CH_2OH$        (c)  $CH_3CHCH_2CH_3$
                                          OH

    (b)  $CH_3CH_2CH_2OH$            (d)      $CH_3$
                                          $CH_3CHCH_2OH$

14. The addition of water to 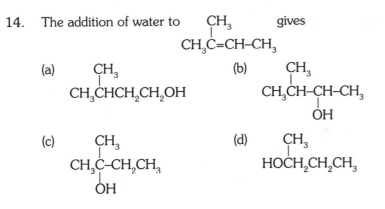 gives

(a)    $CH_3$
       $CH_3CHCH_2CH_2OH$

(b)    $CH_3$
       $CH_3CH-CH-CH_3$
       $OH$

(c)    $CH_3$
       $CH_3C-CH_2CH_3$
       $OH$

(d)    $CH_3$
       $HOCH_2CH_2CH_3$

# ANSWERS

## ANSWERS TO DRILL EXERCISES

### I.    Exercises on Structural Features

1.  (a)  A, B, D, and E
    (b)  A, C, and D   (The term "straight chain" cannot apply to rings.  Among open-chain compounds, as long as the atoms C, O, N, or S are all in a continuous sequence with or without double or triple bonds, the chain is "straight."  If the atoms O, N, or S are appended as part of substituents, the chain is still straight if all the carbons occur in a continuous sequence.)
    (c)  B  (Ring compounds are not classified as branched, either.)
    (d)  A  and B; C and E
    (e)  B  (Only saturated carbons are designated as 1°, 2°, or 3°.)

2.  A  hexane
    B  2,3-dimethylbutane
    E  cyclohexane

3.  E                          F    O

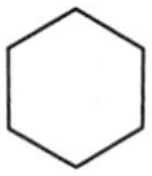

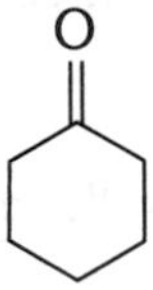

4.  $CH_3CH_2CH_2CH_2CH_3$      pentane

5.  $CH_3$
    $\diagdown$
    $CH{-}CH_2$
    $|\quad\ |$
    $CH_2{-}CH_2$

6.  $Cl$
    $|$
    $CH\quad O$
    $\diagup\quad\diagdown\ \diagup\diagdown$
    $CH_2\quad C$
    $|\qquad\ |$
    $CH\quad CH_2$
    $\diagdown\quad\diagup$
    $CH$

7.  (a)  ethyl      (Actually, ethyl group, but we may omit "group.")

    (b)  isobutyl    (Any four-carbon alkyl group *must* be one of the four butyl groups. The two derived from isobutane are either the isobutyl or the *t*-butyl group.)

    (c)  isopropyl   (Any three-carbon alkyl group *must* be one of the two propyl groups - propyl or isopropyl.)

    (d)  butyl      (Straighten out the chain and you have a group based on butane, not isobutane. The two butyl groups based on butane are the butyl and the *sec*-butyl groups.)

    (e)  isobutyl    (Compare b and c.)

    (f)  *t*-butyl    (It's the only butyl group where the unused bond  is at a 3° carbon.)

    (g)  *t*-butyl    (This is simply f rotated through part of a circle.)

    (h)  isopropyl   (A $C_3$-alkyl group; it must be either propyl or isopropyl.)

    (i)  isobutyl    (Compare b, e, and i.)

    (j)  isopropyl   (Compare c, h, and j.)

    (k)  *sec*-butyl   (It's the only butyl group where the unused bond is at a 2° carbon.)

    (l)  *sec*-butyl   (Compare with k.)

    (m)  *sec*-butyl   (Carefully compare k, l, and m. After chain straightening, the chain is straight in all three and the free bond comes from a 2° carbon.)

    (n)  propyl

    (o)  *sec*-butyl   (Compare with m and the accompanying note.)

    (p)  butyl

    (q)  *t*-butyl    (Just f tipped upside down. They have to be the same. Do you become someone else if you stand on your head?)

    (r)  *sec*-butyl

8.  (a)  butane
    (b)  propane
    (c)  butane
    (d)  isobutane or 2-methylpropane
    (e)  propane

## II.  Summarizing the Chemical Properties of Alkenes

1.  an alkane
2.  an alcohol
3.  a polymer
4.  carbon dioxide and water

## III.  Drill Exercise on the Addition of Hydrogen to Alkenes

1.
$$\underset{\underset{CH_3CHCH_2CH_3}{|}}{CH_3}$$

2.  $CH_3CH_2CH_2CH_3$

3.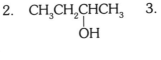

4.
$$\underset{\underset{CH_3CHCH_2CH_3}{|}}{CH_3}$$

5.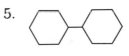

## IV.  Drill Exercise on the Addition of H$_2$O to Alkenes

1.  $\underset{\underset{OH}{|}}{CH_3CHCH_2CH_3}$

2.  $\underset{\underset{OH}{|}}{CH_3CH_2CHCH_3}$

3.  $\underset{\underset{OH}{|}}{\underset{CH_3CCH_2CH_3}{\overset{\overset{CH_3}{|}}{}}}$

4. 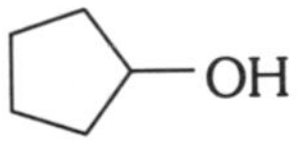  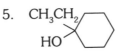—OH

5.  $CH_3CH_2$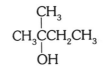
    HO

## ANSWERS TO SELF-TESTING QUESTIONS

### Completion

1.  alkane; butane; butane
2.  alkene; propene
3.  (a)  none
    (b)  $\underset{\underset{OH}{|}}{CH_3CHCH_3}$
    (c)  $CH_3CH_3$
    (d)  none

(e)  none

(f)  none

4.  $Cl_2C=CH_2$

5.  (a)  $CH_2=CH_2$

(b)  $CH_3CH_2CH_2CH_2CH_3$ (may be written as $CH_3(CH_2)_3CH_3$)

(c) 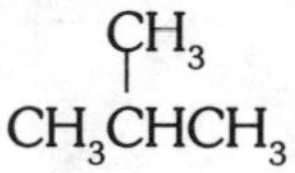  $\underset{\underset{CH_3CHCH_3}{|}}{CH_3}$

(d)  $\underset{\underset{CH_3CHCH_3}{|}}{}$ (may be written as 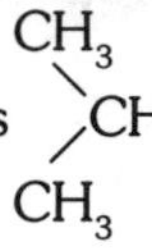  $\underset{\underset{CH_3}{}}{\overset{CH_3}{\diagdown}}CH-$  or as   $(CH_3)_2CH-)$

(e) 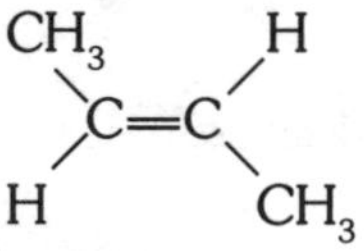  $\underset{H}{\overset{CH_3}{\diagdown}}C=C\underset{CH_3}{\overset{H}{\diagup}}$

6.  (a)  propene                          (c)  2-methylpropene

(b)  ethane                           (d)  cyclopentane

7.  (a)  2-methylbutane                   (c)  3-methyl-1-butene

(b)  2,4-dimethylpentane              (d)  1-bromo-2-methylpropane

## Multiple-Choice

1.  b, d

2.  a, d

3.  b, d

4.  a

5.  b

6.  c

7.  d

8   b

9.  b

10.  c

11.  b, c, d

12.  b

13.  c

14.  c

# CHAPTER 11

# ALCOHOLS, THIOALCOHOLS, ETHERS, AND AMINES

## OBJECTIVES

After you have studied this chapter, and worked the problems you should be able to do the following.

1. *Give the general class structures for alcohols, ethers, thioalcohols, disulfides, and amines.*

2. *Look at the structure of a molecule containing several functional groups and pick out any that are studied in this chapter.*

3. *Write structures and common names for the $C_1$ to $C_4$ compounds in the families of alcohols and amines. (If assigned, write IUPAC names for alcohols.)*

4.  *Tell from the structure of an alcohol if it is 1°, 2°, or 3°.*

5.  *Make a drawing that illustrates hydrogen bonds between molecules in an alcohol, an alcohol dissolved in water, an amine, and an amine dissolved in water.*

6.  *Explain why alcohols have higher solubilities in water and higher boiling points than compounds of similar formula weights in the alkane family or the ether family.*

7.  *Write equations (not necessarily balanced) that are specific illustrations of each of the following kinds of reactions.*

    (a)  *dehydration of an alcohol to an alkene*

    (b)  *conversion of an alcohol into an ether*

    (c)  *oxidation of a 1° alcohol to an aldehyde*

    (d)  *oxidation of a 1° alcohol to a carboxylic acid*

    (e)  *oxidation of a 2° alcohol to a ketone*

    (f)  *oxidation of a mercaptan to a disulfide*

    (g)  *reduction of a disulfide to a mercaptan*

    (h)  *reaction of an amine with aqueous acid*

    (i)  *reaction of an amine salt with aqueous alkali*

    (j)  *reaction of a phenol with aqueous sodium hydroxide*

8.  *Write the structure(s) of expected organic product(s), if any, or a reaction of a member of any family discussed in this chapter with.*

    (a)  $H_2SO_4$ *(and heat)*

    (b)  *(O) - a chemical oxidizing agent*

    (c)  *(H) - a chemical reducing agent*

    (d)  $H_3O^+$ *- dilute aqueous acid at room temperature*

    (e)  $OH^-$ *- dilute aqueous alkali at room temperature*

    *(In other words, be able to complete equations like those found in problems 11.42 and 11.43 in the text.)*

9.  *Name the functional groups and name the groups to which they are changed by action by the following kinds of reagents:*

    (a)  *oxidizing agents;*  (b)  *reducing agents;*  (c)  *dilute acid;*

    (d)  *dilute alkali;*  (e)  *concentrated sulfuric acid (and heat).*

    *(Do this for Chapters 10 and 11 combined.)*

10.  *Recognize the ether function (and realize that we are assuming it undergoes no reactions with the reagents we are studying).*

11.  *Define each term in the Glossary and illustrate or give the structure where applicable.*

Be sure to prepare and learn the one-sentence statements of chemical properties of alcohols, thioalcohols, disulfides, ethers, amines and protonated amines.  Also be sure to prepare the 5 x 8" reaction summary cards and the cards that list the functional groups affected by various kinds of inorganic reactants.

## GLOSSARY

**Alcohol.**  Any organic compound having the –OH group attached to a saturated carbon atom.

**Alcohol Group.**  The –OH group when it is attached to a saturated carbon.

**Alkaloid.**  A physiologically active, heterocyclic amine isolated from plants.

**Amide.**  A compound with a carbonyl-nitrogen bond:
$$\underset{\phantom{x}}{-\overset{\displaystyle O}{\overset{\|}{C}}-\overset{|}{N}-}$$

**Amine.**  Any organic compound in which a trivalent nitrogen atom is attached to one, two, or three carbons of either the alkyl group type or a benzene ring.

**Amine Salt.**  Any organic compound whose molecules have a positively charged, tetravalent, protonated nitrogen atom, as in $RNH_3^+$, $R_2NH_2^+$, or $R_3NH^+$.

**Disulfide.**  Any organic compound having the –S–S– group.

**Ether.**  Any organic compound possessing an oxygen attached to two carbon atoms, *neither* of which is a carbonyl carbon; R–O–R', Aryl–O–R, or Aryl–O–Aryl.

**Glycol.**  A dihydric alcohol.  The two –OH groups are usually (but not necessarily) on adjacent carbons.

**Mercaptan.**  Any organic compound having an –SH group attached to a saturated carbon atom.

**Phenol.**  Any organic compound having the –OH group attached to a benzene ring or to a similar system.

**Primary Alcohol.**  An alcohol in which the –OH group is held by a primary carbon (a carbon with one other carbon joined directly to it); $R–CH_2–OH$.

**Secondary Alcohol.**  An alcohol in which the –OH group is held by a secondary carbon (a carbon with two other carbons joined directly to it); $R_2CH–OH$.

**Tertiary Alcohol.**  An alcohol in which the –OH group is held by a tertiary carbon (a carbon with three other carbons joined directly to it); $R_3C–OH$.

**Thioalcohol.**  A compound containing the –SH group.

## KEY MOLECULAR "MAP SIGNS"

We have likened the functional groups to "map signs" that enable us to "read" the structural formulas of complicated systems.  The principal map signs in this chapter are the following.

| Key Molecular "Map Signs" in Organic Molecules | What to Expect When This Functional Group is Present |
|---|---|
| $-\overset{\displaystyle |}{\underset{\displaystyle |}{C}}-O-H$<br><br>Alcohol group | **Influence on physical properties:**<br>– The –OH group is a good hydrogen-bond donor and hydrogen-bond acceptor.<br>– Molecules with the –OH group tend to be much more polar (e.g., higher boiling points) and much more soluble in water than those without any group that can participate in hydrogen bonding (formula weights being about the same).<br><br>**Influence on chemical properties:**<br>A molecule with the alcohol group is vulnerable to<br>– dehydrating agents (acids + heat); either double bonds are introduced or ethers are made.<br>– oxidizing agents (those that can pull out the pieces of the element hydrogen):<br>1° alcohols ⟶ aldehydes<br>2° alcohols ⟶ ketones<br>3° alcohols ⟶ (no reaction) |

(The list for alcohols will be completed in Chapters 12 and 13.

| OH<br>⬡<br>Phenol system | A phenolic –OH group is a weak acid, but strong enough to neutralize the hydroxide ion. |

–C–O–C–

Ether group

(To be a simple ether, the carbons shown here must hold either Hs or Rs.)

**Influence on physical properties:**

The ether's oxygen can accept hydrogen bonds; ethers therefore are slightly more soluble than alkanes; slightly more polar.

**Influence on chemical properties:**

None. (As far as our study goes, the simple ether group is not attacked by water, dilute acids or bases, or oxidizing or reducing agents. We have to be able to recognize it, but then we ignore it.)

---

–S–H
Thioalcohol group

Easy oxidation to –S–S–; that is the only property that concerns us, because we need to understand it in relation to proteins.

---

–S–S–
Disulfide

Easy reduction to 2 –S–H; again, that is the only property that concerns us, for we need to understand it in relation to proteins.

---

–N:

Amino Group

**Influence on physical properties:**

– If present as $R-NH_2$ or $R_2NH$, amino groups can both accept and donate hydrogen bonds to water, amines, or alcohols.

– If present as $R_3N$, amino groups can only accept hydrogen bonds; are soluble in water much as alcohols are.

**Influence on chemical properties:**

– The presence of an amino group makes the molecule a proton acceptor, a Brønsted base.

– The easy room-temperature changes:

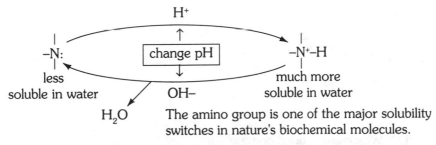

less soluble in water

much more soluble in water

The amino group is one of the major solubility switches in nature's biochemical molecules.

(The list for amines will be completed in Chapter 13.)

## DRILL EXERCISES

### I.    EXERCISES IN HYDROGEN BONDS

We define an *H-bond donor* as a molecule with a δ+ on a hydrogen attached to oxygen or nitrogen (–O–H or –N–H).

An *H-bond acceptor* is a molecule with a δ- on an oxygen or a tri-substituted nitrogen in any functional group.

H-bond donors can establish H-bonds not only to their own kind but also to water molecules and to any other H-bond acceptors.  H-bond donors are invariably H-bond acceptors.

Some molecules, such as ethers, ketones, aldehydes, and esters, can only be H-bond acceptors.  They have oxygens (or nitrogens, as in examples not yet studied), but they do not have –OH or –NH (i.e., a hydrogen with a δ+ and held by oxygen or nitrogen).

Examine these structures and answer the questions that follow.

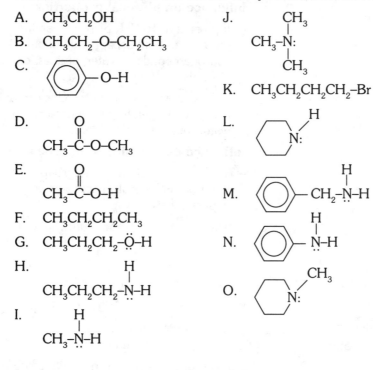

A.   $CH_3CH_2OH$

B.   $CH_3CH_2\text{–}O\text{–}CH_2CH_3$

C.   ⬡—O–H

D.   $CH_3\text{–}\overset{\overset{\textstyle O}{\|}}{C}\text{–}O\text{–}CH_3$

E.   $CH_3\text{–}\overset{\overset{\textstyle O}{\|}}{C}\text{–}O\text{–}H$

F.   $CH_3CH_2CH_2CH_3$

G.   $CH_3CH_2CH_2\text{–}\ddot{O}\text{–}H$

H.   $CH_3CH_2CH_2\text{–}\overset{\overset{\textstyle H}{|}}{\underset{\cdot\cdot}{N}}\text{–}H$

I.   $CH_3\text{–}\overset{\overset{\textstyle H}{|}}{\underset{\cdot\cdot}{N}}\text{–}H$

J.   $CH_3\text{–}\overset{\overset{\textstyle CH_3}{|}}{\underset{\underset{\textstyle CH_3}{|}}{N}}{:}$

K.   $CH_3CH_2CH_2CH_2\text{–}Br$

L.   ⬡N: with H

M.   ⬡—$CH_2\text{–}\overset{\overset{\textstyle H}{|}}{N}\text{–}H$

N.   ⬡—$\overset{\overset{\textstyle H}{|}}{\underset{\cdot\cdot}{N}}\text{–}H$

O.   ⬡N: with CH₃

1.   Which are H-bond donors? _____

2.  Which are H-bond acceptors?                                   _____

3.  Which are H-bond acceptors *only*?                            _____

4.  Which would be completely insoluble in water?                 _____

5.  Which would be more soluble in water, A or B?                 _____

6.  Which would be more soluble in water, C or E?                 _____

7.  Which will have a higher boiling point, G or H?               _____

## II.    DRILL IN WRITING THE PRODUCTS OF DEHYDRATION OF ALCOHOLS

After you have studied Example 11.1 in the text and have tried Practice Exercise 3, you might feel the need for further drill. Write the alkene products (if any) of the dehydration of the following alcohols.

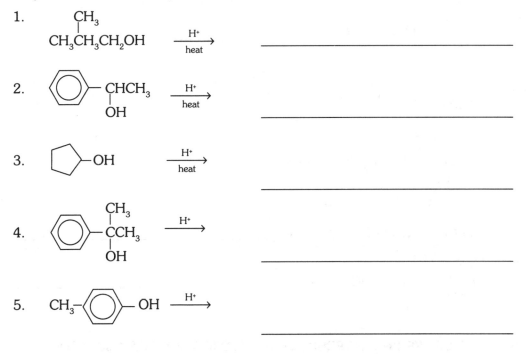

1.     $\underset{\displaystyle CH_3CH_2CH_2OH}{\overset{\displaystyle CH_3}{|}}$   $\xrightarrow[\text{heat}]{H^+}$   _____

2.     [benzene ring]—$\underset{OH}{\overset{}{CHCH_3}}$   $\xrightarrow[\text{heat}]{H^+}$   

_____

3.     [cyclopentane ring]—OH   $\xrightarrow[\text{heat}]{H^+}$   

_____

4.     [benzene ring]—$\underset{OH}{\overset{CH_3}{\underset{|}{\overset{|}{C}CH_3}}}$   $\xrightarrow{H^+}$   

_____

5.     $CH_3$—[benzene ring]—OH   $\xrightarrow{H^+}$

_____

### III.    DRILL IN WRITING THE PRODUCTS OF THE OXIDATION OF ALCOHOLS

Examples 11.2 and 11.3 plus Practice Exercises 4 and 5 are the places to begin this study.  For more drill, write the products of the oxidation of the following alcohols. If oxidation in the sense being studied cannot occur, then write "no reaction."  If the initial product can be oxidized further, then write the next product, too.

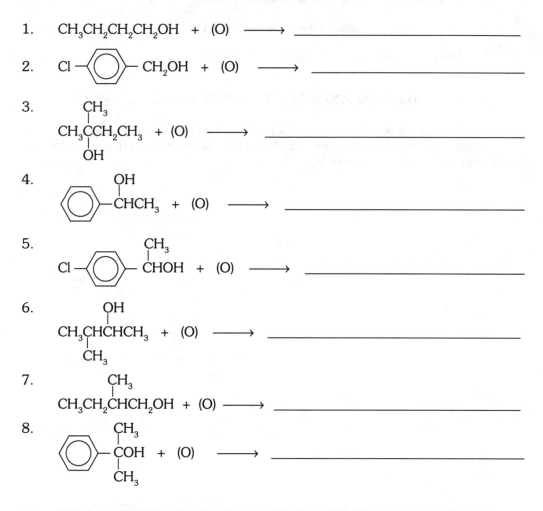

1.    $CH_3CH_2CH_2CH_2OH$  +  (O)  $\longrightarrow$  _____

2.    Cl—⬡—$CH_2OH$  +  (O)  $\longrightarrow$  _____

3.

$$\begin{array}{c} CH_3 \\ | \\ CH_3CCH_2CH_3 \\ | \\ OH \end{array}$$  +  (O)  $\longrightarrow$  _____

4.

$$\begin{array}{c} OH \\ | \\ \text{⬡—}CHCH_3 \end{array}$$  +  (O)  $\longrightarrow$  _____

5.

$$\begin{array}{c} CH_3 \\ | \\ Cl\text{—⬡—}CHOH \end{array}$$  +  (O)  $\longrightarrow$  _____

6.

$$\begin{array}{c} OH \\ | \\ CH_3CHCHCH_3 \\ | \\ CH_3 \end{array}$$  +  (O)  $\longrightarrow$  _____

7.

$$\begin{array}{c} CH_3 \\ | \\ CH_3CH_2CHCH_2OH \end{array}$$  +  (O)  $\longrightarrow$  _____

8.

$$\begin{array}{c} CH_3 \\ | \\ \text{⬡—}COH \\ | \\ CH_3 \end{array}$$  +  (O)  $\longrightarrow$  _____

### IV.    DRILL IN WRITING THE PRODUCT OF THE OXIDATION OF A THIOALCOHOL

Example 11.4 in the text shows how to do this kind of exercise.  For practice, write the products of the oxidation of the following thioalcohols.

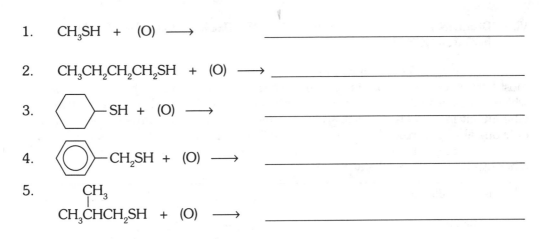

1.    $CH_3SH$  +  (O)  $\longrightarrow$    _____

2.    $CH_3CH_2CH_2CH_2SH$  +  (O)  $\longrightarrow$ _____

3.    ⬡—SH  +  (O)  $\longrightarrow$    _____

4.    ⬡—$CH_2SH$  +  (O)  $\longrightarrow$    _____

5.         $CH_3$
            |
       $CH_3CHCH_2SH$  +  (O)  $\longrightarrow$    _____

## V.    DRILL IN WRITING THE PRODUCT OF THE REDUCTION OF A DISULFIDE

Example 11.5 in the text shows how to do this. Remember that disulfides aren't always symmetrical, like R–S–S–R. Those that we'll encounter in biochemistry usually are not; they're like R–S–S–R'. The reduction of this kind gives two different thioalcohols, so both of them have to be written.

1.                    $CH_3$
                       |
       $CH_3$–S–S–$CHCH_3$  +  2(H)  $\longrightarrow$ _____

2.    ⬡— $CH_2$–S–S–$CH_2CH_3$  +  2(H)  $\longrightarrow$

       _____

3.    $CH_3CH_2OCH_2CH_2$–S–$SCH_3$  +  2(H)  $\longrightarrow$

       _____

4.        S–S
         ╱     ╲
       $CH_2$      $CH_2$  +  2(H)  $\longrightarrow$
         ╲     ╱
         $CH_2$–$CH_2$                _____

5.    ⬡—S–S—⬡  +  2(H)  $\longrightarrow$

       _____

## VI.    DRILL IN WRITING THE STRUCTURES OF ETHERS THAT CAN BE MADE FROM ALCOHOLS

Example 11.6 shows how this is done. In order to make this kind of exercise most helpful for applications in the next chapter, we will include in the drill examples in which you'll construct a structure of an unsymmetrical ether, like R-O-R', from two different given alcohols. We will also make this a review of the names of alcohols. If only one alcohol is named, then the question is what *symmetrical* ether can be made from this alcohol? If two alcohols are named, then the question is what *unsymmetrical* ether can be made from the two?

1.   isobutyl alcohol                            _____

2.   *t*-butyl alcohol                            _____

3.   4-methylcyclohexanol                   _____

4.   isopropyl alcohol and methyl alcohol _____

5.   cyclopentanol and *sec*-butyl alcohol _____

## VII.    DRILL ON WRITING THE PRODUCTS OF THE REACTIONS OF AMINES WITH STRONG ACIDS.

Write the structures of the organic cations that form when each of the following amines reacts with something like hydrochloric acid (which is really a reaction with $H_3O^+$)

1.   $CH_3CH_2CH_2NH_2$                            _____

2.   ⬡— NH–CH₃                                      _____

3.   CH₃CH₂NCH₂CH₃
         |
         CH₃                                      _____

4.   ⬠NH                                             _____

5.   $CH_3NHCH_2CH_2NHCH_3$ (and 2HCl) _____

## VII.  DRILL IN WRITING THE PRODUCTS OF THE REACTIONS OF PROTONATED AMINES WITH STRONG, AQUEOUS BASE

Write the products of the deprotonation of the following cations.

1.  $CH_3CH_2NH_3^+$ _____

2.  $CH_3CH_2\overset{+}{N}HCH_3$
    $\quad\quad\quad\ \ |$
    $\quad\quad\quad CH_3$ _____

3.  $\quad\quad\quad\quad CH_3$
    $\quad\quad\quad\quad\ |$
    benzene$-\overset{+}{N}HCH_3$ _____

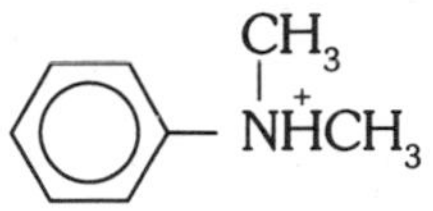

4.  $\quad\quad\quad\quad\quad O$
    $\quad\quad\quad\quad\quad ||$
    $\overset{+}{N}H_3CH_2CH_2CCH_3$ _____

5.  $\overset{+}{N}H_3CHCO_2^-$
    $\quad\ \ |$
    $\quad\ CH_3$ _____

# SELF-TESTING QUESTIONS

### Completion

1.  Review how alkyl groups are named and how their names are used in naming alcohols, amines, ethers, and mercaptans.  Test your knowledge by completing the following:

    Give common names for each:

    (1)  $CH_3CH_2CH_2OH$ _____

    (2)  $CH_3OH$ _____

    (3)  $\quad\quad CH_3$
    $\quad\quad\ |$
    $CH_3C-OH$
    $\quad\quad\ |$
    $\quad\quad CH_3$ _____

    (4)  $(CH_3)_2CH-OH$ _____

(5)  $CH_3NHCH_3$    _____

(6)  $CH_3CH_2OCH_2CH_3$    _____

(7)  $CH_3SH$    _____

2.  Review the chemical reactions of the functional groups of this chapter and test your knowledge by writing the structure(s) of the principal organic product(s) that would be expected to form in each case.  If no reaction will occur, write "none." (The other stipulations in problem 11.42 in the text, also apply.)

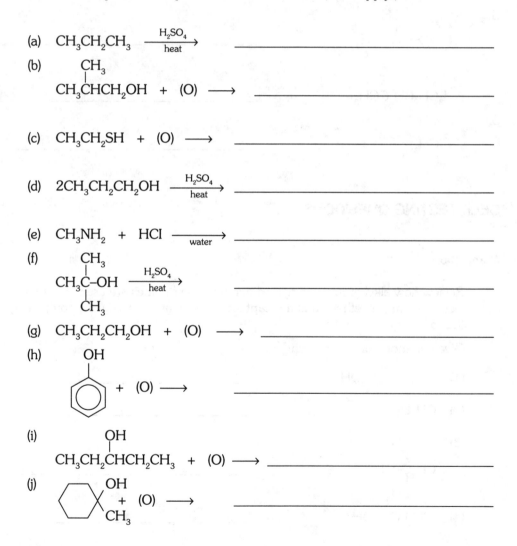

(a)  $CH_3CH_2CH_3$ $\xrightarrow[\text{heat}]{H_2SO_4}$    _____

(b)      $CH_3$
         |
     $CH_3CHCH_2OH$  +  (O)  $\longrightarrow$    _____

(c)  $CH_3CH_2SH$  +  (O)  $\longrightarrow$    _____

(d)  $2CH_3CH_2CH_2OH$ $\xrightarrow[\text{heat}]{H_2SO_4}$    _____

(e)  $CH_3NH_2$  +  HCl $\xrightarrow[\text{water}]{}$    _____

(f)      $CH_3$
         |
     $CH_3C-OH$ $\xrightarrow[\text{heat}]{H_2SO_4}$    _____
         |
         $CH_3$

(g)  $CH_3CH_2CH_2OH$  +  (O)  $\longrightarrow$    _____

(h)      OH

         +  (O)  $\longrightarrow$    _____

(i)        OH
           |
     $CH_3CH_2CHCH_2CH_3$  +  (O)  $\longrightarrow$    _____

(j)        OH

           +  (O)  $\longrightarrow$    _____
         $CH_3$

(k)   $CH_3CH_2CH_2CH_3 + H_2O \xrightarrow[\text{heat}]{H^+}$ _____

(l)   $CH_3OH + (O) \longrightarrow$     _____

(m)   $HOCH_2CH_2CH_3 \xrightarrow[\text{heat}]{H_2SO_4}$     _____

(n)   $CH_2=CHCH_2OH + H_2 \xrightarrow[\text{pressure}]{\text{Ni, heat}}$ _____

(o)   $CH_3CH_2-S-S-CH_2CH_3 \xrightarrow[\text{agent}]{\text{reducing}}$ _____

(p)   $CH_3CH_2-O-CH_2-CH_2-\overset{\overset{\displaystyle OH}{|}}{C}HCH_3 + (O) \longrightarrow$

_____

(q)   $CH_3\underset{\underset{\displaystyle NH_3^+}{|}}{C}HCH_3 + NaOH \xrightarrow[\text{water}]{}$ _____

3.    Write the structure of the organic compound that could be used to synthesize each of the following in one step.  Then write the equation, including the reagent and conditions, in the manner we have followed in our study.

(a)   $CH_3CH_3$

_____

(b)        $CH_3CH_2\overset{\overset{\displaystyle O}{\|}}{C}H$

_____

(c)   $CH_3CH_2OH$

_____

(d)   $CH_3CH_2CH_2OCH_2CH_2CH_3$

_____

(e)   $CH_3CH_2S-SCH_2CH_3$

_____

**Multiple-Choice**

1. The oxidation of $CH_3$–$\overset{\overset{\displaystyle OH}{|}}{C}H$–$CH_2$–$CH_3$ could be made to produce:

    (a)  $CH_3$–$\overset{\overset{\displaystyle OH}{|}}{C}H$–$O$–$CH_2$–$CH_3$

    (c)  $CH_3$–$CH_2$–$\overset{\overset{\displaystyle O}{||}}{C}$–$CH_3$

    (b)  $H$–$\overset{\overset{\displaystyle O}{||}}{C}$–$CH_2CH_2CH_3$

    (d)  $HO$–$\overset{\overset{\displaystyle O}{||}}{C}CH_2CH_2CH_3$

2. The *best* explanation for the solubility of glycerol in water is

    (a)  glycerol is a small molecule

    (b)  glycerol molecules are polar

    (c)  glycerol molecules can donate and accept hydrogen bonds to and from water molecules in the solution

    (d)  glycerol's ions are well-solvated by water

    $HO$–$CH_2$–$\overset{\overset{\displaystyle }{|}}{C}H$–$CH_2$–$OH$
    $\qquad\qquad\;\; OH$

    Glycerol

3. The substance whose structure is $CH_3$–$OH$ is known as

    (a)  wood alcohol

    (c)  methanol

    (b)  grain alcohol

    (d)  potable alcohol

4. The substance whose structure is $CH_3CH_2OCH_2CH_3$ is

    (a)  a common fuel

    (c)  an older anesthetic

    (b)  a common antifreeze

    (d)  an eye irritant in smog

5. The compound whose structure is $CH_3CH_2CH_2NH_2$ is called

    (a)  methylethylamine

    (c)  1-aminopropane

    (b)  propylamine

    (d)  *n*-butylamine

6. The reaction of $CH_3$–$\overset{\overset{\displaystyle }{|}}{\underset{\underset{\displaystyle CH_3}{|}}{N}}H_2^+$ with aqueous sodium hydroxide at

    room temperature will give

    (a)  $CH_3NHCH_3$

    (c)  $CH_3$–$\overset{\overset{\displaystyle }{}}{\underset{\underset{\displaystyle CH_3}{|}}{N}}{}^-$

    (b)  $CH_3\underset{\underset{\displaystyle CH_3}{|}}{N}$–$OH$

    (d)  $CH_3$–$\underset{\underset{\displaystyle CH_3}{|}}{N}$–$CH_3$

7.  The presence of the –NH$_2$ group in an organic compound makes the compound

    (a)  a good proton donor      (c)  more soluble in water

    (b)  less soluble in water      (d)  a good proton acceptor

8.  What is the best representation for hydrogen bonding in methylamine?

    (a)
    $$\begin{array}{cc} \text{H} & \text{H} \\ | & | \\ \text{CH}_3\text{–N–H}\cdots\text{CH}_3\text{–N–H} \end{array}$$

    (b)
    $$\begin{array}{cc} \text{H} & \text{H} \\ | & | \\ \text{CH}_2\text{–N–H}\cdots\text{H–N–CH}_3 \end{array}$$

    (c)
    $$\begin{array}{c} \text{HH} \\ \diagup \;| \\ \text{CH}_3\text{–N}\cdots\text{N–CH}_3 \\ \diagdown \;| \\ \text{HH} \end{array}$$

    (d)
    $$\begin{array}{cc} \text{H} & \text{H} \\ | & | \\ \text{CH}_3\text{–N–H}\cdots\text{N–CH}_3 \\ & | \\ & \text{H} \end{array}$$

Multiple-choice questions 9 through 11 refer to this structure:

$$\underbrace{\text{CH}_3\text{–O–CH}_2}_{①}\text{–CH}_2\text{–}\overset{\overset{\displaystyle \text{OH}\Big\}\,②}{|}}{\text{CH}}\text{–CH}_2\underbrace{\text{CH=CH}_2}_{③}$$

9.  The group labeled ① is

    (a)  an easily hydrolyzed group

    (b)  an easily oxidized group

    (c)  an easily reduced group

    (d)  a generally unreactive group

10.  The group labeled ② could be

    (a)  oxidized to a ketone

    (b)  oxidized to an aldehyde

    (c)  involved in an acid-catalyzed dehydration

    (d)  reduced to a ketone

11. The group labeled ③ could be
    (a) made to react with dilute sodium hydroxide
    (b) made to add a water molecule (if an acid catalyst were available)
    (c) reduced to a carbon-carbon single bond by hydrogen (with catalyst, heat, and pressure)
    (d) involved in hydrogen bonding

12. If the following compounds were arranged in the order of their increasing boiling points, that order would be

    A. $CH_3CH_3$     B.   $CH_3OH$     C.   $CH_3NH_2$

    (a)  A < B < C                    (c)  C < A < B
    (b)  A < C < B                    (d)  B < A < C

13. If the following compounds were arranged in the order of their increasing solubility in water, that order would be

    A. $CH_3CH_2CH_2CH_2CH_2OH$     B.   $CH_3CH_2CH_2CH_2CH_2Cl$

    C.   $CH_3CH_2CH_2CH_2CH_2NH_2$

    (a)  A < B < C                    (c)  B < A < C
    (b)  C < B < A                    (d)  B < C < A

14. Which compound could neutralize aqueous sodium hydroxide?
    (a)  $CH_3OH$                     (c)  $CH_3S$-$S$-$CH_3$
    (b)  $CH_3NH_2$                   (d)  $CH_3NH_3^+Cl^-$

Multiple-Choice Questions 15 to 18 refer to this compound.

$$\overset{②}{\overset{\frown}{NH_2}}$$
$$\underset{①}{\underbrace{CH_3-O-CH_2}}-\underset{③}{\underbrace{CH-CH=CH-CH_2}}-\underset{④}{\underbrace{S-S-CH_3}}$$

15. The group at ① would
    (a)  react with NaOH(aq)
    (b)  react with HCl(aq)
    (c)  be reduced by catalytic hydrogenation
    (d)  be oxidizable by mild reagents
    (e)  none of these

16. The group at ② would

    (a)   neutralize aqueous HCl

    (b)   accept H-bonds from water

    (c)   react with $H_2$

    (d)   none of these

17. The group at ③ would

    (a)   add hydrogen

    (b)   donate hydrogen bonds

    (c)   add water

    (d)   none of these

18. The group at ④ would

    (a)   be easily reduced

    (b)   react with HCl(aq)

    (c)   neutralize NaOH(aq)

    (d)   react with water

# ANSWERS

## ANSWERS TO DRILL EXERCISES

### I.    Exercises in Hydrogen Bonds

1. A, C, E, G, H, I, L, M, and N
2. A, B, C, D, E, G, H, I, J, L, M, N, and O
3. B, D, J, and O
4. F (an alkane) and K
5. A    (It has a lower formula weight and can both donate and accept hydrogen bonds to water, the solvent.  B can only accept hydrogen bonds from water.)
6. E    (It is less hydrocarbonlike, has a lower formula weight, and has two hydrogen bonding sites, the two oxygens.  Structure C, on the other hand, has only one hydrogen bond accepting site.)
7. G    (Its molecules can form stronger hydrogen bonds than those of H.)

## II. Drill in Writing the Products of the Dehydration of Alcohols

1. 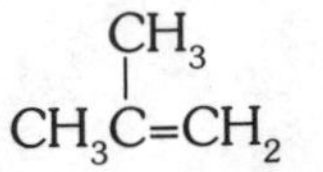

2. 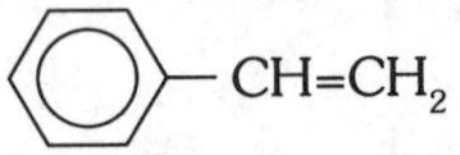

3. 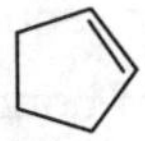

4. 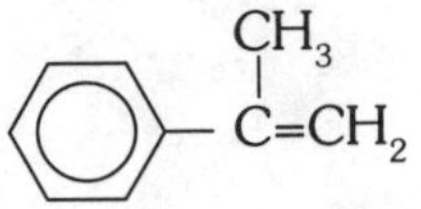

5. no reaction

## III. Drill in Writing the Products of the Oxidation of Alcohols

1. $CH_3CH_2CH_2CHO \xrightarrow{\text{more (O)}} CH_3CH_2CH_2CO_2H$

2. 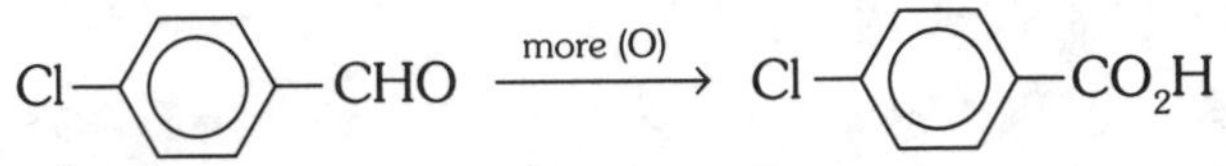

3. no reaction

4. 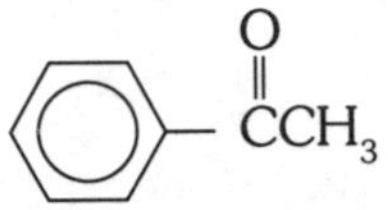

5. 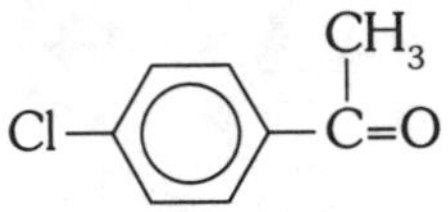

6. 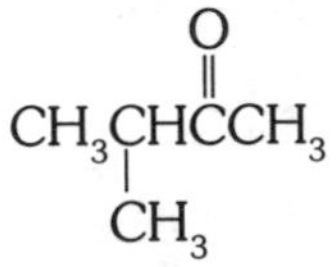

7. 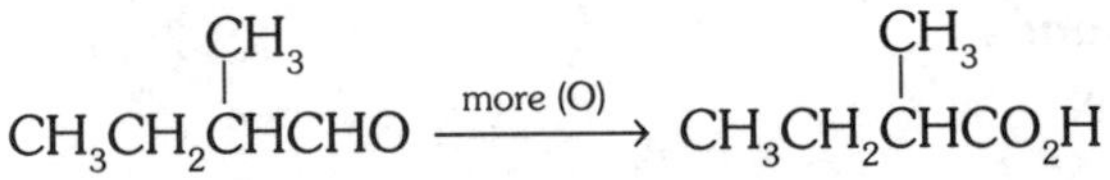

8. no reaction

## IV. Drill in Writing the Product of the Oxidation of a Thioalcohol

1. $CH_3\text{–}S\text{–}S\text{–}CH_3$

2. $CH_3CH_2CH_2CH_2\text{–}S\text{–}S\text{–}CH_2CH_2CH_2CH_3$

3. 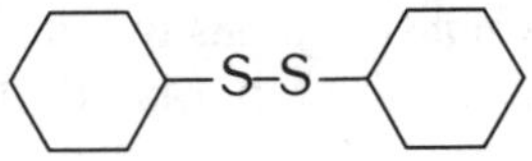

4. 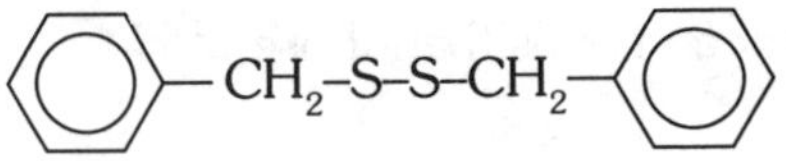

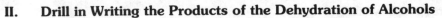

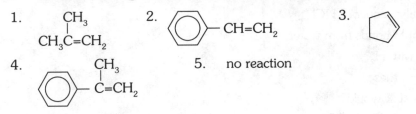

5.

$$\underset{CH_3CHCH_2-S-S-CH_2CHCH_3}{\overset{\overset{\displaystyle CH_3}{|}\qquad\qquad\overset{\displaystyle CH_3}{|}}{}}$$

## V.   Drill in Writing the Product of the Reduction of a Disulfide

1.   $CH_3SH + HSCHCH_3$
   $\qquad\qquad\qquad\quad |$
   $\qquad\qquad\qquad\ CH_3$

2.   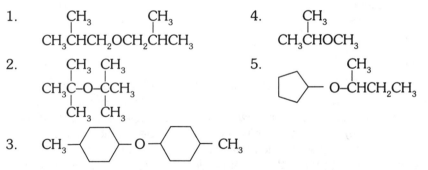$-CH_2SH + HSCH_2CH_3$

3.   $CH_3CH_2OCH_2CH_2SH + HSCH_3$

4.   $HSCH_2CH_2CH_2CH_2SH$

5.   2 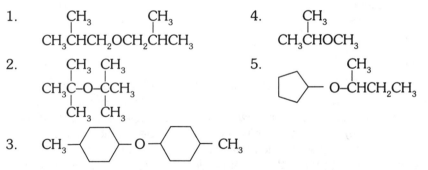$-SH$

## VI.   Drill in Writing the Structures of Ethers That Can Be Made from Alcohols

1.
$$\underset{CH_3CHCH_2OCH_2CHCH_3}{\overset{\overset{\displaystyle CH_3}{|}\qquad\qquad\overset{\displaystyle CH_3}{|}}{}}$$

2.
$$CH_3\overset{\overset{\displaystyle CH_3}{|}}{\underset{\underset{\displaystyle CH_3}{|}}{C}}-O-\overset{\overset{\displaystyle CH_3}{|}}{\underset{\underset{\displaystyle CH_3}{|}}{C}}CH_3$$

3.   $CH_3-$⬡$-O-$⬡$-CH_3$

4.
$$\underset{CH_3CHOCH_3}{\overset{\overset{\displaystyle CH_3}{|}}{}}$$

5.   ⬠$-O-\overset{\overset{\displaystyle CH_3}{|}}{C}HCH_2CH_3$

## VII.   Drill in Writing Products of the Reactions of Amines with Strong Acids

1.   $CH_3CH_2CH_2NH_3^+$

2.   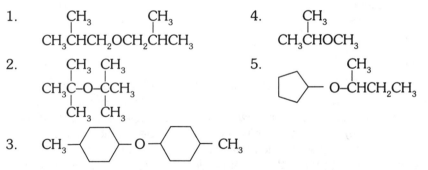$-\overset{+}{N}H_2CH_3$

3.   $CH_3CH_2\overset{+}{N}HCH_2CH_3$
   $\qquad\qquad\ |$
   $\qquad\qquad\ CH_3$

4.   ⬠$\overset{+}{N}H_2$

5.   $CH_3\overset{+}{N}H_2CH_2CH_2\overset{+}{N}H_2CH_3$

## VIII. Drill in Writing the Products of the Reactions of Protonated Amines with Strong, Aqueous Base

1.   $CH_3CH_2NH_2$

2.   $CH_3CH_2NCH_3$
   $\qquad\qquad |$
   $\qquad\qquad CH_3$

3.   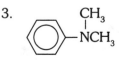$-NCH_3$
   $\qquad\qquad\qquad\quad\overset{\overset{\displaystyle CH_3}{|}}{}$

4. 

$$NH_2CH_2CH_2\overset{\overset{\displaystyle O}{\|}}{C}CH_3$$

5. 

$$\underset{\underset{\displaystyle CH_3}{|}}{NH_2CHCO_2^-}$$

## ANSWERS TO SELF-TESTING QUESTIONS

### Completion

1. (1)  propyl alcohol
   (2)  methyl alcohol
   (3)  *t*-butyl alcohol
   (4)  isopropyl alcohol
   (5)  dimethylamine
   (6)  diethyl ether
   (7)  methyl mercaptan

2. (a)  none
   (b)  
$$\underset{\underset{\displaystyle CH_3CHCH=O}{}}{\overset{\overset{\displaystyle CH_3}{|}}{}}$$

   (c)  $CH_3CH_2-S-S-CH_2CH_3$ (Some students are frustrated by this reaction. Noting that the coefficient 2 does not stand before the structure of the reactant $CH_3CH_2SH$, they ask how this product can form.  The answer is simply that you balance the equation *after* you have written products.  The coefficient 1 before the reactant doesn't stand just for one molecule, anyway.  It also represents a whole mole, surely enough for 1/2 mole of product to form.  The coefficient 2 appeared in the next problem because we agreed to use this as a special signal for handling the options when the same substance can go to two different products.)

   (d)  $CH_3CH_2CH_2-O-CH_2CH_2CH_3$

   (e)  $CH_3NH_3Cl^-$

   (f)  
$$\underset{\underset{\displaystyle CH_2=C-CH_3}{}}{\overset{\overset{\displaystyle CH_3}{|}}{\ _+}}$$

   (g)  $CH_3CH_2CH=O$

   (h) 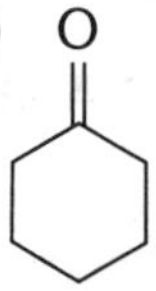  O

(i)     $$CH_3CH_2\overset{\overset{\displaystyle O}{\|}}{C}CH_2CH_3$$

(j)  none

(k)  none

(l)  $CH_2=O$

(m)  $CH_2=CHCH_3$

(n)  $CH_3CH_2CH_2OH$

(o)  $CH_3CH_2SH$

(p)     $$CH_3CH_2-O-CH_2CH_2\overset{\overset{\displaystyle O}{\|}}{C}CH_3$$

(q)  $CH_3\underset{\underset{\displaystyle NH_2}{|}}{C}HCH_3$

3.  (a)  $CH_2=CH_2 + H_2 \xrightarrow[\text{heat, pressure}]{\text{Ni}} CH_3CH_3$

(b)     $$CH_3CH_2CH_2OH \xrightarrow{(O)} CH_3CH_2\overset{\overset{\displaystyle O}{\|}}{C}H + H_2O$$

(c)  $CH_2=CH_2 + H_2O \xrightarrow{H^+} CH_3CH_2OH$

(d)  $2CH_3CH_2CH_2OH \xrightarrow[\text{heat}]{H_2SO_4} CH_3CH_2CH_2OCH_2CH_2CH_3 + H_2O$

(e)  $2CH_3CH_2SH \xrightarrow{(O)} C_3CH_2S-SCH_2CH_3 + H_2O$

## Multiple-Choice

1.  c
2.  c
3.  a, c
4.  c
5.  b, c
6.  a
7.  c, d
8.  d
9.  d
10.  a, c
11.  b, c
12.  b   (The alkane, A, cannot form hydrogen bonds with anything; the amino group sets up weaker hydrogen bonds than does the alcohol group.)

13.  d    (Hydrogen bonding is the clue again.)
14.  d
15.  e
16.  a, b
17.  a, c
18.  a

# CHAPTER 12

# ALDEHYDES AND KETONES

## OBJECTIVES

After studying the chapter and working the exercises in it, you should be able to do the following. Pay particular attention to objectives 1, 5, and 8 through 13. They are very important to our future needs.

1.  *Recognize the following families in the structures of compounds: aldehydes, ketones, carboxylic acids, hemiacetals, hemiketals, acetals, and ketals.*

2.  *Give the common names and structures for aldehydes, ketones, and acids through $C_4$.*

3.  *Compare the physical properties of aldehydes and ketones with those of alcohols or hydrocarbons of comparable formula weights.*

4.  *Write specific examples of reactions for the synthesis of simple aldehydes and ketones from alcohols.*

5.  *Write specific examples of reactions for the oxidation of aldehydes.*

6.  *Describe what one does and sees in a positive Tollens' Test.*

7. *Describe what one does and sees in a positive Benedict's Test.*

8. *Illustrate specific reactions for the reduction of aldehydes and ketones.*

9. *Give specific equations for the addition of an alcohol to an aldehyde (or ketone) to form a hemiacetal (or a hemiketal).*

10. *Write the structures of the original alcohol and aldehyde (or ketone) from the structure of a hemiacetal (or hemiketal).*

11. *Give specific examples of reactions forming acetals or ketals.*

12. *Give specific examples of reactions showing the hydrolysis of acetals or ketals.*

13. *Give definitions of the terms in the Glossary and provide illustrations where applicable.*

Be sure to prepare and learn the one-sentence statements of chemical properties of aldehydes, ketones, hemiacetals, hemiketals, acetals, and ketals. Also, be sure to prepare the 5 x 8" reaction summary cards, and the cards that list the functional groups affected by various kinds of inorganic reactants.

## GLOSSARY

**Acetal.**   Any organic compound in which two ether linkages extend from one carbon:

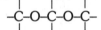

**Aldehyde.**   An organic compound with the general formula RCH=O.

**Aldehyde Group.**   −CH=O

**Benedict's Reagent.**   The metallo-organic complex of the $Cu^{2+}$ ion and the citrate ion in a solution that is basic.

**Benedict's Test.**   A simple test tube procedure using Benedict's reagent to test for the presence of any of three systems of functional groups:

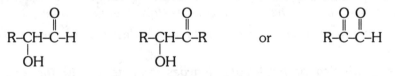

In a positive test, the intensely blue solution changes into a nearly colorless solution and there is a simultaneous separation of a reddish precipitate of copper(I) oxide, $Cu_2O$.  It is a common test for reducing sugars such as glucose.

**Carbonyl Group.**   The carbon-oxygen double bond:  C=O

**Hemiacetal.**   Any organic compound in which an −OH group and an ether linkage come together at the same carbon:

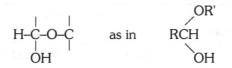

**Hemiketal.** A substance whose molecules have the system:

**Ketal.** A substance whose molecules have the system:

$R_2C(OR')_2$. It is made from a ketone ($R_2C=O$) and two alcohols (HOR').

**Keto Group.** The carbonyl group when it occurs in a ketone.

**Ketone.** Any compound with $C-\overset{\overset{\displaystyle O}{\|}}{C}-C$ in its molecules.

**Tollens' Reagent.** A solution of the diammine complex of the silver ion, $Ag(NH_3)_2^+$, in water that is slightly alkaline.

**Tollens' Test.** A simple test tube test using Tollens' reagent to detect an aldehyde. A positive Tollens' test consists of the appearance of metallic silver, usually in the form of a silver mirror in the test tube.

## KEY MOLECULAR "MAP SIGNS"

| Key Molecular "Map Signs" in Organic Molecules | What to Expect When This Functional Group Is Present |
|---|---|
| $-\overset{\overset{\displaystyle O}{\|}}{C}-H$ <br><br> Aldehyde group | **Influence on physical properties:** <br> The aldehyde group is moderately polar; it can accept H–bonds. <br><br> **Influence on chemical properties:** <br> – One of the most easily oxidized groups: <br>     Changes to a carboxyl group. Gives Tollens' test. <br> – Can be reduced to a 1° alcohol group. <br> – Adds an alcohol molecule to form a hemiacetal. <br> – Can be converted into an acetal system. |

| Key Molecular "Map Signs" in Organic Molecules | What to Expect When This Functional Group Is Present |
|---|---|
| $$\overset{\displaystyle O}{\underset{\displaystyle \phantom{}}{\overset{\parallel}{R-C-R}}}$$ Ketone | **Influence on physical properties:** Same as the aldehyde group **Influence on chemical properties:** – Strongly resists oxidation (unlike the aldehydes). – Can be reduced to a 2° alcohol group. – Adds an alcohol to form a hemiketal (although not as readily as an aldehyde gives the hemiacetal) – Can be converted into a ketal system. |
| $$\underset{\displaystyle O-R'}{\overset{\displaystyle OH}{R-C-H}}$$ Hemiacetal $$\underset{\displaystyle O-R'}{\overset{\displaystyle OH}{R-C-R}}$$ Hemiketal | **Two properties of importance:** – Both the hemiacetal and the hemiketal systems are unstable; they exist in equilibrium with the aldehyde or ketone and the alcohol (R'OH) that formed them. – Both systems can be changed to the acetal or ketal system by a reaction with another alcohol molecule. |
| $$\underset{\displaystyle O-R'}{\overset{\displaystyle O-R'}{R-C-H}}$$ Acetal $$\underset{\displaystyle O-R'}{\overset{\displaystyle O-R'}{R-C-R}}$$ Ketal | **One important property:** – Both the acetal and the ketal systems react with water when an acid catalyst is present, but they do not react in the presence of a basic catalyst; in so doing, they revert back to the original aldehyde $\left(\overset{O}{\overset{\parallel}{R-C-H}}\right)$ or ketone $\left(\overset{O}{\overset{\parallel}{R-C-R}}\right)$ and alcohol (2R'OH). |

## SELF-TESTING QUESTIONS

### Completion

1.  Write the full structure of each compound on the line above its condenses structure.

(a)  _____
     $CH_3CHO$

(b)  _____
     $CH_3CHOCH_3$
        |
        $OH$

(c)  _____
     $CH_3CH_2OH$

(d)  _____
     $CH_3CHOCH_3$
        |
        $OCH_3$

2.  What functional groups have we studied that will be attacked by oxidizing agents? (Give the name of the product, too.)

(a)  In this chapter     _____

     _____

(b)  In previous chapters _____

     _____

     _____

     _____

3.  What functional groups have we studied that will be attacked by reducing agents? (Give the name of the product, too.)

(a)  In this chapter     _____

     _____

(b)  In earlier chapters _____

     _____

     _____

4.    By means of short statements, summarize the chemical reactions of alcohols
      (from both this and earlier chapters).

      **Example:**  Alcohols can be dehydrated to form alkenes.
      _____

      _____

      _____

      _____

      _____

      _____

      _____

      _____

      _____

5.    Examine the structures shown in question one, parts a through d.  Answer the
      following questions by writing the condensed structures of both the reactants and
      the products.

      (a)   Which of the structures in question one will give a positive test with Tollens'
            Reagent?                                    _____

      (b)   Which can be hydrolyzed?                    _____

      (c)   Which can be oxidized under basic conditions?

                                                        _____

      (d)   Which can be reduced?                       _____

6.    Write the structure(s) of the principal *organic* product(s) that could be expected to
      form in each case.  If no reaction occurs, write "none."

      (a)   $CH_3CH=O$  $\xrightarrow{(O)}$            _____

      (b)   $CH_3CH=O + HOCH_3 \rightleftharpoons$      _____

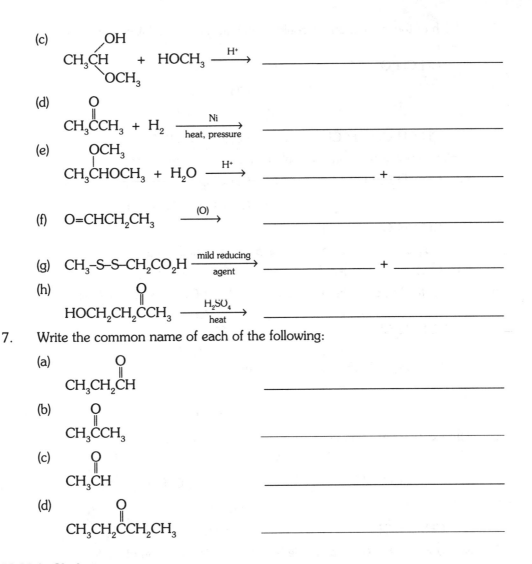

(c)

$$CH_3CH\begin{matrix} OH \\ OCH_3 \end{matrix} + HOCH_3 \xrightarrow{H^+}$$ _____

(d)

$$CH_3\overset{O}{\underset{\|}{C}}CH_3 + H_2 \xrightarrow[\text{heat, pressure}]{Ni}$$ _____

(e)

$$CH_3\overset{OCH_3}{\underset{|}{C}}HOCH_3 + H_2O \xrightarrow{H^+}$$ _____ + _____

(f)   $$O=CHCH_2CH_3 \xrightarrow{(O)}$$ _____

(g)   $$CH_3-S-S-CH_2CO_2H \xrightarrow[\text{agent}]{\text{mild reducing}}$$ _____ + _____

(h)

$$HOCH_2CH_2\overset{O}{\underset{\|}{C}}CH_3 \xrightarrow[\text{heat}]{H_2SO_4}$$ _____

7.   Write the common name of each of the following:

(a)

$$CH_3CH_2\overset{O}{\underset{\|}{C}}H$$   _____

(b)

$$CH_3\overset{O}{\underset{\|}{C}}CH_3$$   _____

(c)

$$CH_3\overset{O}{\underset{\|}{C}}H$$   _____

(d)

$$CH_3CH_2\overset{O}{\underset{\|}{C}}CH_2CH_3$$   _____

## Multiple-Choice

1.   Which of these compounds could be hydrolyzed the most easily (assuming acid catalysis)?

(a)  $CH_3CH_2OCH_2CH_3$          (c)  $CH_3OCH_2CH_2OCH_3$

(b)  $CH_3OCH_2OCH_3$             (d)  $CH_3CH_2CH_2CH_2OH$

2. Which of these compounds could be the most easily oxidized under mild conditions?

(a) $CH_3CH_2OCH_2CH_3$

(c)
$$CH_3CH_2\overset{\overset{\displaystyle O}{\|}}{C}CH_2CH_3$$

(b)
$$CH_3CH_2CH_2CH_2\overset{\overset{\displaystyle O}{\|}}{C}H$$

(d)
$$CH_3CH_2O\overset{\overset{\displaystyle CH_3}{|}}{C}HOCH_2CH_3$$

3. Which of these compounds would be most soluble in water?

(a) $CH_3CH_2CH_2CH_2OH$

(c) $CH_3CH_2CH_2CH_2CH_3$

(b)
$$CH_3CH_2CH_2CH_2\overset{\overset{\displaystyle O}{\|}}{C}H$$

(d) $CH_3CH_2CH_2OCH_2CH_3$

4. The substance that precipitates in a positive Benedict's Test is

(a) CuO    (b) Ag    (c) $Cu_2O$    (d) $Ag^+$

5. The acetal that could be hydrolyzed to $CH_3CH_2OH$ and $CH_3CHO$ is

(a)
$$CH_3CH_2CH\overset{\displaystyle \diagup OH}{\diagdown O-CH_3}$$

(c)
$$CH_3CH\overset{\displaystyle \diagup OH}{\diagdown O-CH_2CH_3}$$

(b)
$$CH_3CH_2O\overset{\overset{\displaystyle }{|}}{C}HCH_3$$
$$\ \ \ \ \ \ \ \ \ OCH_2CH_3$$

(d) none of these

6. The alcohol that could be oxidized to $CH_3\overset{\overset{\displaystyle CH_3}{|}}{C}H-\overset{\overset{\displaystyle O}{\|}}{C}H$ is

(a)
$$CH_3\overset{\overset{\displaystyle CH_3}{|}}{C}H-\overset{\overset{\displaystyle O}{\|}}{C}-OH$$

(c)
$$CH_3\overset{\overset{\displaystyle CH_3}{|}}{C}HCH_2CH_2OH$$

(b)
$$CH_3\overset{\overset{\displaystyle CH_3}{|}}{C}H-OH$$

(d)
$$CH_3\overset{}{C}H-CH_2OH$$
$$\ \ \ \ \ \ \ \underset{\underset{\displaystyle CH_3}{|}}{}$$

7. Isopropyl alcohol could be made by the catalytic reduction of

(a) acetone

(c) methyl ethyl ketone

(b) propionaldehyde

(d) acetaldehyde

8. The oxidation of $CH_3CH_2CH_2\overset{\overset{\displaystyle O}{\|}}{C}H$ would give

(a) isobutyraldehyde

(c) butyraldehyde

(b) butyric acid

(d) 1-butanol

9.  What choice contains the best description of the functional group(s) present in this structure?

(a)  a hemiacetal  (b)  an acetal  (c)  a hemiketal  (d)  a ketal

10. If a hydride donor were used, which of the following systems would be able to accept it?

(a)  $CH_3CH_2CH_2CH_3$

(b)  $CH_3CH_2CH_2OH$

(c)  $CH_3OCH_2CH_3$

(d)
$$CH_3CH_2CH_2\overset{\displaystyle O}{\overset{\displaystyle \|}{C}}H$$

# ANSWERS

### ANSWERS TO SELF-TESTING QUESTIONS

**Completion**

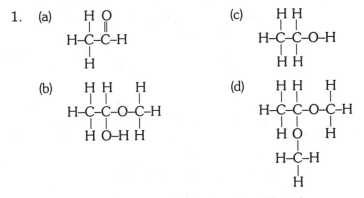

2.  (a)  Aldehydes are oxidized to carboxylic acids.

    (b)  1° alcohols are oxidized to aldehydes or to carboxylic acids.

    2° alcohols are oxidized to ketones.

    Mercaptans are oxidized to disulfides.

3.  (a)  Aldehydes are hydrogenated to 1° alcohols.

    Ketones are hydrogenated to 2° alcohols.

(b)    Alkenes are hydrogenated to alkanes.

Disulfides are reduced to mercaptans.

4.    Alcohols can be dehydrated to form ethers.

1° alcohols can be oxidized to aldehydes, and thence to carboxylic acids.

2° alcohols can be oxidized to ketones.

Alcohols add to aldehydes or ketones to form hemiacetals or hemiketals (both of which are generally too unstable to be isolated).

Alcohols react with hemiacetals or hemiketals to form acetals or ketals.

5.    (a)

$$CH_3CHO \longrightarrow CH_3\overset{\overset{\displaystyle O}{\|}}{C}-OH$$

The hemiacetal, b, will also give a positive test because, in solution, it exists in equilibrium with acetaldehyde and methyl alcohol; acetaldehyde reacts with Tollens' Reagent.

(b)    "To be hydrolyzed" means to undergo a chemical reaction with water.  Only the acetal, d, reacts with water:

$$CH_3\underset{\underset{\displaystyle OCH_3}{|}}{C}HOCH_3 + H_2O \xrightarrow{H^+} CH_3CHO + 2CH_3OH$$

(c)    The aldehyde, a, the hemiacetal, b, and the alcohol, c, can be oxidized under basic conditions.  The alcohol is oxidized to acetaldehyde or to acetic acid, depending on other conditions.  The oxidation of a and b was discussed in the answer to part a of this question.  The acetal is stable in a base and will not hydrolyze; therefore, it cannot break down into oxidizable compounds.

(d)    The aldehyde, a, can be reduced to $CH_3CH_2OH$.  The hemiacetal, b, will also react because it is present in equilibrium with its parent aldehyde ($CH_3CHO$) and alcohol ($CH_3OH$).  As with a, the aldehyde will be taken out by the reducing agent and changed to ethyl alcohol.

6.    (a)    $CH_3CO_2H$          (e)    $CH_3CH=O + 2CH_3OH$

(b)    $CH_3\underset{\underset{\displaystyle OH}{|}}{C}HOCH_3$          (f)    $CH_3CH_2CO_2H$

(c)    $CH_3\underset{\diagdown OCH_3}{\overset{\diagup OCH_3}{C}}H$   $(+H_2O)$          (g)    $CH_3SH + HSCH_2CO_2H$

(d)    $CH_3\underset{\underset{\displaystyle OH}{|}}{C}HCH_3$          (h)    $CH_2=CH\overset{\overset{\displaystyle O}{\|}}{C}CH_3$

7.  (a)  propionaldehyde

    (b)  acetone

    (c)  acetaldehyde

    (d)  diethyl ketone

**Multiple-Choice**

1.  b (an acetal)   (a is an ether; c is a di-ether and not an acetal; and d is an alcohol.)

2.  b (an aldehyde) (a is an ether; c is a ketone; and d is an acetal.  The oxidizing reagent is neutral or basic, as in Tollens' Test or Benedict's Test; therefore, the acetal will hold together.  However, if the reagent is acidic, the acetal will hydrolyze to produce some acetaldehyde (and ethyl alcohol) and the aldehyde will be oxidized.

3.  a (an alcohol)  (It is the only one that can both accept and donate hydrogen bonds.)

4.  c

5.  b

6.  d

7.  a

8.  b

9.  a

10. d

# CHAPTER 13

# CARBOXYLIC ACIDS AND THEIR DERIVATIVES

## OBJECTIVES

After you have studied this chapter and worked the exercises and problems in it, you should be able to do all of the following. Objectives one through five are very important for understanding the applications of this chapter in biochemistry and in our study of the molecular basis of life.

1.  *Recognize the structural features of: the carboxylic acid group, the carboxylate ion group, the ester group and ester bond, the amide group and amide bond.*

2.  *By examining a structure of a substance, determine the probable ability of the substance to neutralize either a base or an acid or to be hydrolyzed, saponified, or esterified.*

3.  Write equations that are specific examples of the formation of:

    (a)  an acid from an alcohol or an aldehyde

    (b)  a carboxylic acid salt from the acid

    (c)  a carboxylic acid from its salt

    (d)  an ester from an alcohol and a carboxylic acid

    (e)  an alcohol and an acid from an ester

    (f)  an alcohol and the salt of a carboxylic acid from an ester

    (g)  an amine and an acid from an amide

4.  Write the structure of the amide that can be made from an acid and ammonia or an amine.

5.  Explain through equations and discussion how the carboxyl group may be used as a "solubility switch."

6.  Write the structural features common to phosphate, diphosphate, and triphosphate esters.

7.  Define the terms in the Glossary, and give illustrations where applicable.

Reaction summary cards should be made for acids, anions of acids, esters, and amides. The partly completed cards for alcohols and amines can now be completed. A list of sentences that summarize chemical facts should be prepared and learned. The lists of reactions organized by key reagents (e.g., acids, bases, water and so forth) should be brought up to date.

## GLOSSARY

**Acid Derivative.**   Any organic compound that can be made from an organic acid or that can be changed back to an organic acid by hydrolysis.

**Acyl Group.**

**Acyl Group Transfer Reaction.**   Any reaction in which an acyl group transfers from a donor to an acceptor.

**Amide.**   Any organic compound having a carbonyl-nitrogen system:

$$
\underset{|}{\overset{\overset{\textstyle O}{\|}}{-C}}-\overset{..}{N}-
$$

**Amide Bond.**   The carbonyl-to-nitrogen bond in an amide

**Carboxylate Ion.**

$$R-\overset{\overset{\displaystyle O}{\|}}{C}-O^-$$

**Carboxylic Acid.**

A compound with the $-\overset{\overset{\displaystyle O}{\|}}{C}-OH$ group (sometimes written as $CO_2H$ or as $COOH$).

**Ester.**   Any organic compound having a carbonyl-oxygen-carbon system:

$$-\overset{\overset{\displaystyle O}{\|}}{C}-O-\overset{\displaystyle |}{\underset{\displaystyle |}{C}}-$$

**Ester Bond.**   The carbonyl-to-oxygen bond in an ester.

**Esterification.**   The formation of an ester through the reaction of an acid with an alcohol.  The acid may be an inorganic oxy-acid (e.g., mono-, di-, or triphosphoric acid) or an organic acid (e.g., carboxylic acid.)

**Saponification.**   The reaction of an ester with aqueous sodium or potassium hydroxide to give an alcohol and the salt of an acid.

## KEY MOLECULAR "MAP SIGNS"

| Key Molecular "Map Signs" in Organic Molecules | What to Expect When This Functional Group Is Present |
| --- | --- |
| $-\overset{\overset{\displaystyle O}{\|}}{C}-O-H$ <br><br> Carboxyl group <br><br> [Often written as $CO_2H$ or $COOH$] | **Influence on physical properties:** <br><br> The carboxyl group is a very polar group; it can both donate and accept H–bonds. <br><br> **Influence on chemical properties:** <br><br> – Can neutralize $OH^-$ (or $HCO_3^-$ or $CO_3^{2-}$); in so doing, the group becomes $-CO_2^-$, the carboxylate ion discussed next. <br><br> – Can be changed into an ester by reacting with an alcohol when a mineral acid catalyst is present. <br><br> – Can be changed into an amide. |

| Key Molecular "Map Signs" in Organic Molecules | What to Expect When This Functional Group Is Present |
|---|---|
| O‖ −C−O⁻  Carboxylate ion | **Influence on physical properties:**<br><br>− One of the most effective groups at bringing long hydrocarbon chains into solution.<br><br>− All salts of the carboxylic acids are solids at room temperature.<br><br>**Influence on chemical properties:**<br><br>− Aqueous solutions will test slightly basic.<br><br>− Can neutralize mineral acids (and revert to carboxyl group) |
| O‖ −C−O−C−  Ester | **Influence on physical properties:**<br><br>The ester group is a moderately polar group; it can accept H− bonds but cannot donate them.<br><br>**Influence on chemical properties:**<br><br>− Can be hydrolyzed (to the carboxylic acid and alcohol).<br><br>− Can be saponified (to the carboxylate ion and alcohol). |
| O‖ −C−N−  Amide | **Influence on physical properties:**<br><br>Amides have a very polar group (particularly if at least one hydrogen is attached to the nitrogen); virtually all amides are solids at room temperature.<br><br>**Influence on chemical properties:**<br><br>− Amides are not basic; they are not proton acceptors in the way that makes amines basic.<br><br>− Can be hydrolyzed by aqueous acids or bases, if heated, to give the carboxylic acid and the amine (or ammonia). |

| Key Molecular "Map Signs" in Organic Molecules | What to Expect When This Functional Group Is Present |
|---|---|
| 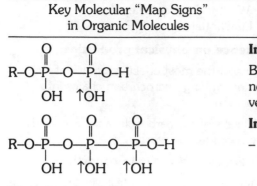<br><br>Di- and triphosphate esters | **Influence on physical properties:**<br><br>Both esters exist at the pH of body fluids as negatively charged ions; therefore they are very soluble in water.<br><br>**Influence on chemical properties:**<br><br>– Can neutralize OH⁻ (to the extent they begin with –OH groups on phosphorus and have not been neutralized).<br><br>– In absence of a catalyst, they only react very slowly with water at the pH of body fluids.<br><br>– In the presence of the appropriate enzymes, they react rapidly with the alcohol (and amino) groups of biochemicals and suffer breakage of the bonds indicated by the arrows (↑). |

## Additional Reactions of Functional Groups Introduced in Earlier Chapters

|  |  |
|---|---|
| $-\overset{\textstyle\vert}{\underset{\textstyle\vert}{C}}-O-H$<br>Alcohol (or phenol) | – Forms esters with carboxylic acids. |
| $-\overset{\textstyle\vert}{N}-H$<br>Ammonia or any amine with at least one H on N | – Can be converted into amides of carboxylic acids. |

# DRILL EXERCISES

## I.    EXERCISES IN STRUCTURES AND NAMES

1.    To make sure that you understand the condensed structures that are often used with carbonyl compounds, write out the following condensed structures as full structures.

(a)  $CH_3CO_2H$

(b)  $HO_2CCH_2CH_3$

(c)  $CH_3CHO$

(d)  $CH_3CH_2OH$

(e)  $CH_3CO_2CH_3$

(f)  $CH_3CONH_2$

(g)  $HOOCCH_3$

2.  It is important to be able to recognize quickly the presence of functional groups in structures; otherwise, the chemical and physical properties of such structures cannot be "read" (in the sense of "reading" a map with knowledge of "map signs").  Study each of the following structures and assign them to their correct families.  Some will have more than one functional group; name them all.

**Examples:**

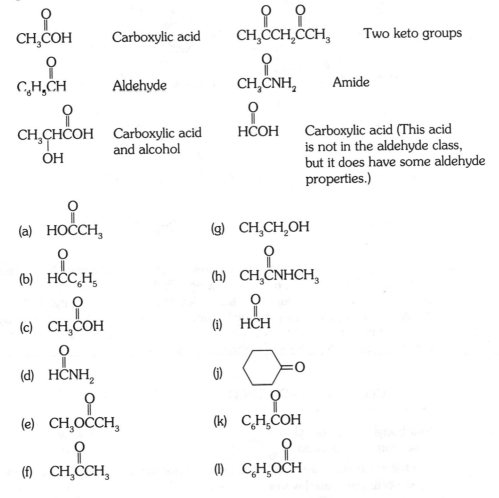

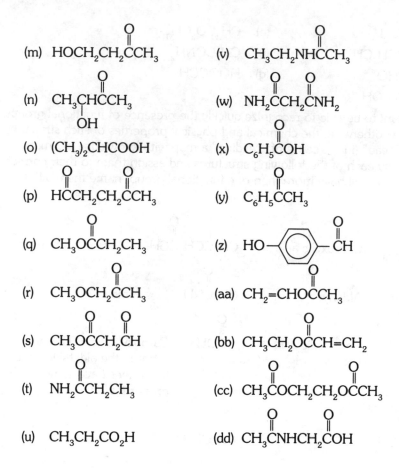

(m)  HOCH$_2$CH$_2$CCH$_3$

(n)  CH$_3$CHCCH$_3$
         |
         OH

(o)  (CH$_3$)$_2$CHCOOH

(p)  HCCH$_2$CH$_2$CCH$_3$

(q)  CH$_3$OCCH$_2$CH$_3$

(r)  CH$_3$OCH$_2$CCH$_3$

(s)  CH$_3$OCCH$_2$CH

(t)  NH$_2$CCH$_2$CH$_3$

(u)  CH$_3$CH$_2$CO$_2$H

(v)  CH$_3$CH$_2$NHCCH$_3$

(w)  NH$_2$CCH$_2$CNH$_2$

(x)  C$_6$H$_5$COH

(y)  C$_6$H$_5$CCH$_3$

(z)  HO—⟨◯⟩—CH

(aa) CH$_2$=CHOCCH$_3$

(bb) CH$_3$CH$_2$OCCH=CH$_2$

(cc) CH$_3$COCH$_2$CH$_2$OCCH$_3$

(dd) CH$_3$CNHCH$_2$COH

3.  If after studying the ways to name esters, amides, and salts discussed in the text, you continue to have trouble, try the following. Prefixes in the common names for esters, amides, and acid salts (and aldehydes) relate to the carbonyl portion of the structure. You may need some drill simply in recognizing what the carbonyl portion is. It is that part of the structure that contains the carbonyl group plus whatever else consisting solely of carbon and hydrogen.

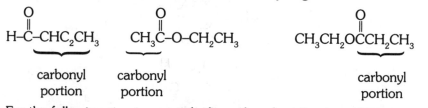

    carbonyl      carbonyl                          carbonyl
    portion       portion                           portion

For the following structures, circle the carbonyl portion in each structure, then write the prefix associated with it.

**Examples:**

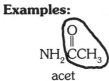

NH₂CCH₃
acet

CH₃CH₂CH₂C OCH₂CH₂CH₂CH₃
butyr

Note that the oxygen that has only single bonds and the R–group attached to it are not included.

(a)  O
     ‖
     HCOCH₃

(b)        O
           ‖
     CH₃CH₂CO⁻Na⁺

(c)          O
             ‖
     CH₃CH₂OCH

(d)          O
             ‖
     CH₃CH₂OCCH₃

(e)          O
             ‖
     CH₃CH₂COCH₃

(f)              O
                 ‖
     CH₃CH₂CH₂COCH₂CH₃

4.  Complete the following table according to the example given.

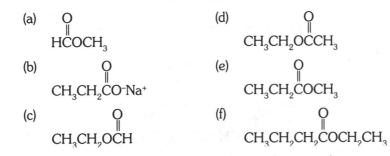

| Structure | Family | Common Name |
|---|---|---|
| CH₃COCH₃ (with C=O) | ester | methyl acetate |
| (a) (CH₃)₂CHOCCH₃ (with C=O) | _____ | _____ |
| (b) CH₃CH₂CNH₂ (with C=O) | _____ | _____ |
| (c) CH₃OCCH₂CH₂CH₃ (with C=O) | _____ | _____ |
| (d) (CH₃)₂CHCH₂OCCH₂CH₂CH₃ (with C=O) | _____ | _____ |
| (e) H–CCH₂CH₂CH₃ (with C=O) | _____ | _____ |

5.    Write the common names for each.

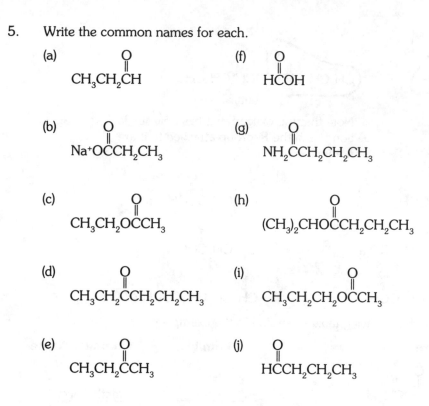

(a)
$$CH_3CH_2CH$$ with O double bonded (aldehyde)

(f)
$$HCOH$$ with O

(b)
$$Na^+OCCH_2CH_3$$ with O

(g)
$$NH_2CCH_2CH_2CH_3$$ with O

(c)
$$CH_3CH_2OCCH_3$$ with O

(h)
$$(CH_3)_2CHOCCH_2CH_2CH_3$$ with O

(d)
$$CH_3CH_2CCH_2CH_2CH_3$$ with O

(i)
$$CH_3CH_2CH_2OCCH_3$$ with O

(e)
$$CH_3CH_2CCH_3$$ with O

(j)
$$HCCH_2CH_2CH_3$$ with O

## II.    DRILL ON THE REACTIONS OF CARBOXYLIC ACIDS WITH STRONG BASES

Write the structures of the salts that form in the following situations.  Assume in every case that the aqueous base is being used at room temperature.  You may write the structures showing the electrical charges (e.g., $CH_3CO_2Na^+$) or not (e.g., $CH_3CO_2Na$), but you should never draw a line between the two parts of the salts (e.g., $CH_3CO_2$–Na) because a line means a <u>covalent</u> bond, which is not present here.

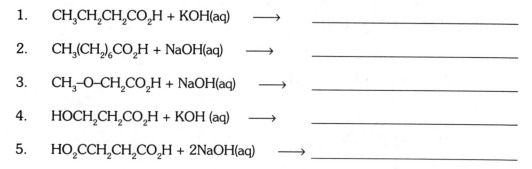

1.    $CH_3CH_2CH_2CO_2H$ + KOH(aq)  $\longrightarrow$  _____

2.    $CH_3(CH_2)_6CO_2H$ + NaOH(aq)  $\longrightarrow$  _____

3.    $CH_3$–O–$CH_2CO_2H$ + NaOH(aq)  $\longrightarrow$  _____

4.    $HOCH_2CH_2CO_2H$ + KOH (aq)  $\longrightarrow$  _____

5.    $HO_2CCH_2CH_2CO_2H$ + 2NaOH(aq)  $\longrightarrow$  _____

## III.    DRILL ON THE REACTIONS OF CARBOXYLIC ACID SALTS WITH STRONG ACIDS

Write the structures of the acids that form in the following situations.  Assume in every case that the aqueous acid is being used at room temperature.

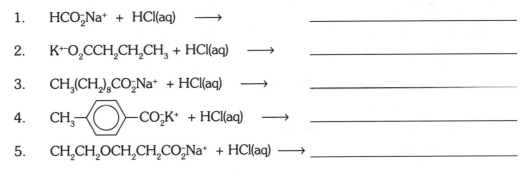

1.    $HCO_2^-Na^+$ + $HCl(aq)$  $\longrightarrow$    _____

2.    $K^+{}^-O_2CCH_2CH_2CH_3$ + $HCl(aq)$  $\longrightarrow$    _____

3.    $CH_3(CH_2)_8CO_2^-Na^+$ + $HCl(aq)$  $\longrightarrow$    _____

4.    $CH_3$—⬡—$CO_2^-K^+$ + $HCl(aq)$  $\longrightarrow$    _____

5.    $CH_2CH_2OCH_2CH_2CO_2^-Na^+$ + $HCl(aq)$ $\longrightarrow$ _____

## IV.    DRILL ON WRITING THE STRUCTURES OF ESTERS THAT CAN FORM FROM GIVEN ACIDS AND ALCOHOLS

Example 13.1 in the text provides the pattern.  If after doing Practice Exercises 3 and 4 you feel the need for more practice, try the following.  Write the structures of the esters that can form between the following pairs of compounds.  This will also serve as a review of the names of acids and alcohols.  On the lines above the names write the structures of the reactants, and then form the structures of the products.

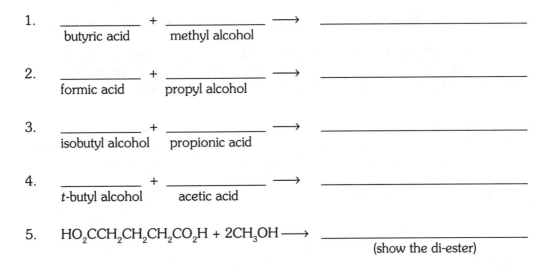

1.    _____ + _____ $\longrightarrow$ _____
         butyric acid        methyl alcohol

2.    _____ + _____ $\longrightarrow$ _____
         formic acid         propyl alcohol

3.    _____ + _____ $\longrightarrow$ _____
      isobutyl alcohol    propionic acid

4.    _____ + _____ $\longrightarrow$ _____
      *t*-butyl alcohol        acetic acid

5.    $HO_2CCH_2CH_2CH_2CO_2H$ + $2CH_3OH \longrightarrow$ _____
                                                            (show the di-ester)

## V.     DRILL ON WRITING THE PRODUCTS OF THE HYDROLYSIS OF ESTERS

Example 13.2 shows how to do this.  For more drill beyond Practice Exercise 5, try the following.  Write the structures of the products of the hydrolysis of the following esters.

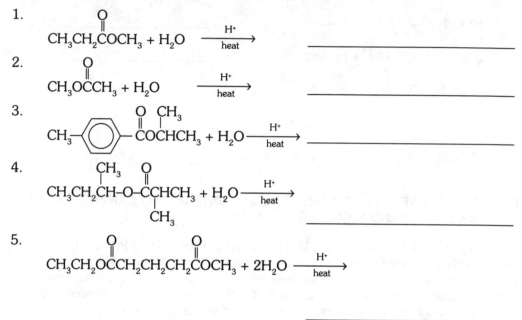

1.

$$CH_3CH_2\overset{\overset{\displaystyle O}{\|}}{C}OCH_3 + H_2O \xrightarrow[\text{heat}]{H^+}$$ _____

2.

$$CH_3O\overset{\overset{\displaystyle O}{\|}}{C}CH_3 + H_2O \xrightarrow[\text{heat}]{H^+}$$ _____

3.

$$CH_3-\!\!\bigcirc\!\!-\overset{\overset{\displaystyle O}{\|}}{C}O\overset{\overset{\displaystyle CH_3}{|}}{C}HCH_3 + H_2O \xrightarrow[\text{heat}]{H^+}$$ _____

4.

$$CH_3CH_2\overset{\overset{\displaystyle CH_3}{|}}{C}H-O-\overset{\overset{\displaystyle O}{\|}}{C}\overset{\underset{\displaystyle CH_3}{|}}{C}HCH_3 + H_2O \xrightarrow[\text{heat}]{H^+}$$ _____

5.

$$CH_3CH_2O\overset{\overset{\displaystyle O}{\|}}{C}CH_2CH_2CH_2\overset{\overset{\displaystyle O}{\|}}{C}OCH_3 + 2H_2O \xrightarrow[\text{heat}]{H^+}$$

_____

## VI.     DRILL IN WRITING THE PRODUCTS OF THE SAPONIFICATION OF ESTERS

Example 13.3 in the text describes how to do this kind of exercise.  For additional drill, write the structures of the products of the complete saponification of the esters of the preceding drill exercise.  Use NaOH(aq) as the saponifying agent.

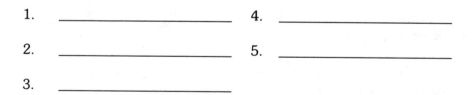

1. _____     4. _____

2. _____     5. _____

3. _____

## VII.    DRILL IN WRITING THE STRUCTURES OF THE AMIDES THAT CAN BE MADE (DIRECTLY OR INDIRECTLY) FROM GIVEN CARBOXYLIC ACIDS AND AMINES

See Example 13.4 for a discussion of how to work this kind of problem. The following will give extra opportunities to drill yourself. Write the structure of the amides that can be made from the given acids and amines (or ammonia).

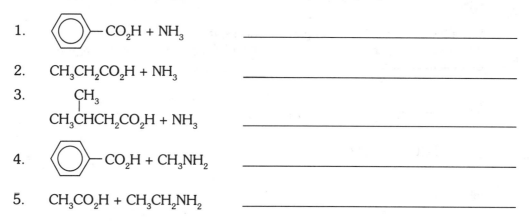

1. ⬡—$CO_2H$ + $NH_3$ _____

2. $CH_3CH_2CO_2H$ + $NH_3$ _____

3. $\overset{\displaystyle CH_3}{\underset{|}{CH_3CHCH_2CO_2H}}$ + $NH_3$ _____

4. ⬡—$CO_2H$ + $CH_3NH_2$ _____

5. $CH_3CO_2H$ + $CH_3CH_2NH_2$ _____

## VIII.    DRILL IN WRITING THE PRODUCTS OF THE HYDROLYSIS OF AMIDES

Example 13.5 in the text discusses how this kind of problem can be worked. For more practice, do the following. Write the structures of the products that can form by the hydrolysis of the following amides. Show the acids as acids, not as their salts. Similarly, show the amines as amines, not in protonated forms.

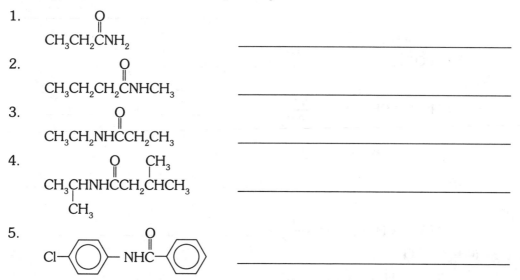

1. $CH_3CH_2\overset{\displaystyle O}{\overset{\|}{C}}NH_2$ _____

2. $CH_3CH_2CH_2\overset{\displaystyle O}{\overset{\|}{C}}NHCH_3$ _____

3. $CH_3CH_2NH\overset{\displaystyle O}{\overset{\|}{C}}CH_2CH_3$ _____

4. $CH_3\underset{|}{CHNH}\overset{\displaystyle O}{\overset{\|}{C}}CH_2\underset{|}{CHCH_3}$ _____
   $\quad\;\;CH_3 \qquad\qquad CH_3$

5. Cl—⬡—$NH\overset{\displaystyle O}{\overset{\|}{C}}$—⬡ _____

## SELF-TESTING QUESTIONS

Use the following questions as your own final examination for the chapter.   As a review, go back to the chapter objectives and find out if you can do them.

### Completion

1.    Write the structure(s) of the organic product(s) that would form in each reaction. If no reaction occurs, write "none."  Some of the reactions will involve a review of earlier chapters.

(a)    $CH_3CO_2H + NaOH \xrightarrow[water]{}$ _____

(b)    $CH_3CO_2H + CH_3OH \xrightarrow[heat]{H^+}$ _____

(c)    $CH_3CH_2CO_2H + NH_3 \xrightarrow[heat]{}$ _____

(d)    $\underset{NH_2\overset{\displaystyle O}{\overset{\|}{C}}CH_3}{} + H_2O \xrightarrow[heat]{}$ _____ + _____

(e)    $CH_3O\overset{\displaystyle O}{\overset{\|}{C}}CH_3 + H_2O \xrightarrow[heat]{H^+}$ _____ + _____

(f)    $\underset{\underset{OH}{|}}{CH_3CHCO_2H} \xrightarrow{(O)}$ _____

(g)    $CH_3O\overset{\displaystyle O}{\overset{\|}{C}}CH_2CH_2\overset{\displaystyle O}{\overset{\|}{C}}OCH_3 \xrightarrow[heat]{H_2O,\ H^+}$ _____ + _____

(h)    $CH_3OCH_2CH_2\overset{\displaystyle O}{\overset{\|}{C}}OCH_3 \xrightarrow[heat]{H_2O,\ H^+}$ _____ + _____

(i)    $\underset{\underset{CH_3}{|}}{NH_2CH_2\overset{\displaystyle O}{\overset{\|}{C}}NHCH\overset{\displaystyle O}{\overset{\|}{C}}OH} \xrightarrow[heat]{H_2O}$ _____ + _____

(j)    $CH_3S\text{-}S\text{-}CH_2\text{-}\overset{\displaystyle O}{\overset{\|}{C}}\text{-}O^-Na^+ \xrightarrow[water]{HCl}$ _____

(k)    $CH_3O\overset{\displaystyle O}{\overset{\|}{C}}CH_3 + NaOH \xrightarrow[heat]{}$ _____ + _____

2.    Which of the reactions in question 1 illustrate

(a)   esterification _____   (b)   saponification _____?

## Multiple-Choice

1.    Organic functional groups that are hydrolyzed by water (usually in the presence of an acid catalyst and heat) are

(a)   ethers                    (d)   disulfides

(b)   acetals                   (e)   esters

(c)   amides                    (f)   carboxylic acids

2.    Organic functional groups that are rather easily oxidized are

(a)   alkanes                   (d)   ketones

(b)   aromatic hydrocarbons     (e)   aldehydes

(c)   mercaptans                (f)   amides

3.    Organic functional groups that are good proton acceptors are

(a)   amino groups             (d)   carboxylate ions

(b)   amides                   (e)   aromatic hydrocarbons

(c)   alkanes                  (f)   mercaptans

4.    Organic functional groups that are good proton donors are

(a)   amino groups             (d)   substituted ammonium ions

(b)   amides                   (e)   carboxylic acids

(c)   alkanes                  (f)   alcohols

5.    Which compound is the most acidic?

(a)   $CH_3CH_2OH$             (c)   $CH_3\overset{\displaystyle O}{\overset{\|}{C}}-NH_2$

(b)   $CH_3CH=O$               (d)   $CH_3CO_2H$

6.    Which compound is the most basic?

(a)   $CH_3CH_2OH$             (c)   $CH_3CH_2NH_3^+Cl^-$

(b)   $CH_3CH_2NH_2$           (d)   $CH_3\overset{\displaystyle O}{\overset{\|}{C}}-NH_2$

7.  In the following structure, the numbered arrows point toward functional groups. What are the numbers of the arrows pointing toward groups readily attacked by relatively mild oxidizing agents?

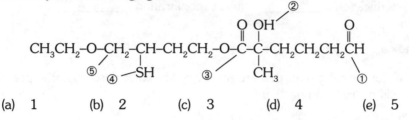

    (a)  1          (b)  2          (c)  3          (d)  4          (e)  5

8.  In the structure of question 7, which groups, if any, are subject to acid–catalyzed hydrolysis?

    (a)  1     (b)  2     (c)  3     (d)  4     (e)  5     (f)  none

9.  In the structure of question 7, which groups, if any, will neutralize aqueous sodium hydroxide at room temperature?

    (a)  1     (b)  2     (c)  3     (d)  5          (e)  none of these

10. The action of 1 mole of water (containing a trace of acid catalyst) on 1 mole of compound Y produces 1 mole of $CH_3CO_2H$ and 1 mole of $CH_3CH_2OH$. The structure of Y is

    (a)  $CH_3CH_2OCH_2CH_3$

    (c)  $CH_3CH-O-CH_2CH_3$
                 |
                 $OH$

    (b)
    $$CH_3CH_2O\overset{\displaystyle O}{\overset{\|}{C}}CH_3$$

    (d)
    $$CH_3\overset{\displaystyle O}{\overset{\|}{C}}-O-\overset{\displaystyle O}{\overset{\|}{C}}CH_2CH_3$$

11.
    The compound $CH_3O-CH_2\overset{\displaystyle O}{\overset{\|}{C}}-NH_2$ could be made by heating a mixture of

    (a)
    $CH_3OH$ and $HOCH_2\overset{\displaystyle O}{\overset{\|}{C}}NH_2$

    (b)
    $CH_3OCH_2OH$ and $H-\overset{\displaystyle O}{\overset{\|}{C}}-NH_2$

    (c)
    $CH_3OCH_2\overset{\displaystyle O}{\overset{\|}{C}}H$ and $NH_3$

    (d)
    $CH_3OCH_2\overset{\displaystyle O}{\overset{\|}{C}}OH$ and $NH_3$

12.  The esterification of propionic acid by ethyl alcohol would produce

(a)
$$CH_3CH_2CH_2O\overset{O}{\underset{\|}{C}}CH_3$$

(c)
$$CH_3CH_2\overset{O}{\underset{\|}{C}}CH_2CH_3$$

(b)
$$CH_3CH_2\overset{O}{\underset{\|}{C}}OCH_2CH_3$$

(d)
$$CH_3CH_2\overset{OH}{\underset{\underset{H}{|}}{C}}OCH_2CH_3$$

13.
The saponification of $CH_3CH_2CH_2O–CH_2CH_2\overset{O}{\underset{\|}{C}}–O–CH_3$ by sodium hydroxide would produce

(a)  $CH_3CH_2CH_2OH + HOCH_2CH_2\overset{O}{\underset{\|}{C}}O^-Na^+ + HOCH_3$

(b)  $CH_3CH_2Cl\,I_2–O–CH_2CH_2\overset{O}{\underset{\|}{C}}–O^-Na^+ + HOCH_3$

(c)  $CH_3CH_2CH_2O–Na^+ + HOCH_2CH_2\overset{O}{\underset{\|}{C}}OCH_3$

(d)  $CH_3CH_2CH_2OCH_2CH_2\overset{O}{\underset{\|}{C}}OH + Na^{+-}OCH_3$

# ANSWERS

## ANSWERS TO DRILL EXERCISES

### I.  Exercises in Structures and Names

1.  (a)
$$\underset{\underset{H}{|}}{\overset{\overset{H}{|}}{H-C}}-\overset{O}{\underset{\|}{C}}-O-H$$

(c)
$$\underset{\underset{H}{|}}{\overset{\overset{H}{|}}{H-C}}-\overset{O}{\underset{\|}{C}}-H$$

(b)
$$H-O-\overset{O}{\underset{\|}{C}}-\underset{\underset{H}{|}}{\overset{\overset{H}{|}}{C}}-\underset{\underset{H}{|}}{\overset{\overset{H}{|}}{C}}-H$$

(d)
$$\underset{\underset{H}{|}}{\overset{\overset{H}{|}}{H-C}}-\underset{\underset{H}{|}}{\overset{\overset{H}{|}}{C}}-O-H$$

(e)
```
    H O   H
    | ||  |
  H-C-C-O-C-H
    |     |
    H     H
```

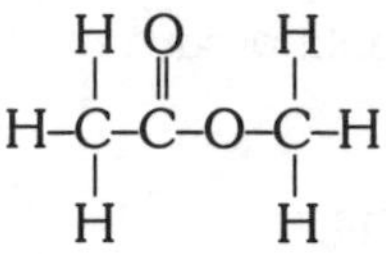

(g)
```
      O H
      || |
  H-O-C-C-H
        |
        H
```

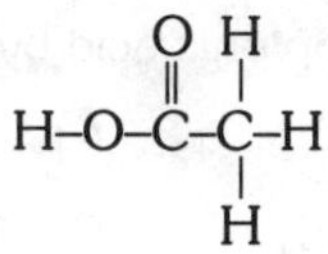

(f)
```
    H O  H
    | || |
  H-C-C-N-H
    |
    H
```

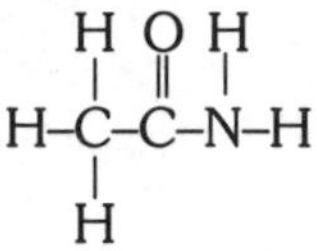

2.  (a)  acid          (k)  acid          (u)  acid
    (b)  aldehyde       (l)  ester         (v)  amide
    (c)  acid           (m)  alcohol, ketone   (w)  diamide
    (d)  amide          (n)  alcohol, ketone   (x)  acid
    (e)  ester          (o)  acid          (y)  ketone
    (f)  ketone         (p)  aldehyde, ketone  (z)  phenol, aldehyde
    (g)  alcohol        (q)  ester         (aa) alkene, ester
    (h)  amide          (r)  ether, ketone    (bb) ester, alkene
    (i)  aldehyde       (s)  ester, aldehyde  (cc) diester
    (j)  ketone         (t)  amide          (dd) amide, acid

3.  (a)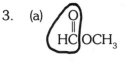
         form

    (d)
```
          O
          ||
  CH₃CH₂O CCH₃
```
         acet

    (b) 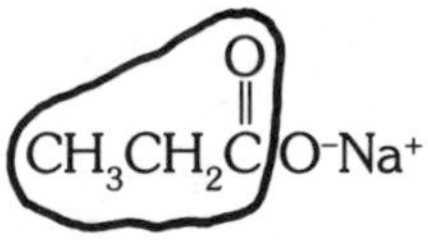
```
          O
          ||
  CH₃CH₂C O⁻Na⁺
```
         propion

    (e)
         propion

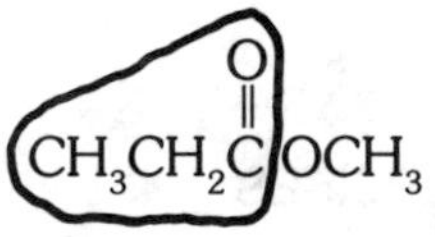

    (c) 
```
          O
          ||
  CH₃CH₂O CH
```
         form

    (f) 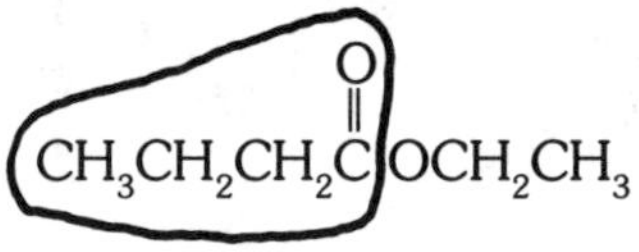
```
              O
              ||
  CH₃CH₂CH₂C OCH₂CH₃
```
         butyr

4.  (a)  ester       isopropyl acetate
    (b)  amide       propionamide
    (c)  ester       methyl butyrate

(d)   ester        isobutyl butyrate

(e)   aldehyde    butyraldehyde

5.   (a)   propionaldehyde

(b)   sodium propionate

(c)   ethyl acetate

(d)   ethyl propyl ketone

(e)   methyl ethyl ketone

(f)   formic acid

(g)   butyramide

(h)   isopropyl butyrate

(i)   propyl acetate

(j)   butyraldehyde

**II.   Drill on the Reactions of Carboxylic Acids with Strong Bases**

1.   $CH_3CH_2CH_2CO_2^-K^+$          4.   $HOCH_2CH_2CO_2^-K^+$

2.   $CH_3(CH_2)_6CO_2^-Na^+$          5.   $Na^+{}^-O_2CCH_2CH_2CO_2^-Na^+$

3.   $CH_3-O-CH_2CO_2^-Na^+$

**III.   Drill on the Reactions of Carboxylic Acid Salts with Strong Acids**

1.   $HCO_2H$

2.   $HO_2CCH_2CH_2CH_3$          4.   $CH_3-\!\!\left\langle\bigcirc\right\rangle\!\!-CO_2H$

3.   $CH_3(CH_2)_8CO_2H$          5.   $CH_3CH_2OCH_2CH_2CO_2H$

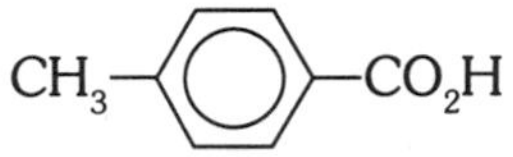

**IV.   Drill on Writing the Structures of Esters That Can Form From Given Acids and Alcohols**

1.   $CH_3CH_2CH_2CO_2CH_3$          4.

2.   $HCO_2CH_2CH_2CH_3$

$$\underset{\underset{CH_3}{|}}{CH_3CO_2\overset{\overset{CH_3}{|}}{C}CH_3}$$

3.

$$CH_3CH_2CO_2CH_2\overset{\overset{CH_3}{|}}{C}HCH_3$$

5.   $CH_3O_2CCH_2CH_2CH_2CO_2CH_3$

**V.   Drill on Writing the Products of the Hydrolysis of Esters**

1.   $CH_3CH_2CO_2H + HOCH_3$

2.   $CH_3OH + HO_2CCH_3$

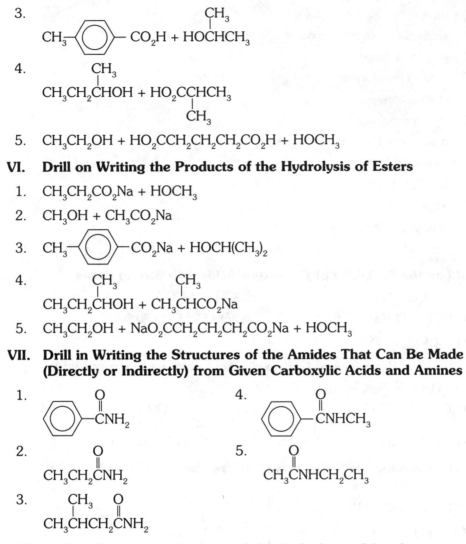

3.   CH$_3$—⟨◯⟩—CO$_2$H + HOCHCH$_3$
         |
         CH$_3$

4.        CH$_3$
          |
     CH$_3$CH$_2$CHOH + HO$_2$CCHCH$_3$
                        |
                        CH$_3$

5.   CH$_3$CH$_2$OH + HO$_2$CCH$_2$CH$_2$CH$_2$CO$_2$H + HOCH$_3$

**VI.   Drill on Writing the Products of the Hydrolysis of Esters**

1.   CH$_3$CH$_2$CO$_2$Na + HOCH$_3$

2.   CH$_3$OH + CH$_3$CO$_2$Na

3.   CH$_3$—⟨◯⟩—CO$_2$Na + HOCH(CH$_3$)$_2$

4.        CH$_3$              CH$_3$
          |                  |
     CH$_3$CH$_2$CHOH + CH$_3$CHCO$_2$Na

5.   CH$_3$CH$_2$OH + NaO$_2$CCH$_2$CH$_2$CH$_2$CO$_2$Na + HOCH$_3$

**VII.   Drill in Writing the Structures of the Amides That Can Be Made
(Directly or Indirectly) from Given Carboxylic Acids and Amines**

1.                  O
                    ‖
     ⟨◯⟩—CNH$_2$

2.              O
                ‖
     CH$_3$CH$_2$CNH$_2$

3.        CH$_3$   O
          |        ‖
     CH$_3$CHCH$_2$CNH$_2$

4.                  O
                    ‖
     ⟨◯⟩—CNHCH$_3$

5.          O
            ‖
     CH$_3$CNHCH$_2$CH$_3$

**VIII.   Drill in Writing the Products of the Hydrolysis of Amides**

1.   CH$_3$CH$_2$CO$_2$H + NH$_3$

2.   CH$_3$CH$_2$CH$_2$CO$_2$H + NH$_2$CH$_3$

3.   CH$_3$CH$_2$NH$_2$ + HO$_2$CCH$_2$CH$_3$

4.                              CH$_3$
                                |
     CH$_3$CHNH$_2$ + HO$_2$CCH$_2$CHCH$_3$
         |
         CH$_3$

5.   Cl—⟨◯⟩—NH$_2$ + HO$_2$C—⟨◯⟩

## ANSWERS TO SELF-TESTING QUESTIONS

### Completion

1. (a) $CH_3CO_2^-Na^+$ (+$H_2O$)

   (b) $CH_3CO_2CH_3$ (+$H_2O$)

   (c) $CH_3CH_2CONH_2$ (+$H_2O$)

   (d) $CH_3CO_2H + NH_3$

   (e) $CH_3OH + CH_3CO_2H$

   (f) $$CH_3\overset{\displaystyle O}{\overset{\|}{C}}CO_2H$$

   (g) $2CH_3OH + HO_2CCH_2CH_2CO_2H$

   (h) $CH_3OCH_2CH_2CO_2H + HOCH_3$ (The ether group is not changed.)

   (i) $NH_2CH_2CO_2H + NH_2\underset{\underset{\displaystyle CH_3}{|}}{C}HCO_2H$

   (j) $CH_3S–SCH_2CO_2H$ (+ NaCl)

   (k) $CH_3OH + CH_3CO_2^-Na^+$

2. (a) b

   (b) k

### Multiple-Choice

1. b, c, and $e$
2. c and $e$
3. a and d
4. d and $e$
5. d
6. b
7. a and d
8. c
9. $e$
10. b
11. d
12. b
13. b

# CHAPTER 14

# CARBOHYDRATES

## OBJECTIVES

After you have studied this chapter and worked the exercises in it, you should be able to do the following.

1.    Describe carbohydrates by their structural features.

2.    Name the three classes of carbohydrates we have studied.

3.    Interpret terms such as "aldose," "hexose," and "aldohexose."

4.    Name the three nutritionally important monosaccharides and give at least one source of each.

5.    Name the three nutritionally important disaccharides, give a source of each, and name the products each gives when hydrolyzed.

6.    Name three polysaccharides made entirely from glucose units and state where each is found in nature.

7.    Write the structures for the α–, β–, and the open forms of glucose. (This would be a minimum goal as far as monosaccharide structures are concerned.)

8.    Look at the cyclic structure of a given monosaccharide and point out its hemiacetal system.

9.    Look at the cyclic structure of a given disaccharide and point out its acetal-oxygen bridge; tell if the disaccharide will give a positive result in Benedict's or Tollens' test.

10.   Give the names of the nutritionally important reducing sugars.

11.   Explain what is meant by "deoxy-".

12.   Write the structure of α– and β– maltose. (This would be a minimum goal as far as disaccharide structures are concerned.)

13.   Write the structure of the repeating unit in amylose (as a minimum goal for illustrating what polysaccharides are like.)

14.   Describe an instance in which two polysaccharides differ only in the orientation (i.e., geometry) of their oxygen bridges.

15.   Describe what one does and sees in the starch-iodine test.

16.   Name the principal components of starch.

17.   Describe the structural relations between amylopectin and amylose. (Do this in words if not by writing the structures.)

18.   With respect to optical isomerism (Section 14.3 in the text, which might be omitted in your course), minimum goals are these.

    (a)   State what physical property a substance must have if it is to be (1) optically active; (2) dextrorotatory; (3) levorotatory.

    (b)   Describe, in words, how D-α-glucose and L-α-glucose are related.

    (c)   Describe what is true about a molecule, structurally, if it is chiral.

    (d)   Explain in general terms the significance of the optical activity of a substance in its reactions in living cells.

19.   Define all of the terms in the Glossary.

## GLOSSARY

**Aldohexose.**   A monosaccharide whose molecules contain six carbons and an aldehyde group.

**Aldose.**   A monosaccharide whose molecules contain an aldehyde group.

**Amylopectin.**   A polymer in which amyloselike units have joined to each other to give a branched polymer of alpha-glucose; it is found in starch.

**Amylose.**   A linear polymer of alpha-glucose; it is found in starch.

**Biochemistry.**   The study of the structures and properties of substances found in living things.

**Blood Sugar.**   The carbohydrates - mostly glucose - that are present in blood.

**Carbohydrate.**   Any naturally occurring substance with the properties of either the polyhydroxyaldehydes or the polyhydroxyketones or any substances that by simple hydrolysis will yield these.

**Chiral.**   Having a handedness such that the mirror image of a structure cannot be superimposed upon the original.

**Chiral Molecule.**   A molecule that has handedness in its molecular structure.

**Chirality.**   The quality of handedness that a molecular structure has that prevents this structure from being superimposable on its mirror image.

**D-Family. L-Family.**   The names of the two optically active families to which substances belong when they are considered solely according to one kind of "handedness" or the other.

**Disaccharide.**   A carbohydrate that can be hydrolyzed into two monosaccharide units.

**Enantiomers.**   Isomers whose molecules are related as object to mirror image but cannot be superimposed.

**Fructose.**   A ketohexose present in honey and one product of the hydrolysis of sucrose. Levulose.

**Galactose.**   An aldohexose that forms, together with glucose, when lactose (milk sugar) is hydrolyzed.

**Glucose.**   An aldohexose widely found as a building block for cellulose, starch, glycogen, dextrin, maltose, sucrose, and lactose; it is the principal carbohydrate found in circulation in the blood. Sometimes it is called *blood sugar* and sometimes, *dextrose*.

**Glycogen.**   The starchlike polymer of alpha-glucose that is an animal's means of storing glucose units.

**Iodine Test.**   A simple test-tube test for starch in which a drop of iodine reagent is added to the solution suspected of containing starch. A positive test consists of the rapid formation of an intensely purple color that may even appear to be black. The *iodine reagent* consists of a dilute solution of iodine in aqueous potassium iodide.

**Ketohexose.**   A monosaccharide whose molecules contain six carbon atoms and have a keto group.

**Ketose.**   A monosaccharide whose molecules contain a keto group. An example is fructose.

**Lactose.**  A disaccharide that can be hydrolyzed to glucose and galactose.  Malt sugar.

**Maltose.**  A disaccharide that can be hydrolyzed to glucose.  Malt sugar.

**Monosaccharide.**  A carbohydrate that cannot be hydrolyzed.

**Optical Activity.**  The ability of a substance to rotate the plane of plane polarized light.

**Optical Isomer.**  One of a set of compounds whose molecules differ only in their handedness.

**Optical Isomerism.**  The existence of isomers whose molecules are chiral and can rotate the plane or plane polarized light.

**Polysaccharide.**  A carbohydrate that can be hydrolyzed into many monosaccharide units.

**Reducing Sugar.**  Any carbohydrate that will give a positive test with Benedict's or Tollens' reagents.

**Simple Sugar.**  Any monosaccharide.

**Starch.**  A naturally occurring mixture of amylose and amylopectin obtained from various plants.

**Substrate.**  A substance whose molecules fit to any enzyme where a reaction is catalyzed.

**Sucrose.**  A disaccharide that can be hydrolyzed to glucose and fructose.  Cane sugar. Beet sugar.


## DRILL EXERCISE


### EXERCISES IN CARBOHYDRATE STRUCTURES AND SYMBOLS

In the simplified structures of the cyclic forms of carbohydrates, the bonds or lines that are parts of the rings (hexagons or pentagons) form a flat surface.  You should imagine that this surface comes out of the page, perpendicular to it.  Bonds or lines that point downward from corners of these rings actually point below the plane of the ring.  Those pointing up from a corner project above the plane.  These relations hold even if we rotate the ring around an imaginary axis going through the center of the ring and perpendicular to the plane of the ring.

Just to make sure that the condensed structural symbols for the monosaccharides are understood, convert this symbol into its full structural formula with the

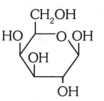

atomic symbols given for all carbon,
hydrogen, and oxygen atoms and with
all bonds represented by lines.

## SELF-TESTING QUESTIONS

### Completion

1.  A carbohydrate whose molecules will react with water to produce two sugar units
    is a _____.

2.  The structural feature involved in the link between two glucose units in maltose is
    the _____.

3.  The hydrolysis of sucrose produces _____ and _____.

4.  The components of starch are called _____ and _____.

5.  The iodine test is used to detect the presence of _____.

6.  Carbohydrates all have high melting points, which means that their molecules are
    strongly attracted to each other in the crystals.  The force of attraction respon-
    sible for this property is the _____.

7.  The names of the two very broad families of optically active compounds, which
    are organized solely on the basis of "handedness," are the _____
    family and the _____ family.

8.  The hydrolysis of lactose produces _____ and _____.

9.  The linear polymer of α-glucose is _____.

10. Starch gives a _____ color with _____ reagent.

11. The technical name for a potential aldehyde group is the _____.

12. The hydrolysis of maltose produces _____.

13.    Because maltose and lactose have the _____ group, they are reducing disaccharides.

14.    The storage form of glucose molecules in animals is _____, which structurally is similar to _____, one of the storage forms of glucose units in plants.

15.    If a polysaccharide can be hydrolyzed into nothing but glucose, it might be any one of the following (make a complete list):

    _____

    _____

## Multiple-Choice

1.    Which is (are) the structure(s) of α-glucose?

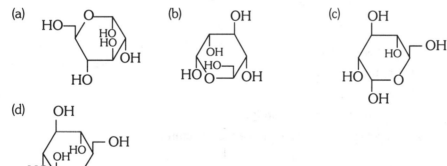

2.    If the letter G is used to represent a glucose unit, then the symbol:

    etc.–O–G–O–G–O–G–O–G–O–G–etc.

    could be a way of showing the basic structural feature of
    (a)  amylopectin      (b)  amylose      (c)  glycogen      (d)  galactose

3.    An example of a reducing carbohydrate is
    (a)  sucrose    (b)  maltose    (c)  cellulose    (d)  galactose

4. An aldohexose would have
   (a) a potential aldehyde group
   (b) five hydroxyl groups in its open form
   (c) six carbons
   (d) one –CH₂OH group

5. The 50:50 mixture of glucose and fructose
   (a) is called dextromaltose
   (b) is obtainable by the hydrolysis of sucrose
   (c) is called invert sugar
   (d) is a disaccharide

6. If the molecules of a substance have the structure:

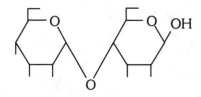

   (a) it is a disaccharide
   (b) it is in the D-family
   (c) it can be hydrolyzed
   (d) it has a β-acetal oxygen bridge

7. If the molecules of a substance have the structure:

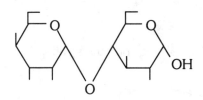

   (a) it can be hydrolyzed into two glucose units
   (b) it can be hydrolyzed into one glucose unit plus another monosaccharide
   (c) it will not give a Benedict's Test
   (d) it will not react with water

8.  If the molecules of a substance have the structure:

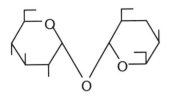

   (a)   it is a disaccharide

   (b)   it will give a positive Benedict's Test

   (c)   it will not be able to react with water

   (d)   it will hydrolyze into two D-galactose units

## ANSWERS

### ANSWERS TO DRILL EXERCISES

**Exercise in Carbohydrate Structures and Symbols**

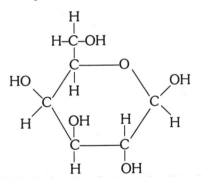

### ANSWERS TO SELF-TESTING QUESTIONS

### Completion

   1.   disaccharide

   2.   $\alpha$-acetal bridge

   3.   fructose, glucose

4. amylose and amylopectin
5. starch
6. hydrogen bond
7. D- and L- (families)
8. glucose, galactose
9. amylose
10. purple, iodine
11. hemiacetal group
12. glucose
13. hemiacetal (or potential aldehyde)
14. glycogen, amylopectin
15. starch, amylose, amylopectin, glycogen, cellulose

## Multiple-Choice

1. a, b, and c
2. b
3. b and d
4. a, b, c, and d
5. b and c

6. a, b, and c
7. b
8. a

# CHAPTER 15

# LIPIDS

## OBJECTIVES

Although all of these objectives are important, one through five constitute minimum goals in our preparation for later chapters. Objectives 11 and 12 are necessary for an understanding of how the membranes of animal cells are organized and how they can hold a cell together while letting substances in and out.

After you have studied this chapter and worked the exercises in it, you should be able to do the following.

1. *Write the structure of a molecule that includes at least one carbon-carbon double bond and is typically found in a triacylglycerol.*

2. *Using that structure, write the products of its reaction with*

   (a) *water - catalyzed by an enzyme (as in digestion);*

   (b) *aqueous sodium hydroxide - saponification; and*

   (c) *hydrogen - hydrogenation with a catalyst, heat, and pressure*

3. Write the names and structures for at least three saturated and three unsaturated fatty acids.

4. Describe the principal structural differences between animal fats and vegetable oils.

5. Explain what polyunsaturated means when used to describe vegetable oils.

6. Look at the structures of two triacylglycerols that have essentially identical formula weights and tell which is the more polyunsaturated.

7. Name two kinds of glycerol-based phospholipids and tell where they may be found in the body.

8. Do the same for two kinds of sphingosine-based lipids.

9. Name several steroids and briefly describe their purposes.

10. Briefly explain why steroids are classified as lipids.

11. Describe the composition and the structure of a biological membrane.

12. Explain how active transport is necessary for certain cellular functions.

13. Define each of the terms in the Glossary.

## GLOSSARY

**Active Transport.**    The movement of a substance through a biological membrane against a concentration gradient and caused by energy-consuming reactions involving parts of the membrane.

**Amphipathic Compound.**    A compound whose molecules have both hydrophilic and hydrophobic groups.

**Fatty Acid.**    Any monocarboxylic acid usually, but not necessarily obtained from a natural product (e.g., by hydrolysis of a fat or oil). Its molecules usually, but not necessarily, have long carbon chains and they have one or more carbon-carbon double bonds.

**Glycolipid.**    A sphingolipid that incorporate a "glycose" unit (e.g., D-glucose or D-galactose) into its molecules instead of a phosphate unit; e.g., the cerebrosides.

**Gradient.**    An unevenness in the value of some physical property throughout a system; e.g., a concentration gradient.

**Hydrophilic Group.**    A structural unit in a molecule that can attract water molecules; e.g., $-OH$, $CO_2^-$, $-\overset{+}{N}H_3$, and $-NH_2$.

**Hydrophobic Group.**    A structural unit in a molecule that has no attraction for water molecules; e.g., alkyl groups.

**Lipid.**   A plant or animal product that is soluble in a nonpolar solvent such as ether, carbon tetrachloride, or benzene.

**Lipid Bilayer.**   An important structural feature of biological membranes characterized by a sheetlike array of two layers of lipid molecules aligned with their hydrophobic "tails" pointing inward so that they are between the "sheets".

**Nonsaponifiable Lipid.**   Any lipid that cannot be saponified (e.g., steroids).

**Phosphoglyceride.**   A phospholipid whose molecules have incorporated a glycerol unit (e.g., plasmalogens and lecithins).

**Phospholipid.**   Complex lipids whose molecules include phosphate ester groups. The principal alcohol unit may be either glycerol or sphingosine. Examples are phosphoglycerides, plasmalogens, and sphingomyelins.

**Plasmalogen.**   One of a family of glycerol-based phospholipids whose molecules have also incorporated an unsaturated fatty alcohol unit; they are important in cell membranes.

**Receptor Molecule.**   A molecule of a protein built into a cell membrane that can accept a molecule of a hormone or a neurotransmitter.

**Saponifiable Lipid.**   Any lipid that can be saponified because its molecules have ester bonds.

**Sphingolipid.**   A complex lipid based on sphingosine, an aminoalcohol. Examples are the sphingomyelins and the cerebrosides.

**Steroids.**   One of the important classes of nonsaponifiable lipids that includes cholesterol as well as many hormones.

**Triacylglycerol.**   A lipid whose molecules are esters formed between glycerol (a trihydric alcohol) and three, usually different, long-chain fatty acids. It is often called *triglyceride* and sometimes simply *glyceride*.

**Triglyceride.**   See **Triacylglycerol**.

**Wax.**   A naturally occurring lipid whose molecules consist of esters formed between long-chain fatty acids and long-chain monohydric alcohols.

## SELF-TESTING QUESTIONS

### Completion

1.  Write the structures of all of the products that would form from the complete digestion of the following lipid.

$$CH_2-O-\overset{\overset{\displaystyle O}{\|}}{C}(CH_2)_7CH=CHCH_2CH=CH(CH_2)_4CH_3$$

$$CH-O-\overset{\overset{\displaystyle O}{\|}}{C}(CH_2)_8CH_3$$

$$CH_2-O-\overset{\overset{\displaystyle O}{\|}}{C}(CH_2)_7CH=CH(CH_2)_7CH_3$$

2.  If the lipid of question 1 had been saponified by sodium hydroxide instead of digested, the fatty acids would have been produced in the form of their

    _____.

3.  Write the structure of the product of the hydrogenation of the lipid in question 1 to a product that would be called a saturated triacylglycerol.

4. A molecule of vegetable oil will normally have more _____ than a molecule of animal fat.

5. The two functional groups in a simple lipid are _____ and _____.

6. An example of a nonsaponifiable lipid is any member of the family of

_____.

7.
$$\overset{O}{\overset{\|}{}}$$
Lipids with the general structure R–O–C–R' (where R and R' are both long-chain) are in the family of _____.

8. The structure of linoleic acid is _____.

9. The acid, other than fatty acids, liberated when complex lipids are hydrolyzed is

_____.

10. A vegetable oil is said to be more _____ than an animal fat (referring to double bonds).

11. The complete hydrolysis (digestion) of the phosphatidycholine would give what products?  (Write their structures and names.)

$$CH_2-O-\overset{O}{\overset{\|}{C}}(CH_2)_7CH=CH(CH_2)_7CH_3$$

$$CH-O-\overset{O}{\overset{\|}{C}}(CH_2)_7CH=CHCH_2CH=CH(CH_2)_4CH_3$$

$$CH_2-O-\overset{O}{\underset{O^-}{\overset{\|}{P}}}-O-CH_2CH_2\overset{+}{N}(CH_3)_3$$

12. Another name for the structure in question 11 is _____.

13. Because the hydrocarbon chains in the structure in question 11 are water-avoiding, they are said to be _____. In the region of the phosphate group, however, the molecule is _____. Overall, the molecule is _____.

14. Because the hydrocarbon chains of the structure in question 11 have double bonds, the lipid would be described as poly-_____.

## Multiple-Choice

1. To reduce the unsaturation in a triacyglycerol, a manufacturer might
   (a) hydrate it          (c) hydrolyze it
   (b) hydrogenate it      (d) dehydrogenate it

2. If a manufacturer hydrogenated most of the alkene double bonds of a vegetable oil, the substance might
   (a) turn rancid
   (b) become a detergent
   (c) become a solid at room temperature
   (d) become a diglyceride

3. In triacylglycerols, aqueous sodium hydroxide attacks
   (a) ester linkages      (c) alkene groups
   (b) ether linkages      (d) alkenelike portions

4. Fatty acids obtained from natural lipids are generally
   (a) of even carbon number
   (b) monocarboxylic
   (c) insoluble in water
   (d) long chain

5. The hydrolysis of a naturally occurring triacylglycerol gives
   (a) glycerol + $RCO_2H$ + $R'CO_2H$ + $R''CO_2H$
   (b) glycerol + $RCO_2^-$ + $R'CO_2^-$ + $R''CO_2^-$
   (c) glycerol + $3RCO_2H$
   (d) glycerol + $3RCO_2^-$

6. The name of a $C_{18}$ acid with three alkene groups is

   (a) stearic acid

   (c) linoleic acid

   (b) oleic acid

   (d) linolenic acid

7. Phospholipids are esters of

   (a) sphingosine only

   (b) either glycerol or sphingosine

   (c) cholesterol only

   (d) glycerol only

8. Phosphoglycerides are

   (a) esters of phosphatidic acids

   (b) cerebrosides

   (c) esters or sphingosine

   (d) glycolipids

9. Cholesterol is classified as a lipid because

   (a) it is an ester or cholesterol

   (b) it dissolves in fat solvents

   (c) it is present in gallstones

   (d) it can be saponified

10. The cell membranes of animals are made of

    (a) lipids

    (c) proteins

    (b) polysaccharides

    (d) cholesterol

11. Molecules of saponifiable lipids have

    (a) hydrophobic groups

    (c) amide groups

    (b) ester groups

    (d) alkene groups

12. The cell membranes of animals are organized

    (a) with hydrophobic groups projecting outward away from the membrane

    (b) as lipid bilayers with imbedded proteins

    (c) with hydrophilic groups projecting inward

    (d) with cellulose molecules lending structural support

# ANSWERS

## ANSWERS TO SELF-TESTING QUESTIONS

### Completion

1.  $HOCH_2CHCH_2OH$ + $CH_3(CH_2)_4CH=CHCH_2CH=CH(CH_2)_7CO_2H$
    $\quad\quad\;\; |$
    $\quad\quad\; OH$ $\quad\quad\quad$ + $CH_3(CH_2)_8CO_2H$

    $\quad\quad\quad\quad\quad\quad$ + $CH_3(CH_2)_7CH=CH(CH_2)_7CO_2H$

2.  sodium salts

3. 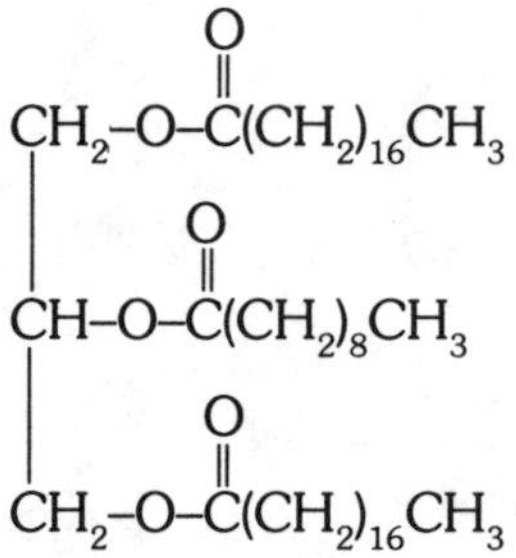
$$
\begin{array}{l}
\quad\quad\quad O \\
\quad\quad\quad \| \\
CH_2-O-C(CH_2)_{16}CH_3 \\
| \\
\quad\quad\quad O \\
\quad\quad\quad \| \\
CH-O-C(CH_2)_8CH_3 \\
| \\
\quad\quad\quad O \\
\quad\quad\quad \| \\
CH_2-O-C(CH_2)_{16}CH_3
\end{array}
$$

4.  carbon-carbon double bonds (or alkene groups or unsaturation)

5.  esters, alkenes

6.  steroids

7.  waxes

8.  $CH_3(CH_2)_4CH=CHCH_2CH=CH(CH_2)_7CO_2H$

9.  phosphoric acid

10. unsaturated

11.

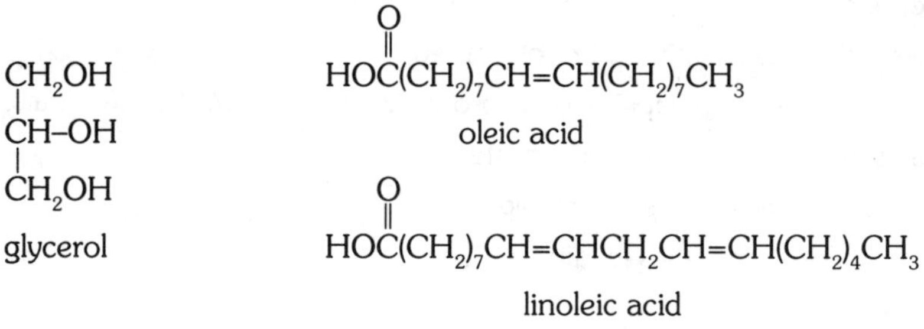

$\quad CH_2OH$ $\quad\quad\quad\quad\quad\quad$ $\overset{\displaystyle O}{\overset{\displaystyle \|}{HOC}}(CH_2)_7CH=CH(CH_2)_7CH_3$

$\quad CH-OH$ $\quad\quad\quad\quad\quad\quad\quad\quad$ oleic acid

$\quad CH_2OH$ $\quad\quad\quad\quad\quad\quad\quad\quad$ $O$

$\quad$ glycerol $\quad\quad\quad\quad\quad\quad$ $\overset{\displaystyle \|}{HOC}(CH_2)_7CH=CHCH_2CH=CH(CH_2)_4CH_3$

$\quad\quad\quad\quad\quad\quad\quad\quad\quad\quad\quad\quad$ linoleic acid

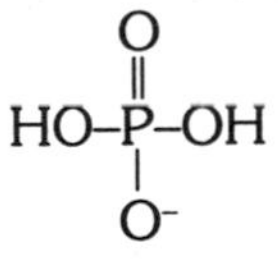

$$\begin{array}{c} O \\ \| \\ HO-P-OH \\ | \\ O^- \end{array}$$

$(H_2PO_4^-$ as well

as $HPO_4^-)$

$HO-CH_2CH_2\overset{+}{N}(CH_3)_3$

choline

dihydrogen-
phosphate ion

12. lecithin

13. hydrophobic; hydrophilic; amphipathic

14. unsaturated

**Multiple-Choice**

1. b

2. c

3. a

4. a, b, c, and d

5. a

6. d

7. b

8. a

9. b

10. a, c, and d

11. a, b, and d

12. b

# CHAPTER 16

# PROTEINS

## OBJECTIVES

These objectives are designed to promote general knowledge about proteins and are important for our understanding of later chapters. When you have completed your study of this chapter and have worked the exercises in it, you should be able to do the following.

1.  *Write the names and structures of five representative amino acids, for example*

    (a)  *glycine - because it is the simplest amino acid;*

    (b)  *alanine - as representative of an amino acid with a hydrocarbon side chain (a hydrophobic group);*

    (c)  *cysteine - because it has the important sulfhydryl group;*

    (d)  *glutamic acid - to represent an amino acid with a side chain $-CO_2H$ (or $-CO_2^-$) group;*

    (e)  *lysine - to represent amino acids with a side chain $-NH_2$ (or $-NH_3^+$) group.*

2.  Based on the amino acids you have learned, write the structure of any di-, tri-, tetra-, or pentapeptide and identify the peptide bonds.

3.  Translate a structure such as gly·ala·glu into a condensed structural formula.

4.  Write the structures that show how disulfide bonds occur in proteins.

5.  Write structures that illustrate how hydrogen bonds and salt bridges participate in the structure of proteins.

6.  Name the four levels of structure in proteins and briefly describe each in terms of the kinds of forces that stabilize it and the kinds of geometric forms it takes.

7.  Explain how hydrophobic and hydrophilic side chains influence the shape of a protein.

8.  Explain the relation between a polypeptide and a protein.

9.  Explain how the solubility of a protein can be changed by changing the pH of its medium.

10. Given the structure of a small polypeptide, write the structures of the products of its digestion.

11. Describe what happens structurally when a protein is denatured.

12. Describe, in general terms, how denaturation affects the properties of a protein.

13. List at least four denaturing agents.

14. Name five fibrous proteins.

15. Name two globular proteins.

16. Define each of the terms in the Glossary.

## GLOSSARY

**Albumins.**  A family of globular proteins that tend to dissolve in water and that in blood contribute to the blood's colloidal osmotic pressure and aid in the transport of metal ions, fatty acids, cholesterol, triacylglycerols and other water-insoluble substances.

**Amino Acid.**  An organic compound whose molecules have both an amino group and a carboxylic group.  In all of the amino acids involved in protein structure, these two groups are attached to the same carbon; they are, in other words, alpha-amino acids.

**Amino Acid Residue.**    The group $-NH-CH-\overset{\overset{\displaystyle O}{\displaystyle \|}}{C}-$ in a polypeptide that is contributed
by one amino acid.

**Collagen.**    The fibrous protein of connective tissue; it changes to gelatin in boiling water.

**Denaturation.**    In protein chemistry, an event whereby some agent - either physical or chemical - causes a protein molecule to lose its natural shape and form without actually hydrolyzing the protein.  Denaturation is usually accompanied by an irreversible loss of the protein's solubility and its biological function.  An agent that can do this is called a *denaturing agent*.

**Dipeptide.**    A compound with one amide (peptide) bond that can be hydrolyzed to two alpha-amino acids.

**Dipolar Ion.**    An ion that has one positively charged site and one negatively charged site (e.g., amino acids).

**Elastin.**    The fibrous protein of tendons and arteries.

**Fibrin.**    The fibrous protein of a blood clot formed from fibrinogen during clotting.

**Fibrous Proteins.**    Water-insoluble proteins found in fibrous tissue.

**Globular Proteins.**    Water-soluble proteins found in blood.

**Globulins.**    Globular proteins in the blood that include γ-globulin.

**α-Helix.**    One of the kinds of secondary structures of a polypeptide in which the molecules are coiled like a spring.

**Hemoglobin (HHb).**    The oxygen carrier in blood found in red blood cells (erythrocytes); a complex between heme (a red, iron-containing pigment) and globin (a protein).

**Isoelectric.**    A condition of a molecule having equal numbers of (+) and (−) charged sites.

**Isoelectric Point (pI).**    The pH of a medium in which a given amino acid or protein is in an isoelectric condition.

**Keratin.**    The fibrous protein of hair, fur, fingernails, hooves, and so forth; it is generally richer in cystine units (and therefore richer in sulfur) than other proteins.

**Myosins.**    Proteins in contractile muscles.

**Peptide Bond.**    The amide linkage in a protein; a carbonyl-to-nitrogen bond,

$$-\overset{\overset{\displaystyle O}{\displaystyle \|}}{C}-\overset{\displaystyle |}{N}-$$

**pI.**  See **Isoelectric Point**.

**β-Pleated Sheet.**  One of the kinds of secondary structure of a polypeptide in which the molecules are aligned side by side in a sheetlike array with the sheet partly pleated.

**Polypeptide.**  A polyamide made from naturally occurring α-amino acids and having the repeating skeletal unit:

**Primary Structure.**  In protein chemistry, the basic feature of all polypeptide molecules; repeating nitrogen-carbon-carbonyl units joined together by peptide bonds.

**Prosthetic Group.**  In protein and enzyme chemistry, a nonprotein molecule joined to a polypeptide to make some biologically active substance.

**Protein.**  A naturally occurring macromolecular substance made up wholly or mostly of polypeptide molecules.

**Quaternary Structure.**  An aggregation of two or more polypeptide strands each having its own primary, secondary, and tertiary structures.

**Salt Bridge.**  A force of attraction between (+) and (−) sites on polypeptide molecules.

**Secondary Structure.**  A shape such as the α-helix adopted by a polypeptide molecule as the result of bonding forces other than the peptide bond; e.g., the salt bridge and the hydrogen bond.

**Side Chain.**  In protein chemistry, one of several organic groups that may be attached to the alpha position of an amino acid.

**Tertiary Structure.**  The shape of a polypeptide molecule arising from the further folding or coiling of its secondary structure.

**Triple Helix.**  The quaternary structure of tropocollagen in which three polypeptide chains are twisted together. These triple helices wrap around each other in a cablelike array to form the basic structural feature of collagen.

## SELF-TESTING QUESTIONS

### Completion

1.  In their dipolar ionic forms, all amino acids have the same basic unit that (without side chains) has the structure: _____

2.  Objective 1 asks you to know the structures of five representative amino acids. The best way to learn them is to learn their side chains, because these are always affixed to the basic unit you wrote in question 1. On the lines provided, write the structures of these side chains.

    _____
    glycine side chain

    _____
    alanine side chain

    _____
    cysteine side chain

    _____
    glutamic acid side chain

    _____
    lysine side chain

3.  Write the structure of the dipeptide that could form from two glycine units. _____

4.  In terms of the three-letter symbols for amino acids, the dipeptide in question 3 would be represented as _____.

5.  Write the structure of a tripeptide that could be made from alanine, glutamic acid, and cysteine. Let alanine be the N-terminal residue and cysteine the C-terminal residue.

6.  In terms of the three-letter symbols, the tripeptide in question 5 has the formula _____ and it contains _____ peptide bonds.

7.  Write the structure of the tetrapeptide: ala·gly·lys·cys

8.  Write the structure of cysteine in its dipolar ionic form.

   _____

9.  If the following polypeptide were subjected to mild reducing conditions, what would form?  Write the structure(s) using the three-letter symbols.

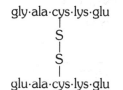

   _____

10. In a discussion of how a polypeptide coils or otherwise assumes some geometric shape, we are talking about what structural level?

   _____

11. Three non-covalent forces that can determine the shape that will be adopted by a

   polypeptide are _____

   _____

12. When we speak of an $\alpha$-helix itself undergoing folding or twisting, we are talking about what level of protein structure?

   _____

13. Which level of protein structure is not necessarily attacked by denaturating agents?

   _____

14. The side chains of which of the five amino acids in Objective 1 are the most susceptible to being altered by a change in the pH or the medium? (Name the amino acids.)

   _____

15. When we speak of two or more polypeptides becoming associated in some way in a gigantic protein system, we are talking about which level of protein structure?

   _____

**Multiple-Choice**

1. The partial hydrolysis of a protein produces a number of amino acids together with several dipeptides. Which of the following fragments would *not* be present?

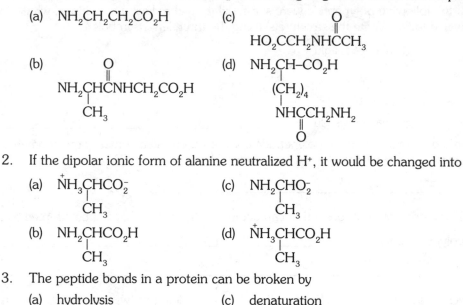

   (a)  $NH_2CH_2CH_2CO_2H$

   (b)

   (c)

   (d)

2. If the dipolar ionic form of alanine neutralized $H^+$, it would be changed into

   (a)  $\overset{+}{N}H_3CHCO_2^-$
        |
        $CH_3$

   (c)  $NH_2CHO_2^-$
        |
        $CH_3$

   (b)  $NH_2CHCO_2H$
        |
        $CH_3$

   (d)  $\overset{+}{N}H_3CHCO_2H$
        |
        $CH_3$

3. The peptide bonds in a protein can be broken by

   (a)  hydrolysis

   (b)  buffer action

   (c)  denaturation

   (d)  hydration

4. Two proteins might be joined together by the action of a mild oxidizing agent if among their amino acid units there was present

   (a)  glutamic acid

   (b)  cysteine

   (c)  glycine

   (d)  lysine

5. Ions of heavy metals (e.g., $Hg^{2+}$ or $Pb^{2+}$) denature proteins by combining with

   (a)  –SH groups

   (b)  –$CH_3$ groups

   (c)  peptide bonds

   (d)  the protein backbones

6. In very strongly acidic solutions, the molecules of alanine would mostly be in what form?

   (a)  $NH_2CHCO_2H$
        |
        $CH_3$

   (c)  $NH_2CHCO_2^-$
        |
        $CH_3$

   (b)  $\overset{+}{N}H_3CHCO_2H$
        |
        $CH_3$

   (d)  $\overset{+}{N}H_3CHCO_2^-$
        |
        $CH_3$

7. In their dipolar ionic forms, amino acids are
   (a) electrically neutral
   (b) weak acids
   (c) weak bases
   (d) solids at room temperature

8. The level of protein structure that has amide bonds is the
   (a) primary   (b) secondary   (c) tertiary   (d) quaternary

9. If two molecules of alanine were joined as a dipeptide
   (a) two isomeric dipeptides are possible
   (b) one dipeptide is possible
   (c) the structural symbol would be ala·ala
   (d) one peptide bond would be present

10. When proteins are digested

    (a)   $\overset{\displaystyle O}{\overset{\|}{-C}}$–NH– units become $-CO_2H + NH_2-$ units

    (b) –SH groups become –S–S– groups

    (c) –S–S– groups become –SH groups

    (d)   $-CO_2H$ and $NH_2$ units join to become $-\overset{\displaystyle O}{\overset{\|}{C}}-NH-$ units

11. If somehow the protein ⌇⌇⌇⌇ were
    with NH$_3^+$ and CO$_2^-$

    changed to ⌇⌇⌇⌇ the protein would be
    with NH$_3^+$ and CO$_2$H

    (a) less soluble in water
    (b) more soluble in water
    (c) at its isoelectric point
    (d) digested

12. In the pentapeptide gly·ala·lys·cys·gly, the N-terminal unit is
    (a) glycine   (b) glutamic acid   (c) lysine   (d) $-NH_2$

13. If the conventions for writing the structures of proteins with the three-letter symbols of amino acids are properly obeyed, the symbol for this tripeptide would be

$$HOCCH_2NHCCHNHCCHNH_2$$

with O, O, O carbonyls and CH$_3$, CH$_2$SH substituents

    (a) gly·ala·cys
    (b) cys·ala·gly
    (c) ala·gly·cys
    (d) ala·cys·gly

14. The principal non-covalent force in protein structure is the
    (a) salt bridge              (c) helix
    (b) disulfide link           (d) hydrogen bond

15. An important secondary structural feature in proteins is the
    (a) salt bridge              (c) disulfide link
    (b) hydrogen bond            (d) $\alpha$-helix

16. Nonprotein molecules that often are associated with proteins are called
    (a) prosthetic groups        (c) side chains
    (b) zwitterions              (d) 3° structures

17. The principal protein in the walls of blood vessels is
    (a) keratin    (b) myosin    (c) elastin    (d) fibrin

18. An important protein in contractile muscle is
    (a) keratin    (b) myosin    (c) elastin    (d) fibrin

19. A protein in hair is
    (a) keratin    (b) myosin    (c) elastin    (d) fibrin

20. One important hemoprotein is
    (a) myoglobin  (b) $\gamma$-globulin  (c) casein  (d) nucleoprotein

# ANSWERS

## ANSWERS TO SELF-TESTING QUESTIONS

### Completion

1.

$$\overset{+}{N}H_3-\underset{|}{C}H-\overset{\overset{O}{\|}}{C}-O^-$$

2.  glycine  alanine  cysteine      glutamic acid              lysine

    $-H$     $-CH_3$   $-CH_2SH$   $-CH_2CH_2CO_2H$    $-CH_2CH_2CH_2CH_2NH_2$

3.

$$NH_2-CH_2\overset{\overset{O}{\|}}{C}-NH-CH_2\overset{\overset{O}{\|}}{C}-OH$$

    (Better yet, the dipolar ionic form:    $\overset{+}{N}H_3-CH_2\overset{\overset{O}{\|}}{C}-NH-CH_2\overset{\overset{O}{\|}}{C}-O^-$)

4.   gly·gly

5.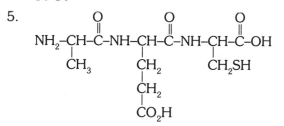

6.   ala·glu·cys, two

7.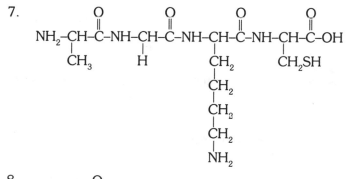

8.

9.   gly·ala·cys·lys·glu
     gly·ala·cys·lys·glu  (Two identical molecules would form.)

10.  the secondary structure

11.  salt bridges, hydrogen bonds, and the water-avoiding (or attracting) responses of hydrophobic (or hydrophilic) groups

12.  the tertiary structure

13.  the primary structure

14.  lysine and glutamic acid

15.  the quaternary structure

**Multiple-Choice**

1.   a, c, and d     (Choice b is a dipeptide, ala·gly, and could be one of the several dipeptides formed. Choice a is not an alpha-amino acid. While choice c has an amide bond, the unit on the right is not an amino acid but rather an acetyl unit. Choice d involves lysine but not in a form found in proteins; the amide bond involves the side chain and not the alpha-amino group.)

2.  d
3.  a
4.  b
5.  a
6.  b
7.  a, b, c, and d
8.  a
9.  b, c, and d
10. a
11. b (it no longer is isoelectric.)
12. a
13. b (The N-terminal unit is cys, and the conventions tell us to write the N-terminal unit on the left.)
14. d
15. d (All of the others *stabilize* secondary structures.)
16. a
17. c
18. b
19. a
20. a

# CHAPTER 17

# ENZYMES, HORMONES, AND NEUROTRANSMITTERS

## OBJECTIVES

Objectives 1 through 8 are mostly concerned with the basic vocabulary of enzymes; 9 and 10 are about how enzymes work in general.  The heart of this chapter is the study of the ways in which the body controls which reactions go, which shut down, which accelerate, and which go slower.  The different chemical and physiological functions of enzymes, hormones, and neurotransmitters are "must" topics of the chapter.

After you have studied this chapter and worked its Exercises and Review Questions you should be able to do the following.

1. *Describe the composition of enzymes and the kinds of cofactors in general terms.*

2.  *Describe what enzymes do.*

3.  *Explain how enzymes possess "specificity."*

4.  *Explain why enzymes are sensitive to denaturing conditions and pH.*

5.  *Describe in general terms how certain B vitamins are vital to some enzymes.*

6.  *Recognize from the name of a substance whether it is an enzyme.*

7.  *Tell what an enzyme's substrate (or type of reaction) is from its name.*

8.  *Define and give an example of an isoenzyme and describe how an analysis of serum isoenzymes can be used in medical diagnosis.*

9.  *Describe the "lock and key" mechanism of enzyme action.*

10. *Describe the "induced fit" model for enzyme action.*

11. *Explain how feedback regulation of an enzyme works.*

12. *Explain how substrate molecules can activate an enzyme with more than one active site.*

13. *Describe what effectors do and give the names of two.*

14. *Explain how calcium ion is involved in enzyme activation.*

15. *Describe the sequence of events that follow the nerve signal for a heart muscle contraction and that involve ions, channels, and effectors.*

16. *Describe what a zymogen is and how it is charged into an enzyme.*

17. *Describe in general terms what happens when a blot clot forms.*

18. *Describe what plasminogen is and how it is involved in dissolving a blot clot.*

19. *Explain what calcium channel blockers do and why they are used.*

20. *Explain how some poisons work.*

21. *Name three kinds of poisons that act as irreversible enzyme inhibitors.*

22. *Explain in general terms what kind of substance is used to measure the concentration of an enzyme in some body fluid.*

23. *Name the enzymes whose serum levels are measured in (a) viral hepatitis and (b) myocardial infarction.*

24. *Name the two kinds of primary chemical messengers.*

25. *Without necessarily reproducing molecular structures, describe what cyclic nucleotides are.*

26. *List the steps in the overall process that occurs when cyclic AMP is involved after a signal releases a hormone or a neurotransmitter.*

27. *Describe how a hormone finds its target cells and recognizes them.*

28. *In general terms, explain how neurotransmitters work and why they must eventually be deactivated.*

29.  Describe in general terms what the following do.

(a)  acetylcholine and choline acetyltransferase

(b)  norepinephrine

(c)  monoamine oxidases

(d)  antidepressant drugs

(e)  dopamine

(f)  drugs for schizophrenia, like Thorazine® and Haldol®

(g)  GABA

(h)  mild tranquilizers, like Valium® and Librium®

(i)  enkaphalins and endorphins

30.  Define each of the terms in the Glossary.

## GLOSSARY

**Active Site.**  The portion of an enzyme molecule most directly responsible for the catalytic effect of the enzyme.

**Agonist.**  A compound whose molecules can bind to a receptor on a cell membrane and cause a response by a cell.

**Antagonist.**  A compound that can bind to a membrane receptor but not cause any response by the cell.

**Antimetabolite.**  A substance that inhibits the growth of bacteria.

**Apoenzyme.**  The part of an enzyme made entirely of polypeptide(s).

**Binding Site.**  A part of an enzyme molecule whose function is to hold the substrate and position it over the active sites.

**Coenzyme.**  An organic compound needed to make a complete enzyme from an apoenzyme.

**Cofactor.**  A nonprotein substance essential either to the completion of the synthesis of some enzymes or to their ability to function.

**Effector.**  A nonsubstrate molecule or ion that activates an enzyme.

**Enzyme-Substrate Complex.**  The combination of an enzyme plus a substrate that exists while catalytic action is occurring.

**Feedback Inhibition.**  The competitive inhibition of an enzyme by a product of its own action.

**Homeostasis.**  The response of an organism to a stimulus such that the organism is restored to its pre-stimulated state.

**Hormone.**   A primary chemical messenger made by an endocrine gland and released by the gland in response to some stimulus and carried by the blood to a target organ where the hormone's molecules trigger a particular response.

**Induced Fit Theory.**   Some enzymes can be induced by their substrates to undergo changes in shape that let the enzyme-substrate complex form.

**Inhibitor.**   A substance that interacts with an enzyme to prevent its acting as a catalyst.

**Isoenzymes.**   Slightly different structural versions of basically the same enzyme produced by different tissues.

**Lock-and-Key Theory.**   The specificity of an enzyme for its substrate is caused by the need for the substrate molecule to fit to the enzyme surface much as a key can fit and turn only one tumbler lock.

**Monoamine Oxidase.**   An enzyme that catalyzes the partial oxidation of amines, particularly amines involved as neurotransmitters or drugs of the nervous system.

**Neurotransmitter.**   A substance released by one nerve cell to carry a signal to the next nerve cell.

**Poison.**   A substance that reacts in some way in the body to cause changes in metabolism that threaten health or life.

**Substrate.**   A substance whose reaction is catalyzed by an enzyme.

**Target Cell.**   A cell at which a hormone molecule finds a site where it acts.

**Vitamin.**   An organic substance satisfying all these criteria: it is required by an organism, but the organism cannot synthesize it; it must be acquired through the organism's diet and is normally present in trace amounts in various foods; it is not a carbohydrate, a lipid, or a protein; it is required for normal growth and health; and its absence leads to a "deficiency" disease.


## SELF-TESTING QUESTIONS

### Completion

1.   An organic compound that is needed, in some cases, to complete an enzyme is called a _____.

2.   The specific site on a large enzyme molecule where a substrate experiences the catalytic activity of the enzyme is called the _____.

3.  An enzyme might lose its catalytic ability through the action of heat or a change in pH because an enzyme is made up mostly of a _____.

4.  The temporary union of an enzyme with the compound on which it acts is called the _____.

5.  The theory that accounts for the unusual specificity of an enzyme is the _____ theory.

6.  Coenzymes can be made in the body, but in many cases it is essential that the body be supplied with _____ via the diet because they are a necessary part of many coenzyme molecules.

7.  Some metabolic sequences are shut down when molecules of their final product combine with and inactivate an _____ for one of the early steps in the sequence.

8.  An enzyme that catalyzes the hydrolysis of an ester link is called an _____; and an enzyme that helps in transferring a phosphate group is called a _____.

9.  The enzyme *sucrase* catalyzes the hydrolysis of _____.

10. Some enzymes can remove a pair of electrons from a substrate; because of this particular kind of action, they are called _____.

11. What are the *names* of the vitamins needed to make each of these coenzymes?

    $NAD^+$            _____

    FMN               _____

    thiamine pyrophosphate        _____

12. When the activity of an enzyme is suppressed by its binding at its active site a molecule of one of the products of the action of the enzyme, the regulation is called _____.

13. An enzyme that is subject to feedback inhibition participates in a general, self-regulating process called _____.

14. An enzyme with more than one active site can be activated by its

    _____.

15. An enzyme that initially forms with its active site covered up by a fragment of its own _____ is called a _____.

16. Plasminogen is involved in _____ a blood clot, but first it must be converted to _____ by _____. Plasminogen is an example of a _____.

17. Calmodulin and _____ are examples of enzyme _____ but before they can work they must first bind _____.

18. In heart muscle contraction, a nerve signal opens _____ in cell membranes. When the signal stops, _____ are removed from the cell by _____.

19. The number of available calcium channels in heart muscle cells can be reduced by drugs called _____.

20. Compounds that inhibit normal metabolism in disease-causing bacteria have the general name of _____.

21. By using the _____ for an enzyme as an analytical reagent, the concentration of the enzyme in some body fluid can be measured.

22. In diseases or injuries of the liver, one of the liver enzymes with the symbol _____ escapes into the _____ along with a lower level of another liver enzyme with the symbol _____. The ratio of these two is typically _____ in victims of viral
    (higher or lower)

hepatitis than in healthy individuals.

23. In a myocardial infarction, three enzymes with the short symbols of _____, _____, and _____ leak from heart tissue into general circulation. One of these can be found in the form of three isoenzymes which have the symbols _____ if present in skeletal muscle, _____ when found in heart muscle, and _____ when present in brain tissue. A procedure called _____ can be used to separate these isoenzymes from each other in a sample of blood.

24. In an infarct, the enzyme lactate dehydrogenase, symbolized as _____, appears in the blood, but there are _____ isoenzymes for it. Normally, the first two of these isoenzymes appear in relative concentrations in which the level of the first is less than that of the second. This relative relationship is inverted in an infarct, and the phenomenon is called an _____.

25. The general names for the two kinds of primary chemical messengers are _____ and _____. Endocrine glands make the _____ and _____ make the other kind of chemical messenger.

26. Many primary chemical messengers function at _____ cells by activating a membrane-bound enzyme called _____ cyclase. This enzyme then catalyzes the conversion of _____ to _____. This then activates an _____ inside the cell. To halt the process and switch the signal off, another enzyme called _____ hydrolyzes _____.

27. Hormones exert their influence in a variety of ways. Adrenaline, for example, is an _____ activator. Insulin and human growth hormone affect the _____ of the membranes of their target cells. Some sex hormones function by activating _____ which then direct the synthesis of _____. Prostaglandins are called _____ because they usually work right where they _____.

28.   Chemical communication from one neuron to another occurs across a very narrow, fluid-filled space called a _____, and the general name for substances that move across this space carrying a "message" is _____.

29.   A substance whose molecules, by binding to a receptor, cause a change to occur in the cell is an _____. A substance that can bind to the same receptor but cause no change in the cell is called the _____.

30.   A complex forms between a receptor in the neuron and the _____ that works to activate _____. This enzyme then catalyzes the formation of _____ in the membrane of the neuron.

31.   To switch a nerve signal off, enzyme-catalyzed reactions have to degrade or otherwise deactivate molecules of _____.

32.   Choline acetyltransferase catalyzes the hydrolysis of the neurotransmitter called _____. Nerve gases work by _____, and the botulinus bacillus works by _____. Local anesthetics like nupercaine and procaine block pain signals by _____ _____.

33.   Several antidepressants work by interfering with the normal use of a neurotransmitter named _____. The deactivation of excess or unused amounts of this substance is catalyzed by enzymes called the _____ symbolized as _____. If this enzyme is inhibited by one kind of antidepressant then the signal carried by the neurotransmitter is _____.
                                                                          (kept up or blocked)

34.   The neurotransmitter called _____ is believed to be involved with schizophrenia.  Drugs that bind to receptors for this neurotransmitter inhibit its _____ to these receptors.  Symptoms similar to those of schizophrenia can be induced by the abuse of drugs called _____ that work by triggering the _____ of this neurotransmitter.
                       (release or decomposition)

35. The full name of GABA is _____. If the
   work of this neurotransmitter is enhanced, the result is that nerve signals are

   _____.
   (accelerated or inhibited)

36. Two families of neurotransmitters that include powerful painkillers are named the

   _____ and the _____.

## Multiple-Choice

1. The site on an enzyme that is able to accept the corresponding substrate is called
   the
   (a)  allosteric site         (c)  catalytic site
   (b)  binding site            (d)  effector site

2. A substance with the name *protease* is
   (a)  a prostaglandin         (c)  an enzyme
   (b)  a coenzyme              (d)  a hormone

3. An enzyme for the reaction; $A + B \rightleftharpoons C + D$
   (a)  increases the quantity of C that forms
   (b)  increases the quantity of A that remains unreacted
   (c)  shifts the equilibrium to favor the right side
   (d)  accelerates the establishment of the equilibrium

4. The catalytic activity of an enzyme will generally
   (a)  increase with increasing temperature
   (b)  increase with decreasing temperature
   (c)  be unaffected by temperature
   (d)  be greatest in a particular temperature range

5. Several coenzymes
   (a)  activate apoenzymes
   (b)  are made from various B-vitamins
   (c)  catalyze hydrolysis reactions
   (d)  are esters of carboxylic acids

6. The symbol $NAD^+$ stands for
   (a)  an enzyme     (b)  a vitamin     (c)  a coenzyme     (d)  an isoenzyme

7. When an enzyme that contains FMN interacts with one having NADH, then which can form?

   (a) $FMNH + NAD^+$            (c) $NADP^+ + FMNH_2$

   (b) $NAD^+ + FMNH_2$         (d) $NAD^+ + H^+ + FMN$

8. Isoenzymes usually differ in

   (a) the substrates they will accept

   (b) their coenzymes

   (c) their apoenzyme portions

   (d) their cofactors

9. The symbol CK stands for

   (a) creatine kinase           (c) cofactor kinase

   (b) carbon-potassium          (d) an isoenzyme

10. A theory that is used to explain enzyme specificity is called

   (a) the theory regulatory enzymes

   (b) cyclic AMP theory

   (c) the induced fit theory

   (d) the lock-and-key theory

11. Consider the following sequence of reactions, each with its own enzyme (En):

$$ A \xrightarrow{En_a} B \xrightarrow{En_b} C \xrightarrow{En_c} D \xrightarrow{En_d} E $$

   If the enzyme, $En_a$, for the conversion of A to B is deactivated by molecules of E, then we have an example of

   (a) feedback inhibition       (c) enzyme poisoning

   (b) gene activation           (d) isoenzyme formation

12. The digestion of proteins involves a compound called trypsinogen. This compound is

   (a) a neurotransmitter        (c) an effector

   (b) a zymogen                 (d) a coenzyme

13. Plasminogen is activated by

   (a) tPA                       (c) its substrate

   (b) $NAD^+$                   (d) $Ca^{2+}$

14. Troponin is activated by

   (a) tPA                       (c) its substrate

   (b) $NAD^+$                   (d) $Ca^{2+}$

15. Calmodulin is an example of

   (a)  a zymogen                  (c)  an effector

   (b)  an enzyme inhibitor        (d)  a neurotransmitter

16. The general principle that makes possible the measurement of just one enzyme out of several that are present in blood serum is that of

   (a)  enzyme specificity         (c)  enzyme inhibition

   (b)  electrophoresis            (d)  gene activation

17. Symptoms of a myocardial infarction include

   (a)  $LD_1$-$LD_2$ flip

   (b)  a rise in the serum level of the CK(MB) isoenzyme

   (c)  a drop in the serum level of GOT

   (d)  a change in the serum level of glucose oxidase

18. Primary chemical messengers are

   (a)  enzymes and hormones

   (b)  hormones and neurotransmitters

   (c)  neurotransmitters and cyclic nucleotides

   (d)  cyclic nucleotides and axons

19. The ability of a hormone to recognize its own target cells depends on a "lock-and-key" fit of the hormone molecule to

   (a)  a receptor protein         (c)  cyclic AMP

   (b)  a neurotransmitter         (d)  adenyl cyclase

20. An example of a hormone that activates an enzyme is

   (a)  adrenalin                  (c)  testosterone

   (b)  insulin                    (d)  human growth hormone

21. Near a receptor of a neuron, adenylate cyclase is activated by

   (a)  a neurotransmitter

   (b)  a receptor

   (c)  a neurotransmitter-receptor complex

   (d)  phosphodiesterase

22. A substance that serves as both a neurotransmitter and a hormone is

   (a)  acetylcholine              (c)  epinephrine

   (b)  choline acetyltransferase  (d)  norepinephrine

23. Norepinephrine that is reabsorbed by the neuron that released it is degraded by the help of

   (a) GABA    (b) MAO    (c) LD    (d) CK(MB)

24. A drug that binds to a neuron's receptor protein

   (a) inhibits the transmission of a nerve signal

   (b) accelerates the activation of adenylate cyclase

   (c) prolongs the receipt of signals from the presynaptic neuron

   (d) inactivates MAO enzymes

25. Amphetamines stimulate presynaptic neurons to release

   (a) L-DOPA    (b) dopamine   (c) GABA    (d) MAO

26. A pain-killer produced in the brain itself is

   (a) substance P                     (c) enkephalin

   (b) GABA                            (d) morphine

## ANSWERS

### ANSWERS TO SELF-TESTING QUESTIONS

**Completion**

1. coenzyme    (Cofactor is not correct because it would include trace elements too. Vitamin as the answer is close, but coenzyme is better here because vitamins usually have to be changed into coenzymes first. Also, remember that just a few vitamins work this way.)

2. active site (or catalytic site)

3. protein (It is more specific than apoenzyme.)

4. enzyme-substrate complex

5. lock-and-key

6. vitamins

7. enzyme

8. esterase; kinase

9. sucrose

10. oxidases

11. nicotinamide (or nicotinic acid, but it is the amide that is used); riboflavin; thiamine

12.   feedback inhibition
13.   homeostasis
14.   substrate
15.   polypeptide chain; zymogen
16.   dissolving; plasmin; tissue plasmogen activator; zymogen
17.   troponin; effectors; calcium ion
18.   calcium channels; calcium ions, active transport
19.   calcium channel blockers
20.   antimetabolite
21.   substrate
22.   GPT; bloodstream; GOT; higher
23.   CK, LD, and GOT; CK(MM); CK(MB); CK(BB); electrophoresis
24.   LD; 5; $LD_1$-$LD_2$ flip
25.   hormones and neurotransmitters; hormones; neurons (nerve cells)
26.   target; adenylate; ATP to cyclic AMP; enzyme; phosphodiesterase; cyclic AMP
27.   enzyme; permeabilities; genes; enzymes; local hormones; are made
28.   synapse; neurotransmitter
29.   agonist; antagonist
30.   neurotransmitter; adenylate cyclase; cyclic AMP
31.   neurotransmitter
32.   acetylcholine; inhibiting choline acetyltransferase; blocking the synthesis of acetylcholine; blocking the receptor protein for acetylcholine
33.   norepinephrine; monoamine oxidases; MAO; kept up
34.   dopamine; binding; amphetamines; release
35.   gamma-aminobutyric acid; inhibited
36.   endorphins and enkephalins

**Multiple-Choice**

| | | | | | |
|---|---|---|---|---|---|
| 1. | b | 11. | a | 21. | c |
| 2. | c | 12. | b | 22. | d |
| 3. | d | 13. | a | 23. | b |
| 4. | d | 14. | d | 24. | a |
| 5. | b | 15. | c | 25. | b |
| 6. | c | 16. | a | 26. | c |
| 7. | b | 17. | a, b | | |
| 8. | c | 18. | b | | |
| 9. | a | 19. | a | | |
| 10. | c, d | 20. | a | | |

# CHAPTER 18

# EXTRACELLULAR FLUIDS OF THE BODY

## OBJECTIVES

Objectives 1 through 4 are concerned with the chemistry of digestion and the digestive juices. The first objective should be considered a minimum goal for this area.

In the professional health care fields, the chemistry of respiration and the chemistry of the blood are areas of vital importance because respiratory problems arise in a number of emergency situations. To deal swiftly and correctly with the emergency, health care personnel must obtain and quickly evaluate several measurements, including the blood pH, its bicarbonate level, and the partial pressures of the blood gases. Early in your career, you will most likely have to learn about the chemistry of respiration and the chemistry of blood in order to increase your professional capabilities. Objectives 5 through 15 deal with these areas (although a study of the chemistry of blood is not complete without a study of the function of the kidneys, the subject of objectives 16 and 17). If you use this occasion for a thorough study, your value as a professional will improve that much more quickly.

After you have studied this chapter and worked the exercises in it, you should be able to do the following.

1. Name the end products of the digestion of carbohydrates, proteins, and lipids.

2. List the major digestive reactions of the mouth, the stomach, and the duodenum.

3. Name the digestive juices and their principal enzymes and zymogens.

4. Describe the functions of bile salts.

5. Describe the functions of the principal components of the blood.

6. Explain how fluids and nutrients exchange at capillary loops.

7. Name two situations in which proteins leave the blood and describe the consequences.

8. Give three ways edema may arise.

9. Explain how oxygen binds cooperatively to hemoglobin.

10. Discuss how the ability of hemoglobin to hold and carry oxygen varies with (a) the pH of the blood, and (b) the partial pressure of carbon dioxide in blood.

11. Describe how differences in partial pressure between alveoli and actively metabolizing tissue aid in the exchange of respiratory gases.

12. Describe how localized changes in pH aid in gas exchange.

13. Describe how $pCO_2$ and $pO_2$ aid in gas exchange.

14. Explain how waste $CO_2$ is transported in blood.

15. Outline how values of blood pH, $pCO_2$, and $[HCO_3^-]$ change from the normal in clinical situations involving metabolic or respiratory acidosis and alkalosis.

16. Describe the role of the kidneys in preventing acidosis.

17. Describe what vasopressin, aldosterone, and renin do.

18. Define the terms in the Glossary.

## GLOSSARY

**Acidosis, (Acidemia).**   A condition defined by a drop in the pH of the blood. *Metabolic acidosis* is acidosis brought about by a defect in some metabolic pathway. *Respiratory acidosis* is caused by a defect in respiration.

**Aldosterone.**   A steroid hormone made in the adrenal cortex that is secreted into the bloodstream when the sodium ion level is low. Its target organ is the kidney where it signals the kidney to leave sodium ions in the bloodstream.

**Alkalosis (Alkalemia).**   A condition defined by a rise in the pH of the blood. *Metabolic alkalosis* is caused by a defect in some metabolic pathway. *Respiratory alkalosis* is brought about by a defect in respiration.

**Bile Salts.**   Steroid-based detergents that emulsify fats and oils in the diet, thereby removing them from coating other food particles so that the latter may be more easily attacked by water.

**Carbaminohemoglobin.**   Hemoglobin that contains chemically bound carbon dioxide.

**Chloride Shift.**   The movement of a chloride ion into a red blood cell when a bicarbonate ion migrates out.

**Digestion.**   The hydrolysis of food molecules that takes place in the digestive tract.

**Electrolytes.**   In blood chemistry, the salts dissolved in blood.

**Extracellular Fluids.**   Body fluids that are outside of cells.

**Hemoglobin.**   The complex protein in red cells that carries oxygen.

**Internal Environment.**   Everything enclosed within an organism, with emphasis on the circulating fluids of the body, their composition, and how that composition as well as body temperature are maintained.

**Interstitial Fluids.**   Fluids in tissue but not inside cells.

**Isohydric Shift.**   The movement of the hydrogen ion that is released from carbonic acid as carbon dioxide enters a red blood cell to react with oxyhemoglobin and release oxygen.

**Oxyhemoglobin.**   Hemoglobin that is bearing oxygen molecules.

**Vasopressin.**   A hormone made in the hypophysis, whose target organ is the kidney, and that regulates the concentrations of dissolved and dispersed materials in the blood by instructing the kidneys to collect urine if the blood is too dilute and not to collect urine if the blood is too concentrated.

**Zymogen.**   A polypeptide that can be changed to an enzyme by the loss of a few amino acid residues or by some other change in its structure.

## SELF-TESTING QUESTIONS

### Completion

Questions 1 through 13 are concerned with the chemistry of digestion.

1.  The first digestive juice to have much effect on proteins in the diet is

    _____.

2.  Some digestive enzymes occur in their respective digestive juices initially in inactive forms called _____.

3.  The proteolytic enzyme active in the stomach is called _____ and its zymogen is _____.

4.  The partially digested material in the stomach that moves into the upper intestinal tract is called _____.

5.  Starch in the diet is acted upon first in the _____ by the enzyme _____ found in _____.

6.  All of the reactions of digestion are classified as

    _____ reactions.

7.  For the most effective digestion of fats and oil in the diet, we depend on the emulsifying action of _____, compounds that are best described as _____ and that are released into the _____ of the digestive tract from an organ called the _____.

8.  A special enzyme in intestinal juice called _____ helps convert trypsinogen to trypsin.

9.  Besides helping to digest proteins in the diet, trypsin also catalyzes the conversion of _____ to chymotrypsin, of _____ to carboxypeptidase, and of _____ to elastase.

10.  The arrival of acidic chyme in the duodenum causes the release of enzyme-rich fluids called _____ and _____.

11.  Another fluid stimulated by the arrival of chyme in the upper intestinal tract is called _____.

12.  The principal fat-digesting enzyme released in pancreatic juice is called

_____.

13.  Complete the following table by writing the names of the end products of the complete digestion of each family of foods given in the first column.

| Food | End Products of Complete Digestion |
|---|---|
| Proteins | _____ |
| Starch | _____ |
| A mixture of lactose, maltose, and sucrose | _____ |
| Triacylglycerols | _____ |

14.  From the standpoint of osmosis and dialysis, which is the more concentrated mixture, blood or interstitial fluid?

_____

15.  In which direction will water naturally have a net tendency to migrate: from a blood capillary into the interstitial compartment or from the interstitial compartment into the bloodstream?

_____

16.  Besides the forces generated by osmosis and dialysis, what force is present in the circulatory system that is especially important on the arterial side of a capillary loop?

_____

17.  Materials are taken away from tissues by both veins and the _____ ducts.

18. If fluids accumulate in some tissue (e.g., the lower limbs) a condition of
    _____ exists.

19. Within the red blood cells, called _____ are molecules of
    _____, which carry oxygen from the lungs to tissues needing
    oxygen.

20. Each molecule of _____ in red blood cells has
    _____ subunits and each subunit can carry
    _____ molecule(s) of oxygen.

21. The reaction whereby oxygen is picked up may be symbolized as follows:

    $HHb + O_2 \rightleftharpoons$ _____ + _____

22. To help force the reaction from left to right, the system acts to neutralize the
    _____ ion.

23. The ion largely responsible for doing this (question 22) has the formula
    _____, and the product of the neutralizing reaction is
    _____ and water.

24. The binding of oxygen to hemoglobin is described as _____,
    which means that the binding of one molecule of oxygen by a molecule of
    hemoglobin changes the system to make it cooperate in the binding of the
    remaining _____ molecules of oxygen.

25. The symbol $HbO_2^-$ is our symbol for _____, and that symbol is
    incorrect to the extent it does not show that each molecule of this substance
    carries _____ molecules of oxygen.

26. At active tissues needing oxygen, ions with the symbol _____ are gener-
    ated which help $HbO_2^-$ release oxygen.

27. The equation for the formation of carbaminohemoglobin is:

    _____ $\rightleftharpoons$ _____

28.  The equation of question 27 is one way that newly formed molecules of carbon dioxide are taken up by the bloodstream. The other chemical change to carbon dioxide that occurs when it enters a red blood cell has the equation (an equilibrium):

_____  $\rightleftharpoons$  _____

29.  The equations (or equilibria) of questions 27 and 28 occur at actively metabolizing tissue, and their operation shifts what equilibrium involving oxygen release?

_____  $\rightleftharpoons$  _____

As you have written this equilibrium, how does it shift at actively metabolizing cells? _____
     (left or right)

30.  To recapitulate, the equilibrium of question 29 shifts to the _____ at cells needing _____. They need it because that have done some chemical work that produces water and _____ which reacts somewhat with the water to give _____ according to the equilibrium of question _____. $CO_2$ also reacts somewhat with hemoglobin according to the equilibrium of question _____. These last two reactions generate _____ ions that aid in forcing the equilibrium of question 29 to the _____ and thereby help in releasing

_____.

31.  The use of the hydrogen ions generated in the equilibria of questions 27 and 28 to shift the equilibrium of 29 is called the _____.

32.  Bicarbonate ions travel back to the lungs in the serum, not inside

_____.

33.  What forces the bicarbonate ions out of the red cell is the migration of _____ ions into the red cell. This exchange is called the

_____.

34. The combined action of H$^+$ and $CO_2$ generated at actively metabolizing cells

    serves to _____ the extent to
    
    (raise or lower)

    which oxygen is held by _____.

35. A hemoprotein at muscles that can bind oxygen more strongly than hemoglobin

    is called _____.

36. An acidosis brought on by an error in metabolism is called

    _____.

37. An acidosis caused by a breakdown in the respiratory centers or by some deterio-

    ration of the lungs is called _____.

38. Hyperventilation is a method used by a system to expel excess

    _____, the loss of which should help

    _____ the pH of the blood.
    
    (raise or lower)

39. Prolonged vomiting may cause metabolic _____.

40. Shallow breathing, or _____, is the way the system

    tries to retain _____ which has the effect of retaining

    _____ acid in the carbonate buffer.  This helps

    _____ the pH of the blood.  Thus, shallow
    
    (raise or lower)

    breathing may be a response to metabolic _____.
    
    (acidosis or alkalosis)

41. If shallow breathing occurs because the respiratory centers are not working, the

    individual cannot efficiently expel _____ and may experience

    respiratory _____.

42. The hyperventilation of a patient in hysterics causes an overremoval

    _____, a loss of the _____ buffer, and a

    _____ in the pH of the blood.  These response result in respiratory _____. Rebreathing exhaled air helps suppress this

because it supplies more _____ to the lungs and thence to the

bloodstream.

43.  If air too enriched in oxygen is given to a patient for too long a period, that

     individual will have trouble getting _____ out of actively

     metabolizing cells.

44.  What organ(s), in effect, removes acid from the blood?

     _____

45.  If the blood pressure drops greatly, the kidneys may not be able to

     _____ the blood because the blood flow through them is

     reduced.

46.  The kidneys respond to a fall in blood pressure by secreting an enzyme called

     _____ into the blood which acts on a circulating proenzyme called

     _____. This proenzyme is changed to

     _____ which helps to generate _____,

     the most powerful _____ known.

47.  The endocrine gland called the _____ is somehow sensitive to

     the _____ pressure of the blood.  If this pressure goes up, it

     means that the concentration of dissolved and dispersed substances in the blood

     is too _____. If this happens, the gland
               (high or low)

     secretes the hormone _____, whose target organs are

     _____.

48.  Secretion of this hormone (in question 47) eventually results in the formation of a

     _____ volume of urine.
          (lesser or greater)

49.  The hormone that helps the bloodstream to conserve its sodium ions is called

     _____.

**Multiple-Choice**

1. The end products of the digestion of milk sugar are
   (a) glucose and galactose    (c) glucose and fructose
   (b) only glucose    (d) glucose and maltose

2. If gastric juice were completely devoid of its hydrochloric acid, this would impair the digestion in the stomach of
   (a) lipids    (c) proteins
   (b) carbohydrates    (d) chyme

3. Removal of the gall bladder would reduce the efficiency of the digestion of
   (a) lipids    (c) proteins
   (b) carbohydrates    (d) fatty acids

4. The enzyme that catalyzes the conversion of trypsinogen to trypsin is
   (a) amylase    (b) pepsin    (c) mucin    (d) enteropeptidase

5. The enzyme in saliva is
   (a) mucin    (b) pepsin    (c) $\alpha$-amylase    (d) amylose

6. Bile contains
   (a) a lipase    (c) a carbohydrase
   (b) a protease    (d) no enzymes

7. Without bile salts, which substances would not be as easily absorbed from the intestinal tract into the bloodstream?
   (a) fat-soluble vitamins    (c) glucose
   (b) water-soluble vitamins    (d) amino acids

8. Plasma and interstitial fluids are most unlike in their concentrations of
   (a) electrolytes    (c) lipids
   (b) proteins    (d) carbohydrates

9. Oxygen is transported in the bloodstream chiefly as
   (a) molecules of $O_2$    (c) oxyhemoglobin ions
   (b) hydronium ions    (d) methemoglobin ions

10. Hemoglobin in blood has a relatively higher ability to hold oxygen molecules in the tissue capillaries of those tissues where
    (a) the pH is dropping    (c) $pCO_2$ is rising
    (b) the pH is rising    (d) $pCO_2$ is low

11.    Hemoglobin more easily accepts its second, third, and fourth molecules of oxygen than it does its first because the first initiates

(a)    a cooperative effect          (c)    an isohydric effect

(b)    a Bohr effect                 (d)    a releasor effect

12.    The equilibrium $HHb + O_2 \rightleftharpoons HbO_2^- + H^+$ will respond to a drop in pH by

(a)    shifting to the right

(b)    shifting to the left

(c)    remaining unchanged

(d)    absorbing more $O_2$

13.    If the equilibria $CO_2 + H_2O \rightleftharpoons H^+ + HCO_3^-$ shifts to the right, then the equilibrium of question 12 will tend to

(a)    shift to the right          (c)    remain unchanged

(b)    shift to the left           (d)    absorb oxygen

14.    The equilibrium $Hb + CO_2 \rightleftharpoons HbCO_2 + H^+$ shifts to the right in a region where

(a)    the ability of HHb to hold oxygen is high

(b)    the $pO_2$ is relatively high

(c)    the pH is relatively low

(d)    the $pCO_2$ is relatively high

15.    When the carbonate buffer system in blood acts, a hydrogen ion is replaced by a water molecule at the expense of a

(a)    hydroxide ion               (c)    bicarbonate ion

(b)    hydronium ion               (d)    calcium ion

16.    Healthy kidneys respond to acidosis by

(a)    putting $H^+$ into urine          (c)    retaining $Na^+$ in blood

(b)    putting $HCO_3^-$ into blood      (d)    removing organic anions

17.    When respiratory centers are healthy, the body can respond to acidosis by

(a)    hyperventilation

(b)    hypoventilation

(c)    retaining $CO_2$

(d)    retaining $H^+$

18. Hyperventilation aids in controlling acidosis by
    (a)    removing $CO_2$ from blood
    (b)    bringing in more $O_2$
    (c)    promoting the chloride shift
    (d)    increasing the $pCO_2$ of blood

19. In severe emphysema, acidosis may develop because
    (a)    hyperventilation cannot be stopped
    (b)    shallow breathing cannot be used
    (c)    oxygen toxicity has become a problem
    (d)    the removal of $CO_2$ from the blood at the lungs is impaired

20. If for any reason hemoglobin molecules leave the lungs not fully saturated with oxygen, the condition is called
    (a)    oxygen affinity        (c)    hypoxia
    (b)    oxygen toxicity        (d)    anoxia

21. If for any reason a tissue cannot get oxygen, the condition is called
    (a)    oxygen affinity        (c)    hypoxia
    (b)    oxygen toxicity        (d)    anoxia

22. The nitrogen wastes present in urine is (are)
    (a)    urea                   (c)    uric acid (and the urate ion)
    (b)    creatine               (d)    ammonia

23. If the osmotic pressure of the blood rises by even as little as 2%,
    (a)    the kidneys release renin
    (b)    the hypophysis releases vasopressin
    (c)    the adrenal cortex releases aldosterone
    (d)    the liver releases fibrinogen

24. The hormone that helps to regulate the level of sodium ions in the blood is
    (a)    angiotensin I          (c)    renin
    (b)    vasopressin            (d)    aldosterone

25. On the arterial side of a capillary loop, the blood pressure is
    (a)    lower than on the venous side
    (b)    equal to that on the venous side
    (c)    lower than the osmotic pressure from the interstitial areas
    (d)    higher than the osmotic pressure from the interstitial areas

## ANSWERS

### ANSWERS TO SELF-TESTING QUESTIONS

#### Completion
1. gastric juice
2. zymogens
3. pepsin, pepsinogen
4. chyme
5. mouth, amylase, saliva
6. hydrolysis (or hydrolytic)
7. bile salts, detergents (or soaps or steroid-based detergents), upper intestine (or duodenum), gall bladder
8. enteropeptidase
9. chymotrypsinogen, procarboxypeptidase, proelastin
10. intestinal juice and pancreatic juice
11. bile
12. pancreatic lipase
13. amino acids

    glucose

    glucose, fructose, galactose

    fatty acids and glycerol
14. blood
15. from the interstitial compartment into the bloodstream
16. simple blood pressure (from the pumping action of the heart)
17. lymph
18. edema
19. erythrocytes, hemoglobin
20. hemoglobin, four, one
21. $HbO_2^- + H^+$
22. hydrogen
23. $HCO_3^-$; carbon dioxide
24. cooperative; three
25. oxyhemoglobin, four
26. $H^+$ ($H_3O^+$ is better)
27. $HHb + CO_2 \rightleftharpoons Hb\text{--}CO_2^- + H^+$
28. $CO_2 + H_2O \rightleftharpoons HCO_3^- + H^+$

29.  $HHB + O_2 \rightleftharpoons HbO_2^- + H^+$
     to the left (or right to left)
30.  left, oxygen; carbon dioxide, bicarbonate ion and hydrogen ion, 28; 27; hydrogen ($H^+$), left, oxygen
31.  isohydric shift
32.  red blood cells (or erythrocytes)
33.  chloride ($Cl^-$); chloride shift
34.  lower, hemoglobin
35.  myglobin
36.  metabolic acidosis
37.  respiratory acidosis
38.  carbon dioxide, raise
39.  alkalosis
40.  hypoventilation, carbon dioxide, carbonic; lower; alkalosis
41.  carbon dioxide, acidosis
42.  carbon dioxide, carbonate, rise; alkalosis; carbon dioxide
43.  carbon dioxide
44.  the kidneys
45.  filter or cleanse
46.  renin, angiotensinogen; angiotensin I, angiotensin II, vasoconstrictor
47.  hypophysis (or pituitary), osmotic; high; vasopressin, the kidneys
48.  lesser than usual
49.  aldosterone

## Multiple-Choice

1.  a
2.  c
3.  a
4.  d
5.  c
6.  d
7.  a
8.  b
9.  c
10. b and d
11. a
12. b
13. b
14. d
15. c
16. a, b, c, and d
17. a
18. a
19. d
20. c
21. d
22. a, b, c, and d
23. b
24. d
25. d

# CHAPTER 19

# MOLECULAR BASIS OF ENERGY FOR LIVING

## OBJECTIVES

After you have studied this chapter and worked the exercises in it, you should be able to do the following.

1.   *Name the two products of digestion that are most frequently used for chemical energy in living systems.*

2.   *Name the end products of the complete catabolism of carbohydrates and simple lipids.*

3.   *Write the structures of the triphosphate and diphosphate networks.*

4.   *Explain the basis for classifying organophosphates as high or low energy.*

5.    Name the principal triphosphate used as an immediate source of chemical energy in cells.

6.    Outline (by a flow sheet) the principal pathways in biochemical energetics between products of digestion and the synthesis of ATP.

7.    Give the symbols for the chief hydride-accepting coenzymes and write equations (in two ways) illustrating their activity.

8.    In general terms, explain what kind of gradient is forced into existence by the operation of the respiratory chain, and where it occurs.

9.    In general terms, describe the connections between this gradient and the synthesis of ATP (using the chemiosmotic theory).

10.    Give the general purpose of the citric acid cycle (Kreb's Cycle).

11.    Write an overall equation for glycolysis - glucose to lactate.

12.    Explain the importance of glycolysis.

13.    Explain why anaerobic glycolysis ends at lactate, not pyruvate.

14.    Give the general purpose of the pentose phosphate pathway of glucose catabolism.

15.    Describe the overall results of one turn of the fatty acid cycle.

16.    Using palmitic acid as an example explain what, in a general sense, becomes of its carbon atoms when it is degraded by the fatty acid cycle.

17.    List the three main fates of amino acids in the body.

18.    Describe the nitrogen pool.

19.    Given the structure of an amino acid, write the structure of what it changes into when it undergoes (a) transamination or (b) oxidative deamination.

20.    List the major possible uses of the chemical energy and the molecular parts of amino acids.

21.    Name the ketone bodies and explain how they are produced in greater than normal amounts when effective insulin is missing.

22.    Explain the relations between (a) ketonemia and ketonuria, (b) ketonemia and ketone breath, and (c) ketonemia and ketoacidosis.

23.    Discuss the step-by-step progression of events from a condition in which effective insulin is lacking to the coma that results if the condition remains untreated.

24.    Define the terms in the Glossary.

## GLOSSARY

**Acetyl Coenzyme A.**  A molecule that can be made from glucose or fatty acids and that the body uses to transfer acetyl groups into the citric acid cycle.

$$CH_3-\overset{\overset{\textstyle O}{\|}}{C}-S-CoA$$

**Adenosine Triphosphate (ATP).**  A high-energy triphosphate ester that provides chemical energy for several metabolic processes as it changes to adenosine diphosphate (ADP) and inorganic phosphate.

**Aerobic Sequence.**  The citric acid cycle as coupled to the respiratory chain and that cannot operate without oxygen.

**Anaerobic Sequence.**  Oxygen-independent catabolism of glucose to lactate.

**ATP Synthase.**  The enzyme in a mitochondrion that catalyzes the formation of ATP from ADP and Pi, and then holds the ATP until a proton flow set up by the respiratory chain occurs.

**Beta Oxidation.**  A synonym for the fatty acid cycle.

**Catabolism.**  Reactions in living things that break molecules down.

**Chemiosmotic Theory.**  The energy for oxidative phosphorylation comes from the establishment by the respiratory chain of a gradient of protons and (+) charges between the outside and the inside of the inner membrane of a mitochondrion.

**Citric Acid Cycle.**  A series of reactions that dismantle acetyl units, sending electrons (and protons) down the respiratory chain and putting the carbon of the acetyl unit into carbon dioxide.

**Deamination, Direct.**  The removal of an amino group from some molecule (e.g., an amino acid) and the formation of ammonia or the ammonium ion.

**Fatty Acid Cycle.**  A series of repeating reactions whereby fatty acids are broken down to give molecules of acetyl coenzyme A for various metabolic needs; it is also called the fatty acid spiral.

**Gluconeogenesis.**  The making of glucose from smaller molecules and ions.

**Glycolysis.**  The breakdown of glucose to pyruvate or lactate.

**Ketoacidosis.**  Acidosis resulting from untreated ketosis.

**Ketone Bodies.**  Acetoacetic acid, $\beta$-hydroxybutyric acid, and acetone.

**Ketonemia.**  A higher than normal level (concentration) of ketone bodies in the bloodstream.

**Ketonuria.**  The appearance of higher than normal concentrations of the ketone bodies in the urine.

**Ketosis.**   A condition denoting the combined occurrence of ketonemia, ketonuria, and detectable odor of acetone on the breath.

**Lipigenesis.**   The synthesis of fatty acids from acetyl units.

**Nitrogen Pool.**   The sum total of nitrogen in all its forms - dissolved amino acids, proteins, nucleic acids, etc. - in an organism.

**Oxidative Deamination.**   The reverse of reductive amination; the change of an amino group to a keto group.

**Oxidative Phosphorylation.**   The synthesis of a high-energy phosphate (e.g., ATP) by a series of reactions involving the respiratory chain and inorganic phosphate.

**Oxygen Debt.**   The condition in a tissue when anaerobic glycolysis has operated and lactate has been excessively produced.

**Pentose Phosphate Pathway.**   The series of reactions in the metabolism of glucose 6-phosphate whereby NADP is made; pentose sugars are intermediates.

**Respiratory Chain.**   A series of reactions (or, in different contexts, the series of enzymes directly involved in the reactions) that dismantle the intermediates of the citric acid cycle, extract electrons from them, and send the electrons along to the last enzyme in the respiratory chain which gives them to oxygen as water forms.

**Respiratory Enzymes.**   Electron-acceptor and electron-donor enzymes of the respiratory chain.

**Transamination.**   The transfer of the amino group from an amino acid to a keto acid such that the keto group is changed to an amino group.

## SELF-TESTING QUESTIONS

### Completion

1.   When glucose is burned in air, the elements present in the glucose emerge in molecules of _____ and _____.

2.   When glucose is carried through both the anaerobic and aerobic sequences of metabolism in the body, its elements emerge in molecules of _____ and _____.

3.   The energy produced by the combustion of glucose appears largely as _____.

4.  Some of the energy produced by the breakdown of glucose in the body appears
    as _____, but of greater importance to the body, it
    also appears in the form of molecules of _____.

5.  The structure of ATP in its electrically neutral, un-ionized form is (complete the
    following):

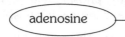

6.  When adenosine triphosphate reacts, provides energy for some chemical change,
    and loses one phosphate, the remainder of its molecule is called

    _____.

7.  The symbol, Pi, stands for a mixture having (give the formulas)
    _____ and _____ as its two
    principal ions.  The exact proportion of these ions in the mixture is largely a
    function of the _____ of the medium.

8.  When proteins of a relaxed muscle interact with ATP, the muscle contracts as
    _____ and _____ are produced.

9.  To put these two substances (question 8) back together again, the body taps
    energy from the _____, a series of reactions whereby
    electrons are taken from some metabolite of the _____ cycle and
    are delivered to _____ to produce water.

10. To fuel the _____ cycle, the body uses acetyl _____.

11. The chief location in the cell for the synthesis of ATP is the
    _____ of a mitochondrion.  When the respiratory
    chain operates, the net effect is to change $MH_2$ and half a molecule of oxygen to
    M and _____ and to generate a gradient of _____ ions and
    _____ charge across the _____ of the mitochondrion.

12. As _____ ions move across the inner membrane of the mitochon-drion at those places where such movement can occur, changes occur in the enzyme _____, that cause it to release _____, whose synthesis it has catalyzed from _____ and Pi.

13. Our symbol for a donor of H:⁻ units to the respiratory chain is _____, and the metabolic pathway that is the richest supplier of them is called the

    _____.

14. The first receptor of H:⁻ in the respiratory chain is an enzyme whose coenzyme has the short symbol of _____; the symbol for its reduced form is _____. This reduced form can pass H:⁻ to another enzyme with a coenzyme having the short symbol of _____ and whose reduced form is _____.

15. When $FMNH_2$ passes H:⁻ to the next enzyme in the chain having the short symbol _____, only the electrons in H:⁻ go to this next enzyme and the hydrogen nuclei, $H^+$, are put _____. In this way, the first "installment" of the _____ gradient that extends from relatively high concentrations of these species on the _____
    (outside or inside)
    of the _____ mitochondrial membrane to lower concentrations on the _____ of this membrane.
    (outside or inside)

16. Following the iron-sulfur enzyme and the enzyme containing coenzyme Q is a series of enzymes collectively called the _____.

17. The final acceptor of the electrons being transferred from one of these enzymes to the next is _____, and water forms.

18. The organic group that is joined to coenzyme A has the structure:

    _____, and the symbol we use for the combination of this with coenzyme A is _____.

19.   The metabolic sequence that degrades acetyl coenzyme A is called

_____. Certain intermediates in this sequence are

_____, and agents causing this kind of chemical change are
            (oxidized or reduced)

enzymes of another sequence called the _____.

20.   A sequence in glucose catabolism that can be run anaerobically is also called

_____. Its end product is _____ which

(when there is enough oxygen available) can lose the pieces of the element

_____ and be thereby oxidized to _____.

This, in turn, can be broken down oxidatively to acetyl _____.

21.   The anaerobic sequence enables a tissue to make a fresh supply of

_____ for energy-demanding processes without oxygen

being immediately available.

22.   When a tissue is operating anaerobically, we say it is running an oxygen

_____, and an accumulation of _____, an

end-product of _____, occurs.

23.   The body's chief source of NADPH is the _____.

24.   A large fraction of the lactate produced in the anaerobic sequence is converted to

_____. The energy for accomplishing this is provided by

sending the smaller fraction of the lactate (or some other metabolite) through the

_____ sequence.

25.   When the acyl group of a fatty acid is catabolized inside a mitochondrion, it is

attached to a coenzyme called _____. The first step in the

oxidation is the loss of _____ from the $\alpha$- and $\beta$-carbon atoms to

give a functional group adjacent to the carbonyl group having the name

_____. In the next step a molecule of _____ adds

to this functional group to give a $\beta$-_____ acyl system. This is

then oxidized in third step to give the _____-group. Finally a

molecule of coenzyme A interacts with this compound to split out _____
and leave an acyl-CoA system having _____ fewer carbon atoms
than it originally had.  Then this shorter version is subjected to the same series of
steps, and the process is called the _____ cycle or
_____ oxidation.

26.    The hydrogen removed from the acyl CoA in the fatty acid cycle to give the $\alpha$,
       $\beta$-unsaturated acyl CoA derivative is accepted by an enzyme having as its coen-
       zyme _____, whose reduced form is symbolized as _____.

27.    The hydrogen removed in the oxidation of the $2^0$ alcohol group in the third step
       of the fatty acid cycle is accepted by an enzyme having the coenzyme
       _____, whose reduced form is symbolized as    _____

28.    The acetyl CoA units produced by the fatty acid cycle enter the metabolic
       pathway having the name _____.

29.    Amino acids and other nitrogenous substances in the body wherever they are,
       make up the _____.

30.    Nitrogen enters the system largely as nitrogen compounds that are products of
       the digestion of _____.

31.    Some intermediates in the catabolism of amino acids are used to make the
       brain's favorite nutrient, _____, by a series of reactions
       called _____.

32.    Some of the intermediates in the catabolism of amino acids can be used to make
       fatty acids by a series of reactions called _____.

33.    If the following kind of reaction occurs during the catabolism of an amino acid:

$$\underset{\text{G--CH--CO}_2^-}{\overset{\overset{\displaystyle NH_3^+}{|}}{}} + \underset{\text{R--C--CO}_2^-}{\overset{\overset{\displaystyle O}{||}}{}} \longrightarrow \underset{\text{G--C--CO}_2^-}{\overset{\overset{\displaystyle O}{||}}{}} + \underset{\text{R--CH--CO}_2^-}{\overset{\overset{\displaystyle NH_3^+}{|}}{}}$$

the amino acid has undergone _____.

34.  If an amino acid undergoes the following reaction:

$$\underset{|}{\overset{NH_3^+}{G-CH-CO_2^-}} + NAD^+ + H_2O \longrightarrow \underset{}{\overset{O}{\underset{||}{G-C-CO_2^-}}} + NADH + NH_4^+ + H^+$$

the amino acid has been involved in a reaction called

_____.

35.  When serine undergoes the following change:

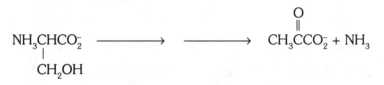

$$\underset{\underset{CH_2OH}{|}}{NH_3CHCO_2^-} \longrightarrow \quad \longrightarrow \quad \overset{O}{\underset{||}{CH_3CCO_2^-}} + NH_3$$

serine                                        pyruvate

the overall reaction is called _____.

36.  Gluconeogenesis under conditions of starvation requires a breakdown of the
body's _____.

37.  When transamination occurs to alanine, the product is called _____,
and it can be used to make acetyl CoA, the "fuel" for the
_____ or as a building block for making
_____ in lipigenesis.

38.  An enhanced rate of fatty acid oxidation in the liver may produce excessive
amounts of _____ in the blood, a condition known
as _____. This will slowly lower the pH of the blood, a
condition known as _____.

39.  One disease that leads to a large increase in the level of ketone bodies in the
blood is _____.

40.  Efforts by the body to eliminate ketone bodies and associated positive ions,
nitrogen wastes, and excess glucose from the bloodstream mean that more
_____ than usual is also removed.

41. If insufficient water is drunk per day, the blood may _____,
its circulation in the brain in sufficient quantities may become more difficult, and
some blood flow may be diverted from the _____
in order to supply the brain.  This diversion only makes it more difficult for the
body to eliminate wastes.

42. As the anions from ketonemia become part of the urine being made, the chief
cation to leave with them has the formula _____.  The
loss of this cation is sometimes called "the loss of _____", but
in reality this ion is not a base.  However, each of these cations that is lost
represents the neutralization of one _____ ion, so the net effect if the loss
of base within the system.

43. A simplified statement of one sequence of events in undetected (and therefore
untreated) diabetes mellitus would be as follows.  The excessive release of
_____ from adipose tissue and
                      (a)
the increased rate of catabolism of fatty acids in the _____
                      (b)
lead to the production of _____ at a rate faster than
                      (c)
can normally be handled.  As a result there is a slow rise in the concentration of
_____ in the bloodstream (a condition
                      (d)
called _____).  Because two of them are acids, the
                      (e)
_____ of the blood slowly drops.
                      (f)

**Multiple-Choice**

1.  One of the substances generated by biochemical energetics is
    (a) $NAD^+$    (b) ATP    (c) vitamin C  (d) hydrogen

2.  Among the substances essential to the aerobic sequence of glucose catabolism is (are)
    (a) $NAD^+$    (b) $O_2$    (c) ATP    (d) ADP

3.  The product of the operation of the respiratory chain that is most needed by the body is
    (a) ATP    (b) $H_2O$    (c) $[H:^- + H^+]$ (d) FAD

4.  Among the accomplishments of the citric acid cycle is (are)
    (a) the synthesis of acetyl coenzyme A
    (b) supplying $H:^-$ and $H^+$ to the respiratory chain
    (c) glycogenesis
    (d) glycolysis

5.  Materials needed to make ATP include (one choice)
    (a) glucose    (b) ADP + Pi  (c) fatty acids  (d) acetyl CoA

6.  The most direct initiator of the synthesis of ATP is the appearance of
    (a) glucose    (c) $NAD^+$
    (b) oxygen    (d) ADP

7.  To complete the following reaction, we should include as a product

    $$ATP + H_2O \longrightarrow \underline{\hspace{3cm}} + Pi$$
    (a) AMP    (b) ADP    (c) $H_2O_2$    (d) $CO_2 + H_2O$

8.  The chief use of acetyl CoA is to provide "fuel" for the
    (a) citric acid cycle    (c) anaerobic sequence
    (b) glycolysis    (d) homeostasis

9.  The chief purpose of the respiratory chain is to
    (a) accept $H:^-$    (c) resupply ATP
    (b) use up acetyl CoA    (d) use $O_2$

10.  Between the respiratory chain and glycolysis in the catabolism of glucose occurs the
    (a) anaerobic sequence    (c) the citric acid cycle
    (b) oxidative phosphorylation    (d) homeostasis

11. The operation of the respiratory chain establishes a gradient within a mitochondrion that involves specifically

   (a) $OH^-$ and $Cl^-$ ions     (c) $NAD^+$ and $NADH_2$

   (b) $H^+$ ions     (d) electron pairs and ATP

12. ATP synthase is activated by $H^+$ ion

   (a) to make ATP     (c) to launch the respiratory chain

   (b) to release ATP     (d) to make ADP

13. The oxidation of $MH_2$ to M and $H_2O$ that begins with the enzyme bearing the coenzyme $NAD^+$ in the respiratory chain can generate a maximum of how many molecules of ATP?

   (a) 2     (b) 3     (c) 4     (d) 6

14. One important source of acetyl CoA is

   (a) pyruvate + CoA–SH + $NAD^+$

   (b) citrate

   (c) the respiratory chain

   (d) oxidative phosphorylation

15. When glucose is catabolized to $CO_2$ and $H_2O$, oxygen atoms from $O_2$ in the air end up as parts of molecules of

   (a) $CO_2$     (b) $H_2O$     (c) ATP     (d) Pi

16. The end product of glycolysis under anaerobic conditions is

   (a) phosphoenolpyruvate     (c) lactate

   (b) pyruvate     (d) acetyl CoA

17. The end product of glycolysis under aerobic conditions is

   (a) phosphoenolpyruvate     (c) lactate

   (b) pyruvate     (d) acetyl CoA

18. If lactate is generated by hard work,

   (a) glycolysis has occurred

   (b) an oxygen debt exists

   (c) a fraction is eventually converted back to glycogen

   (d) the blood will become hyperglycemic

19. One important purpose of the pentose phosphate pathway of glucose catabolism is to make

   (a) ATP     (c) NADH

   (b) NADPH     (d) glucose 1-phosphate

20. The fatty acid cycle
    (a)  produces lactic acid
    (b)  requires the β-form of oxygen
    (c)  involves extensive glycogenolysis
    (d)  produces units of acetyl coenzyme A

21. During each "turn" the fatty acid cycle produces
    (a)  $FADH_2$, NADH, acetyl CoA
    ( b ) $FADH_2$, NADH, acetoacetyl CoA
    (c)  $FADH_2$, NADP, ADP
    (d)  $FADH_2$, NADH, β-hydroxybutyrate

22. Once the substance $R-\overset{\overset{\text{O}}{\|}}{C}-S-CoA$ has been formed, the next step in the fatty acid cycle is
    (a)  the addition of water          (c)  dehydration
    (b)  an attack by CoASH            (d)  dehydrogenation

23. Once the substance $R-CH=CH-\overset{\overset{\text{O}}{\|}}{C}-S-CoA$ has formed in the fatty acid cycle, the next step is
    (a)  the addition of water          (c)  dehydration
    (b)  an attack by CoASH            (d)  dehydrogenation

24. The fatty acid cycle degrades fatty acids by how many carbons per "turn"?
    (a)  1          (b)  2          (c)  3          (d)  4

25. The following reaction is an example of

$$CH_3\overset{\overset{\text{O}}{\|}}{C}CO_2H + HO_2CCH_2CH_2\underset{\underset{\text{NH}_2}{|}}{C}HCO_2H \longrightarrow CH_3\underset{\underset{\text{NH}_2}{|}}{C}HCO_2H + HO_2CCH_2CH_2\overset{\overset{\text{O}}{\|}}{C}CO_2H$$

    (a)  oxidative deamination          (c)  decarboxylation
    (b)  transamination                 (d)  gluconeogenesis

26. The following reaction is an example of

$$CH_3\underset{\underset{\text{NH}_2}{|}}{C}HCO_2H \xrightarrow[\text{H}_2\text{O}]{\text{NAD}^+} \longrightarrow CH_3\overset{\overset{\text{O}}{\|}}{C}CO_2H + NH_3 + NADH$$

    (a)  oxidative deamination          (c)  decarboxylation
    (b)  transamination                 (d)  gluconeogenesis

27. If transamination occurred to

$$CH_3CH_2\overset{\overset{\displaystyle CH_3}{|}}{C}H\overset{\overset{}{}}{C}HCO_2^-, \text{ it would become}$$
$$\underset{\underset{\displaystyle NH_3^+}{|}}{}$$

   (a)        $CH_3CH_2\overset{\overset{\displaystyle CH_3}{|}}{C}HCH_2NH_3^+$

   (c)   $CH_3CH_2\underset{\underset{\displaystyle H_3C}{|}}{C}H\underset{\underset{\displaystyle NH_3^+}{|}}{C}HCO_2^-$

   (b)        $CH_3CH_2\overset{\overset{\displaystyle CH_3}{|}}{C}H\underset{\underset{\displaystyle O}{||}}{C}CO_2^-$

   (d)   $CH_3CH_2\underset{\underset{\displaystyle CH_3}{|}}{C}HCH_2CO_2^-$

28. The end products in the catabolism of proteins in the body are

   (a) amino acids

   (b) nitrogen, water, and carbon dioxide

   (c) ammonia, water, and carbon dioxide

   (d) urea, water, and carbon dioxide

29. Intermediates in the catabolism of amino acids may be used to make

   (a) glucose            (c) ketone bodies

   (b) fatty acids       (d) other amino acids

30. The major site of the catabolism of amino acids is the

   (a) liver             (c) adipose tissue

   (b) kidneys        (d) gall bladder

31. Which one of these structures is not a ketone body?

   (a)   $CH_3\overset{\overset{\displaystyle O}{||}}{C}CH_2\overset{\overset{\displaystyle O}{||}}{C}O^-$

   (c)   $CH_3\overset{\overset{\displaystyle O}{||}}{C}CH_3$

   (b)   $CH_3\overset{\overset{\displaystyle O}{||}}{C}O^-$

   (d)   $CH_3\underset{\underset{\displaystyle HO}{|}}{C}HCH_2\overset{\overset{\displaystyle O}{||}}{C}O^-$

32. Which one of the conditions given is mostly closely linked to acidosis?

   (a) acetone breath       (c) ketonuria

   (b) ketonemia           (d) glycosuria

33. To remove excesses of ketone bodies the kidneys also remove excessive amounts of

   (a) $HCO_3^-$     (b) water      (c) $Cl^-$      (d) $Na^+$

34.  In a ketotic state the system loses its fight against acidosis because the kidneys cannot, at a sufficient rate, resupply the blood with

(a)  $K^+$        (b)  $HCO_3^-$        (c)  $Cl^-$        (d)  $OH^-$

## ANSWERS

### ANSWERS TO SELF-TESTING QUESTIONS

**Completion**

1.  carbon dioxide, water
2.  carbon dioxide, water
3.  heat
4.  heat, ATP

5.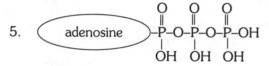

6.  adenosine diphosphate, ADP
7.  $H_2PO_4^-$, $HPO_4^{2-}$; pH
8.  ADP, Pi
9.  respiratory chain, citric acid, oxygen
10. citric acid cycle; CoA
11. inner membrane; $H_2O$; $H^+$; (+) charge; inner membrane
12. $H^+$; ATP synthase; ATP; ADP
13. $MH_2$; citric acid cycle
14. $NAD^+$: NADH; FMN: $FMNH_2$
15. FeS–P; to the outside of the inner membrane of the mitochondrion; $H^+$ ion; outside; inner; inside
16. cytochromes
17. oxygen
18. 

$$\underset{CH_3-C-}{\overset{O}{\|}}; \qquad \underset{CH_3-C-S-CoA}{\overset{O}{\|}}$$

19. citric acid cycle; oxidized; respiratory chain
20. glycolysis; lactate, hydrogen, pyruvate; coenzyme A
21. ATP

22. debt, lactate, glycolysis
23. pentose phosphate pathway of glucose catabolism
24. glucose; aerobic
25. coenzyme A; hydrogen; alkene group; water; hydroxy; keto; acetyl CoA; two; fatty acid; beta
26. FAD; $FADH_2$; respiratory chain
27. $NAD^+$, NADH
28. citric acid cycle
29. nitrogen pool
30. proteins
31. glucose; gluconeogenesis
32. lipigenesis
33. transamination
34. oxidative deamination
35. direct deamination
36. protein
37. pyruvate; citric acid cycle; fatty acids
38. ketone bodies; ketonemia; acidosis (or ketoacidosis)
39. diabetes
40. water
41. thicken (or become more viscous); kidneys
42. $Na^+$; base; $H^+$
43. (a)  fatty acids
    (b)  liver
    (c)  ketone bodies
    (d)  hydrogen ions (or, hydronium ions)
    (e)  acidosis (or, in this case, ketoacidosis)
    (f)  pH

**Multiple-Choice**

| | | | |
|---|---|---|---|
| 1. | b | 18. | a, b, and c |
| 2. | a, b, and d | 19. | b |
| 3. | a | 20. | d |
| 4. | b | 21. | a |
| 5. | b | 22. | d |
| 6. | d | 23. | a |
| 7. | b | 24. | b |
| 8. | a | 25. | b |
| 9. | c | 26. | a |
| 10. | c | 27. | b |
| 11. | b | 28. | d |
| 12. | b | 29. | a, b, c, and d |
| 13. | b | 30. | a |
| 14. | a | 31. | b |
| 15. | b | 32. | b |
| 16. | c | 33. | b and d |
| 17. | b | 34. | b |

# CHAPTER 20

# METABOLISM AND MOLECULE BUILDING

## OBJECTIVES

When you have finished studying the chapter and have answered the Review Questions, you should be able to do the following.

1. *Discuss the factors that influence the blood sugar level, using the technical terms for high and low values.*

2. *List some of the effects of hypoglycemia.*

3. *Describe the influences that the hormones epinephrine, growth hormone, glucagon, insulin, and somatostatin have on glucose metabolism.*

4. *Describe the glucose tolerance test and its purpose.*

5.  State the overall result of gluconeogenesis including (in general terms) its raw materials.

6.  Tell what adipose tissue is and describe its roles in lipid metabolism.

7.  Describe how lipids are transported in the blood.

8.  Describe the overall change in lipigenesis.

9.  List three important fates of acetyl coenzyme A.

10. State the role of amino acids in gluconeogenesis.

11. Give the difference between the essential amino acids and the nonessential amino acids.

## GLOSSARY

**Blood Sugar.**   The carbohydrates, mostly glucose, in blood.

**Blood Sugar Level.**   The concentration of glucose in the blood (usually expressed in units of milligrams per 100 mL).

**Diabetes Mellitus ("Diabetes").**   Clinically defined, a condition in which there is an insufficiency of effective insulin and an impairment of glucose metabolism.

**Epinephrine.**   A hormone of the adrenal medulla that activates the enzymes that the system needs to get glucose released promptly from glycogen.

**Essential Amino Acids.**   The amino acids that must be supplied by the diet because the system cannot make them from other amino acids, carbohydrates, or lipids.

**Glucagon.**   A hormone secreted by the alpha cells of the pancreas in response to a drop in the blood sugar level and that stimulates the liver to release glucose stored as liver glycogen.

**Gluconeogenesis.**   The making of glucose from smaller organic molecules and ions.

**Glucose Tolerance.**   The ability to manage the intake of dietary glucose in a normal manner.

**Glucose Tolerance Test.**   A series of tests of the blood sugar level following the ingestion of a considerable amount of glucose used to obtain information about an individual's glucose tolerance.

**Glucosuria.**   The appearance of glucose in the urine.

**Glycogenesis.**   The synthesis or glycogen.

**Glycogenolysis.**   The hydrolysis of glycogen and the release of glucose units.

**Human Growth Hormone.**   A hormone that stimulates the release of glucagon.

**Hyperglycemia.**   A blood sugar level above normal.

**Hypoglycemia.**   A blood sugar level below normal.

**Insulin.**   A protein hormone made by the pancreas that is normally released in response to a rise in the blood sugar level and is needed by certain cells to help them admit glucose molecules.

**Lipigenesis.**   A series of reactions whereby the body makes fatty acids (and then lipids) from two-carbon acetyl units.

**Lipoprotein Complex.**   A complex of blood proteins and lipids (fatty acids, cholesterol, and cholesterol esters) that carries lipids in the bloodstream.

**Normal Fasting Level.**   In blood chemistry, this usually refers to the level of blood sugar in a healthy patient about four hours after a meal.

**Reductive Amination.**   The conversion of a keto group to an amino group by ammonia and a reducing agent.

**Renal Threshold.**   The concentration of some solute in blood (e.g., glucose) above which some of the substance, removed by the kidneys, is left in the urine.

**Somatostatin.**   A hormone of the hypothalamus that inhibits or slows down the release of glucagon and insulin from the pancreas.

**Transamination.**   The transfer of an amino group of an amino acid to a molecule with a keto group which is then changed to an amino group.

## SELF-TESTING QUESTIONS

### Completion

1.   The synthesis of glycogen can be called _____.

2.   The release of glucose from glycogen is called _____.

3.   A hormone of the adrenal medulla that can help release glucose from glycogen in a hurry is _____.

4.   A hormone of the alpha cells of the pancreas that stimulates the liver to release glucose is _____.

5.   If the _____ rises above normal, a condition of hyperglycemia is said to exist.

6.  The appearance of glucose in urine is called _____.

7.  Hyperinsulinism or starvation may produce a condition of

    _____-emia.

8.  Clinically, an insufficiency of effective insulin is associated with

    _____.

9.  If the kidneys are releasing glucose into the urine, the _____

    for glucose has probably been exceeded.

10. A hormone that is an even better activator of glycogenolysis at the liver than

    epinephrine is called _____.

11. The hormone secreted by the pancreas in response to a *rise* in blood sugar level

    is _____, and it stimulates removal of

    _____ from the bloodstream.

12. The "signal" for the release of insulin is a rise in _____.

13. The ability of the pancreas to respond to glucose is measured by the

    _____ test.

14. The effect of reduced insulin activity while glucagon activity remains normal is to

    reduce the supply of _____ inside cells.  In this situation,

    the cells increase their rate of _____.

15. After triacylglycerols have been completely digested, the products are a mixture

    of _____ and _____.  As these

    migrate out of the digestive tract they are changed into _____.

16. The principal storage site for triacylglycerols is _____, but

    when they are in the bloodstream they are carried as _____.

17. In a 70-kg adult make, there is roughly _____ of

    triacylglycerol, enough to last about _____ if it has to

    serve as the sole source of caloric needs.

18. If glycolysis produces acetyl CoA that the cell doesn't need for energy, this excess acetyl CoA can be made into _____ by a series of reactions called _____.

19. The reducing agent needed for lipigenesis is _____.

20. The liver can make cholesterol from _____.

21. The synthesis of cholesterol in the liver is normally switched off by molecules of

_____.

22. Amino acids that we cannot make from others and that must be present in the diet are classified as _____ amino acids.

23. One of the serious protein deficiency diseases is called _____.

24. If pyruvate underwent the following reaction:

$$CH_3\overset{\overset{\displaystyle O}{\|}}{C}CO_2^- + NH_4^+ + NADPH + H^+ \longrightarrow CH_3\overset{\overset{\displaystyle NH_3^+}{|}}{C}HCO_2^- + NADP^+ + H_2O$$

the change would be called _____.

## Multiple-Choice

1. The concentration of monosaccharides in blood is called
   (a) diabetes mellitus
   (b) the renal threshold
   (c) the blood sugar level
   (d) the normal fasting level

2. Hormones that tend to raise the blood sugar level include
   (a) insulin
   (b) epinephrine
   (c) glucagon
   (d) somatostatin

3. If too much insulin somehow appears in the bloodstream,
   (a) glucosuria will result
   (b) the blood will become hypoglycemic
   (c) the blood will become hyperglycemic
   (d) the renal threshold for glucose will be exceeded

4.  A hormone whose target tissue is the pancreas and that acts to slow down the release of insulin from the pancreas is

    (a) glucagon                    (c) epinephrine

    (b) somatostatin                (d) human growth hormone

5.  The carbon atoms in glucose most directly made by gluconeogenesis come from

    (a) $HCO_3^-$    (b) lactate    (c) acetyl coenzyme    (d) $CO_2$

6.  To help get glucose inside cells of certain tissues, the blood should carry

    (a) epinephrine    (b) insulin    (c) cyclic AMP    (d) NAD

7.  Which concentration of glucose is within the range called the normal fasting level?

    (a) 60 mg/dL  (b)  125 mg/dL    (c)  52 mg/dL (d)  90 mg/dL

8.  The appearance of glucose in the urine is called

    (a) glucosemia                  (c) glucosuria

    (b) hepatoglucose               (d) renal glucose

9.  The very rapid onset of hypoglycemia can be caused by

    (a) a high salt diet

    (b) a high carbohydrate diet

    (c) an overdose of insulin

    (d) an overdose of epinephrine

10. A term that describes the release of glucose from glycogen is

    (a) glycogenolysis              (c) glycogenesis

    (b) glycolysis                  (d) gluconeogenesis

11. If the amount of energy taken into a healthy body in the form of carbohydrates is greater than the body needs for energy, the body will experience

    (a) glucosuria

    (b) a greater rate of running the lipigenesis cycle

    (c) a greater rate of running the fatty acid cycle

    (d) an enhanced rate of glycogenolysis

12. The average, adequately nourished adult male has enough chemical energy in storage as lipids to sustain life for how long?

    (a) 1 day      (b) 1 week    (c) 1 month    (d) 3 months

13. Fatty acids are transported in the bloodstream bound as complexes to molecules of

    (a) protein    (b) triacylglycerol  (c) FFA      (d) cholesterol

14. Fatty acid synthase is a multienzyme complex that makes fatty acids from

    (a)  acetic acid                 (c)  acetoacetic acid

    (b)  acetyl CoA                  (d)  acetone

15. The uses of acetyl CoA in the body include

    (a)  the synthesis of fatty acids

    (b)  the synthesis of cholesterol

    (c)  the synthesis of certain amino acids

    (d)  fuel for the citric acid cycle

16. Glycine is listed as a nonessential amino acid because

    (a)  not very many proteins use or need glycine

    (b)  the body cannot help but get all it needs of this amino acid in the diet

    (c)  the body can make it from nonprotein sources

    (d)  humans happen to have no need for this particular amino acid

17. The nitrogen pool includes

    (a)  enzymes                     (c)  urea

    (b)  amino acids                 (d)  NADPH

18. Alanine could be made by the reductive amination of

    (a)      $\overset{\displaystyle OH}{\underset{\displaystyle |}{}}$
             $CH_3CHCH_3$

    (c)      $\overset{\displaystyle O}{\underset{\displaystyle \parallel}{}}$
             $CH_3CCO_2^-$

    (b)      $\overset{\displaystyle O}{\underset{\displaystyle \parallel}{}}$
             $CH_3CCH_3$

    (d)      $\overset{\displaystyle O}{\underset{\displaystyle \parallel}{}}$
             $CH_3CCH_2CO_2^-$

# ANSWERS

## ANSWERS TO SELF-TESTING QUESTIONS

### Completion
1. glycogenesis
2. glycogenolysis
3. epinephrine
4. glucagon

5. blood sugar level
6. glucosuria
7. hypoglycemia
8. diabetes mellitus
9. renal threshold
10. glucagon
11. insulin; glucose
12. the blood sugar level
13. glucose tolerance
14. glucose; gluconeogenesis
15. fatty acids, glycerol; triacylglycerols
16. adipose tissue; lipoprotein complexes
17. 12 kg; 43 days
18. fatty acids; lipigenesis
19. NADPH
20. acetyl CoA
21. cholesterol
22. essential
23. kwashiorkor
24. reductive amination

## Multiple-Choice

| | | | |
|---|---|---|---|
| 1. | c | 10. | a |
| 2. | b and c | 11. | b |
| 3. | b | 12. | c |
| 4. | b | 13. | a |
| 5. | b | 14. | b |
| 6. | b | 15. | a, b, c, and d |
| 7. | d | 16. | c |
| 8. | c | 17. | a and b |
| 9. | c | 18. | c |

# CHAPTER 21

# NUCLEIC ACIDS

## OBJECTIVES

Heredity is not only involved with substances having certain *functions*; it is also concerned with molecules having particular *structures* that carry out these functions. You are not expected to memorize the structures of the heterocyclic bases, their corresponding nucleotides, or sections of nucleic acid molecules. You should, however, be able to name the hydrolysis products of the nucleic acids. You should also be prepared to give some of the condensed structural representations for these products using the symbols for the phosphate-pentose-phosphate-pentose chain and the letters - A, T, G, C, and U - that represent the side chains.

After you have completed your study of this chapter, you should be able to do the following.

1. *Give the name and the abbreviation for the chemical of an individual gene.*

2. *Give the general name for the monomer unit of a nucleic acid.*

3. *Give the names of the compounds produced when these monomeric units are hydrolyzed.*

4. *Describe in words the two main structural differences between DNA and RNA.*

5. *Using simple symbols, describe the features of a DNA strand.*

6. *Describe what in that structure is the genetic code.*

7. *Describe in words the contribution of Crick and Watson to the chemistry of heredity.*

8. *Describe in words the structure and shape of paired strands of DNA and the forces that stabilize it.*

9. *Explain why, regardless of species, A and T are always found in a ratio of 1:1, and why G and C are found in the same ratio.*

10. *Name and outline the process by which a gene copies itself.*

11. *Describe the functions of the four types of RNA and where they work.*

12. *Using words and simple drawings, explain how a gene specifies a unique amino acid sequence on a polypeptide.*

13. *Using letter symbols - A, T, C, G, and U - explain what a codon and its anticodon are and name the chemicals that bear them.*

14. *In general terms, explain how a virus works in a host cell.*

15. *In general terms, describe some of the things that can be done by recombinant DNA technology.*

16. *Give four examples of diseases related to genetic disorders.*

17. *Define each of the terms in the Glossary.*

## GLOSSARY

**Anticodon.**   A sequence of three adjacent heterocyclic amines located on a molecule of tRNA and complementary to a codon.

**Bases, Heterocyclic.**   The heterocyclic amines obtained by the hydrolysis of nucleic acids.

**Base Pairing.**   In the chemistry of heredity, the attraction by means of hydrogen bonds of pairs of bases - adenine with thymine and guanine with cytosine. (Uracil can substitute for thymine.)

**Chromatin.**   Filaments in a cell nucleus consisting of strands of DNA associated with polypeptides called histones.

**Chromosome.**   Small threadlike bodies within the nucleus of a cell that carry the genes of an organism in a linear array and are visible microscopically during cell division.

**Codon.**  A specific sequence of three adjacent heterocyclic amines in an mRNA strand that constitutes the code "word" for a specific amino acid.

**Deoxyribonucleic Acid (DNA).**  The chemical of a gene; it is a polymer in which the repeating units are contributed by four nucleotides, each of which carries one of four heterocyclic amines - adenine (A), thymine (T), guanine (G), and cytosine (C).

**DNA.**  See **Deoxyribonucleic Acid**.

**Double Helix.**  In the chemistry of heredity, an arrangement of two DNA strands held together by hydrogen bonds between an appropriate pair of heterocyclic amines.  The two strands are directed in opposite ways and are intertwined in a right-handed corkscrew-like helix.

**Exon.**  A segment of a DNA strand that gets expressed eventually in specifying a segment of a polypeptide.

**Gene.**  A hereditary unit carried within a cell on its chromosomes and consisting of DNA or a section of a DNA molecule.

**Genetic Code.**  (1) The relation between an amino acid and particular codons that specify that amino acid during the synthesis of a polypeptide. (2) The sequence of anticodons in a DNA strand (but this usage applies more to *genetic message*).

**Intron.**  A segment of a DNA strand that does not become expressed as a segment of a polypeptide.

**Messenger RNA (mRNA).**  RNA that carries the genetic code for a specific polypeptide out from the cell's nucleus to the cytoplasm.  The mRNA attaches ribosomes to itself.  It then accepts a parade of amino acids, each borne by its own tRNA molecule, and fashions a specific polypeptide.

**Nucleic Acid.**  A polymer of nucleotides in which the repeating units are a phosphate-pentose system and the side chains, attached to the pentose units, are heterocyclic amines.

**Nucleotide.**  A monomer of nucleic acid that consists of a phosphate ester of either ribose or deoxyribose attached, in turn, to one of a small set of heterocyclic amines.

**Primary Transcript RNA (ptRNA).**  RNA made directly at the guidance of DNA and from which messenger RNA (mRNA) is made.

**Recombinant DNA.**  DNA that is transferred into a plasmid which then is reinserted into a bacterium or yeast cell and which can direct the synthesis of a polypeptide not normally made by the organism.

**Replication.**  The reproductive duplication of a DNA double helix prior to mitosis.

**Replisome.**  A small package, mostly protein, that occurs at a site on the nuclear matrix and through which DNA strands loop in and out.

**Ribonucleic Acid (RNA).**   One of a family of polymers in which all of the members have the same kinds of basic structural features but have different functions.  All RNAs are polymers of ribonucleotides; the principal heterocyclic amines are adenine (A), uracil (U), guanine (G), and cytosine (C).

**Ribosomal RNA (rRNA).**   RNA incorporated into cytoplasmic granules called ribosomes.

**Ribosome.**   A complex of rRNA and protein formed into a granule that can become attached to an mRNA strand; polypeptide synthesis is catalyzed on the surface of this particle.

**RNA.**   See **Ribonucleic Acid**.

**Transcription.**   The synthesis of messenger RNA under the direction of DNA.  The genetic message on DNA is transcribed into a sequence of nucleotides on mRNA.

**Transfer RNA (tRNA).**   RNA that serves to carry an amino acid to a specific accep-tor site at a ribosome of an mRNA molecule and there puts the amino acid into place on a growing polypeptide chain.

**Translation.**   In the chemistry of heredity, the synthesis of a polypeptide under the direction of messenger RNA; the genetic code is used to make a sequence of amino acid units on a polypeptide.

**Virus.**   One of a large number of substances consisting of nucleic acid (usually RNA) surrounded by a protein that can enter particular kinds of cells, multiply, and thereby destroy the cell.

## SELF-TESTING QUESTIONS

### Completion

1.   Nucleic acids are polymers whose monomers have the general name of

   _____.  These monomers can be hydrolyzed to give one

   or the other of two sugars named _____ and

   _____, an inorganic _____ ion and

   a set of heterocyclic _____.

2.   Nucleic acids that are made with ribose have the full name of

   _____ which is usually abbreviated _____.  The four

bases usually obtained from this kind of nucleic acid have the names and one-letter symbols of _____ _____ _____ _____ .

3.  Nucleic acids made with deoxyribose have the full name of

    _____ which is usually abbreviated _____ . The four bases obtained from this kind of nucleic acid have the names and one-letter symbols of _____ _____ _____ _____ .

4.  The "backbones" of all nucleic acids contain a chain of diesters of

    _____ and the particular kind of sugar molecule. The bases are attached to the backbone, one base at each _____ unit.

5.  The bases have functional groups and molecular geometries that permit them to form base-pairs with each other by means of _____ bonds. Base A always pairs with either _____ or with _____; G always pairs with _____ .

6.  Two strands of DNA form a twisting _____ according to

    _____ and _____ . Base A of one strand pairs to base

    _____ opposite it on the other strand; and base C pairs to base

    _____ .

7.  The helices are further coiled into _____ .

8.  An individual hereditary unit called a _____ consists of a particular series of triplets of nucleotides in one of the kinds of nucleic acids,

    _____ .

9.  In higher organisms, complete genes come in interrupted sequences of triplets. The interrupting segments in DNA molecules are called _____,

    and the segments that make up a full gene are called _____ .

10. The process whereby a gene becomes reproduced exactly, just prior to cell division, is called _____ . The faithfulness of the copying depends

    on _____ .

11.   When DNA is used to direct the synthesis of a polypeptide, the DNA first directs the synthesis of a specific form of RNA called _____, abbreviated _____. This product is then processed to delete the triplets in its chain that correspond to the _____ segments of DNA. Then the remaining triplets are "knitted" back together to give another form of RNA called _____, abbreviated _____.

12.   Each triplet in mRNA is called a _____, and each is able to specify a particular _____ residue in a completed _____.

13.   The overall process of using DNA to direct the formation of mRNA is called _____.

14.   Following this process is another complicated process called _____, and its final product is a specific _____.

15.   To accomplish this last process, the cell needs two other kinds of RNA. One kind is used to make ribosomes and is called _____, abbreviated _____. The other kind is called _____, abbreviated _____, and its function is to carry _____ to polypeptide assembly sites on _____.

16.   One particular triplet of bases on a tRNA molecule can match a complementary triplet on mRNA. This tRNA triplet is called _____.

17.   Substances that can invade particular "host" cells and take over the genetic machinery are called _____.

18.   The genetic material in a virus is either _____ or _____ but not both.

19.   Most viruses have an "overcoat" made of _____ which includes an _____ capable of catalyzing the formation of an opening in the membrane of the _____ of the virus.

20. Viruses take over the genetic machinery of a host cell in order to make

    _____.

21. A retrovirus is able to use RNA to make _____.

22. The AIDS virus is an example of a _____ and its name

    is _____, abbreviated _____.

23. In reconbinent DNA technology, particles called _____ in a

    bacterium are given new genetic material.

24. Non-bacterial polypeptides that have been made by genetic engineering are

    those of (name four) _____

    _____.

25. Four diseases attributed to defective genes are

    (1) Associated with the overproduction of thick mucus in the lungs:

        _____

    (2) Associated with an impairment in the metabolism of the amino acid

        phenylalanine: _____

    (3) Associated with a blood disorder: _____

    (4) Associated with poor pigmentation of the eyes and the skin:

        _____

## Multiple-Choice

1. What is transmitted from parents to offspring is a complete set of
   (a) DNA molecules          (c) enzymes
   (b) polypeptides           (d) hormones

2. The site of polypeptide synthesis in a cell is
   (a) a chromosome           (c) a ribosome
   (b) a gene                 (d) a nuclear membrane

3.  If DNA were fully hydrolyzed, the products would be

    (a)  ribose                        (c)  phosphoric acid

    (b)  deoxyribose                   (d)  a few heterocyclic amines

4.  Molecules of tRNA differ from molecules of mRNA in being

    (a)  longer                        (c)  shorter

    (b)  triple helices                (d)  inside ribosomes

5.  The overall process of transcription proceeds in which order?

    (a)  exon to intron to polypeptide

    (b)  mRNA to polypeptide

    (c)  DNA to ptRNA to mRNA

    (d)  DNA to rRNA to tRNA to mRNA

6.  DNA segments that appear to be uninvolved in polypeptide synthesis are called

    (a)  introns    (b)  exons    (c)  triplets    (d)  anticodons

7.  Human insulin can be made by bacteria or yeasts by a method called

    (a)  recombinant bacteria         (c)  recombinant plasmids

    (b)  recombinant RNA              (d)  recombinant DNA

8.  Gene-directed polypeptide synthesis proceeds in which order of events?

    (a)  gene to mRNA to ptRNA to tRNA

    (b)  gene to ptRNA to mRNA to polypeptide

    (c)  gene to rRNA to tRNA to ptRNA

    (d)  gene to replicated gene to polypeptide

9.  According to the Crick-Watson theory, the genetic message carried by a gene is related most particularly to

    (a)  the kinds of heterocyclic amines projecting from the phosphate-pentose chain in the gene

    (b)  the sequence in which the heterocyclic amines are lined up along the "backbone" of the gene

    (c)  the absence of one of the –OH groups normally found in RNA

    (d)  the sequence of amino acids in gene molecule

10. The functional groups in adenine are geometrically arranged to enable adenine to pair by hydrogen bonding with

    (a)  adenine    (b)  guanine    (c)  thymine    (d)  uracil

11. If a codon triplet were U–C–G, the anticodon would be

    (a)  A–C–G    (b)  C–G–A    (d)  T–G–C    (d)  A–G–C

# ANSWERS

## ANSWERS TO SELF-TESTING QUESTIONS

### Completion

1. nucleotides; ribose and deoxyribose; phosphate; bases (or amines)
2. ribonucleic acids: RNA

   adenine    A              guanine    G

   uracil     U              cytosine   C
3. deoxyribonucleic acids; DNA

   adenine    A              guanine    G

   thymine    T              cytosine   C
4. phosphoric acid; sugar (or pentose)
5. hydrogen; T or U; C
6. double helix; Crick and Watson; T; G
7. super helices
8. gene; DNA
9. introns; exons
10. replication; base pairing of A to T and of G to C
11. primary transcript RNA; ptRNA; intron; messenger RNA; mRNA
12. codon; amino acid; polypeptide
13. transcription
14. translation; polypeptide
15. ribosomal RNA; rRNA; transfer RNA; tRNA; amino acyl units; a ribosome (or at an mRNA site at a ribosome)
16. an anticodon
17. viruses
18. RNA or DNA
19. protein; enzyme; host cell
20. more virus
21. DNA
22. retrovirus; human immunodeficiency virus; HIV
23. plasmids
24. insulin; growth hormone, interferon, erythropoeitin

25.  (1)  cystic fibrosis
     (2)  PKU (phenylketonuria)
     (3)  sickle cell anemia
     (4)  albinism

## Multiple-Choice

1.  a
2.  c
3.  b, c, d
4.  c
5.  c
6.  a
7.  d
8.  b
9.  b
10.  c, d
11.  b

# CHAPTER 22

# RADIOACTIVITY AND HEALTH

## OBJECTIVES

After you have studied this chapter and have worked the Exercises, Review Questions, and Review Problems in it, you should be able to do the following. The first nine objectives are basic to the study of radioactivity and nuclear chemistry. Objectives 10 through 12 are most important for those of you who may one day work with radioisotopes. Finally, objectives 13 through 15 will help you in reading and studying current issues involving nuclear power and radioactive pollution.

1.  *Name three kinds of atomic radiations from natural sources.*

2.  *Contrast a chemical change with a nuclear change.*

3.  *Balance a nuclear equation when given all but one of the particles involved.*

4.  *Contrast alpha, beta, and gamma rays according to their composition and their relative penetrating power.*

5.  Explain the term "half-life".

6.  Do inverse square law calculations.

7.  Describe the steps one can take to protect oneself from radiations.

8.  Name three ways of detecting radiations.

9.  Describe in general terms why radiations are damaging to cells.

10. Discuss five desirable properties of a radionuclide that is to be used in medical diagnosis.

11. Explain how technetium-99 is used in the diagnosis of disease.

12. Explain how I-131 and I-123 can be used in the diagnosis of disease.

13. Not all of the many units of radiation measurement (Section 22.3) may be assigned in your course. For those that are, name the unit used to describe each of the following (as assigned) and give its symbol (if it has one).

    (a) how radioactive is a sample;

    (b) how intense is an exposure to X rays (or gamma rays);

    (c) how much energy does a particular kind of radiation release in a given mass of tissue (irrespective of the kind of tissue);

    (d) how much energy is associated with a specific radiation.

    (e) how relatively stable is a particular radioactive isotope.

14. Describe the kinds of atomic events that occur in the reactor of a nuclear power plant.

15. Name three radioactive isotopes that can be released into the environment from a nuclear power plant.

16. Define each of the terms given in the Glossary.

## GLOSSARY

**Alpha Particle.**   The nucleus of a helium atom; $^4_2$He.

**Alpha Radiation.**   A stream of high-energy alpha particles (alpha particles are the nuclei of helium atoms with the symbol $^4_2$He); an ionizing radiation.

**Background Radiation.**   Cosmic rays plus the natural atomic radiation emitted by the traces of radioactive isotopes present in soils and rocks.

**Beta Particle.**   The electron; $^0_{-1}$e

**Beta Radiation.**   A stream of high-energy electrons; an ionizing radiation.

**Cosmic Ray.**   Streams of ionizing radiations coming to us from the sun and outer space and consisting mostly of protons but also alpha particles, electrons, and the nuclei of atoms of lower atomic number (up to number 28).

**Curie (Ci).**   A unit of measure for the activity of a radioactive source. $1 \text{ Ci} = 3.70 \times 10^{10}$ disintegrations per second.

**Electron Volt (eV).**   A very small unit of energy often used in describing the energy associated with a particular radiation.

$1 \text{ eV} = 1.6 \times 10^{-19} \text{ joule}$

$1 \text{ eV} = 3.8 \times 10^{-20} \text{ calorie}$

$1000 \text{ eV} = 1 \text{ KeV (1 kiloelectron volt)}$

$1000 \text{ KeV} = 1 \text{ MeV (1 megaelectron volt)}$

**Fission.**   The splitting of the nucleus of a heavy atom approximately in half which is accompanied by the release of one to several neutrons and much energy.

**Gamma Radiation.**   A radiation similar to powerful X rays but formed naturally during radioactive disintegrations.

**Half-Life ($t_{1/2}$).**   The period of time needed for one-half of the atoms in a sample of a particular radioactive isotope to undergo radioactive decay.

**Inverse Square Law.**   The intensity of radiation at some distance from a source varies inversely with the square of the distance.

**Ionizing Radiation.**   Radiation that can create ions from the neutral molecules of a material through which it passes and by which it is eventually absorbed. (Alpha, beta, gamma, X, and cosmic radiations, as well as neutrons beams, are all ionizing radiations.)

**Nuclear Chain Reaction.**   The mechanism of nuclear fission with which one fission event makes enough fission-initiators (neutrons) to trigger more than one additional fission event.

**Nuclear Equation.**   A shorthand symbol of a nuclear transformation such as radioactive decay or transmutation in which the symbols for the particles include their mass numbers and atomic numbers.

**Rad.**   1 rad = 100 ergs of energy absorbed per gram of tissue as a result of receiving ionizing radiations.

**Radiation Sickness.**   The set of symptoms that develop following exposure to heavy doses of ionizing radiations.

**Radiation Sickness.**   The set of symptoms that develop following exposure to heavy doses of ionizing radiations.

**Radical.**   A particle with one or more unpaired electrons.

**Radioactive.**   The property of unstable atomic nuclei whereby they eject alpha, beta, or gamma rays.

**Radioactive Decay.**   The change of a radioactive isotope into another isotope by the emission of alpha rays or beta rays.

**Radioactive Disintegration Series.**   A series of isotopes selected and arranged in such a way that each isotope but the first is produced by radioactive decay from the isotope preceding it in the series.  The last isotope in the series is nonradioactive.

**Radionuclide.**   A radioactive isotope.

**Rem.**   One rem is that quantity of any given radiation that will produce the same effect in man as one roentgen of X rays or gamma rays.

**Roentgen.**   One roentgen is that quality of X rays or gamma rays that will generate ions with an aggregate of $2.1 \times 10^9$ units of charge in 1 mL of dry air at normal pressure and temperature.

**Threshold Exposure.**   The level of exposure to some toxic substance or condition below which no harm is done.

**Transmutation.**   The change of an isotope of one element into an isotope of a different element by the emission of an alpha or a beta ray.

**Tumorogen.**   Any chemical or radiation that can initiate the formation of a tumor.


## SELF-TESTING QUESTIONS

### Completion

#### *Questions About General Aspects of Radioactivity*

1.   A substance that will ruin photographic film even when it is stored at room temperature in an airtight black wrapping is probably _____.

2.   The nuclei of helium atoms are often called _____ particles.

3.   If a substance emits beta radiation, its atoms have lost _____ from their _____, and their atomic number has changed by (how many) _____ unit(s) to a _____ value.
     (higher or lower)

4.   If a substance emits gamma radiation, its nuclei have lost _____ and they have experienced a change in _____ unit(s) of atomic number and _____ unit(s) of atomic mass.

5.  Traces of radionuclides in soils and rocks contribute to our

    _____.

6.  If a radioactive source initially weighing 8 grams had only 4 grams of that isotope left after 20 days, then the _____ of this source is 20 days.

7.  If a radioactive source initially weighing 12 grams had only 3 grams of that isotope left after 8 days, its half-life is _____ days.

8.  If a radionuclide emits an alpha particle, its atomic number changes by

    _____ unit(s) _____ and its atomic
                                           (up or down)

    mass changes by _____ unit(s) _____.
                                                           (up or down)

9.  Radiations that closely resemble gamma radiation are called _____.

10. Because alpha radiation emitted from radioactive sources has enough energy to knock electrons from neutral molecules of air, we say that this radiation is an

    _____.

11. Complete and balance these nuclear equations.

    (a)  $^{210}_{84}\text{Po} \longrightarrow {}^{4}_{2}\text{He} +$         _____

    (b)  $^{212}_{86}\text{Rn} \longrightarrow {}^{208}_{84}\text{Po} +$        _____

    (c)  $^{212}_{82}\text{Pb} \longrightarrow {}^{0}_{-1}\text{e}$              _____

### Questions About Medical and Health Aspects of Radiations

12. Potentially, the most serious damage that a radiation can do to a cell is to hit its

    _____.

13. When a high-energy particle hits a water molecule, the molecule can lose first an electron an then a proton leaving an unstable particle with the formula

    _____.

14. Exposure to high intensity radiations can cause a set of symptoms that collectively are called _____.

15. Four kinds of serious, long-term consequences of exposure even to low levels of ionizing radiations are _____, _____, _____, and _____.

16. The most penetrating radiations are _____ and _____.

17. Two strategies for protecting oneself from exposure to the radiations of some radioactive source are _____ and _____.

18. The inverse square law states that _____.

19. If the intensity of radiation, in arbitrary units, is 25.0 units at a point 5.00 m from a source, then at a point 10.0 m from the source the intensity will be _____ units.

20. If the intensity of radiation, in arbitrary units, is 100 units at 1.00 m from a source, then how many meters from the source would you have to move to reduce the exposure to 1.00 unit?

_____

21. Because there are some naturally occurring radionuclides in such common substances as rocks and soil, we can never escape exposure to _____ radiation.

22. Three radionuclides produced as wastes in the operation of atomic reactors and the tissues they "seek" in the body are

| Radionuclide | Tissue It Seeks |
|---|---|
| _____ | _____ |
| _____ | _____ |
| _____ | _____ |

23. Which is more likely to be used in diagnosis and which in cancer therapy?

X rays of 100 keV energy _____

X rays of 1.3 MeV energy _____

24. When a film badge dosimeter is used, the amount of exposure to radiations is correlated with the degree of _____ on the film.

25. If a nucleus of a molybdenum-98 atom successfully captures a neutron, the product will be which radionuclide?

_____

26. When selecting a radionuclide for use in diagnosis, the radiologist must find one with chemical properties that are compatible with the body, and beside this are five other properties:

    (1) _____

    (2) _____

    (3) _____

    (4) _____

    (5) _____

27. Complete the following table in the manner in which it is now partly filled. The first column contains the names of radioactive isotopes used in medicine and therapy. The second column gives the chemical form in which the isotope is administered. The third column is for a very brief statement describing the purpose(s) of the use of the isotope.

| | Isotope | Form | Purpose(s) |
|---|---|---|---|
| (a) | technetium-99$^m$ | $TcO_4^-$ | _____ |
| (b) | iodine-131 | _____ | _____ |
| (c) | iodine-123 | _____ | _____ |
| (d) | cobalt-60 | _____ | _____ |
| (e) | phosphorus-32 | _____ | _____ |

## *Questions About Units of Radiation Measurements*

28.   Complete the following table relating the unit of measurement used to the kind of measurement taken.

|     | Kind of Measurement | Common Unit Used | SI Unit |
| --- | --- | --- | --- |
| (a) | The energy of the particles in a stream of alpha or beta rays or the energy of gamma rays | _____ | — — — _____ |
| (b) | Exposure to X rays or gamma rays | _____ | — — — _____ |
| (c) | Absorbed dose | _____ | _____ |
| (d) | Activity of a radioactive source | _____ | _____ |

29.   The energy released in tissue when it is exposed to ionizing radiations is called the _____, which is commonly described in units of the

_____.

30.   A radioactive source with as many disintegrations per second as a 0.5-gram sample of radium has an activity, in common units, of _____.

31.   To allow variations in the responses of different types of radiations to different types of tissues, the absorbed dose may be multiplied by various fractions (modi-fying factors or quality factors) to give a number called the _____

_____, which is expressed in units of _____.

32.   If a 0.10-gram sample of some radioactive source is rated as 5.0 microcuries in activity, a 0.20-gram sample would have an activity of _____.

33.   If some radioactive source has a half-life of 15 years, then a 0.2-gram sample of this substance would have a half-life that is _____
(half, the same, or double)

as a 0.1-gram sample.

**Multiple-Choice**

## Questions About General Aspects of Radioactivity

1.  If a radiation from a radioactive element consists of high-energy helium nuclei, the radiation is called

    (a)   an alpha ray          (c)   a gamma ray

    (b)   a beta ray            (d)   cosmic rays

2.  In the nuclear reaction

    $$^{210}_{83}\text{Bi} \longrightarrow {}^{210}_{84}\text{Po} + \underline{\hspace{3cm}}$$

    the other product is

    (a)   an alpha particle     (c)   a neutron

    (b)   a beta particle       (d)   a proton

3.  In the nuclear reaction

    $$^{230}_{90}\text{Th} \longrightarrow {}^{\underline{\phantom{000}}}_{88}\text{Ra} + \text{alpha particle}$$

    the blank line (the mass number of Ra–88) should have the number

    (a)   230        (b)   226        (c)   234        (d)   229

## Questions About Medical and Health Aspects of Radiations

4.  The best material for shielding yourself from ionizing radiation is

    (a)   concrete    (b)   plexiglass    (c)   glass    (d)   lead

5.  If you can move away from a radioactive source so that you have doubled your distance from it, your exposure to its radiations will be reduced by a factor of

    (a)   two        (b)   three        (c)   four        (d)   sixteen

6.  If you were forced to live where you were continuously exposed to a radioactive source, you would be wisest in selecting one with

    (a)   a very short half-life      (c)   gamma-ray emissions

    (b)   a very long half-life       (d)   beta-ray emissions

7.  The principal reason for the danger associated with radioactivity is that the radiations

    (a)   liberate a great amount of energy in tissue

    (b)   cannot be shielded

    (c)   generate ions and radicals in tissue

    (d)   provoke a rapid rise in the white-cell count

## Questions About Units of Radiation Measurements

8.  The unit of the dose equivalent is the

    (a)   curie        (b)   roentgen    (c)   rad        (d)   rem

9.   A radiologist selecting an isotope for use in diagnosis would try to find one that
 (a)   is quickly eliminated by the body
 (b)   has a short half-life
 (c)   can do the job in the least concentration
 (d)   is a gamma emitter only

10.   A unit of the activity of a radioactive source is the
 (a)   curie       (b)   roentgen   (c)   rad         (d)   half-life

11.   If exposure to radiation results in the absorption of 10 joule of energy per kilogram of tissue, then the tissue has absorbed
 (a)   1 rad       (b)   10 Gy      (c)   1 rem       (d)   100 rems

12.   A radioactive source rated at 1 millicurie would be equivalent to
 (a)   100 curie              (c)   1,000 g of radium
 (b)   1 milligram of radium  (d)   1 half-life of 1 millisecond


## ANSWERS


### ANSWERS TO SELF-TESTING QUESTIONS

**Completion**
 1.   radioactive
 2.   alpha
 3.   electrons, nuclei, one, higher
 4.   energy (no particle is lost), no, no
 5.   background radiation
 6.   half-life
 7.   four
 8.   two, down, four, down
 9.   X rays
10.   ionizing radiation
11.   The missing formulas are
 (a)   $^{206}_{82}Pb$     (b)   $^{4}_{2}He$      (c)   $^{212}_{83}Bi$
12.   nucleus
13.   $\cdot\ddot{O}H$

14. radiation sickness
15. cancer, tumor, mutation, birth defect
16. X rays and gamma radiation
17. Use a dense shielding material like lead and get as far from the source as practical.
18. The intensity of radiation is inversely proportional to the square of the distance from the source.
19. 6.25
20. 10.0 m
21. background
22. strontium-90, bone seeker

    iodine-131, thyroid seeker

    cesium-137, general circulation
23. 100 keV - diagnosis

    1.3 MeV - therapy
24. fogging
25. molybdenum-99
26. (1)  short $t_{1/2}$

    (2)  decays to a nonradioactive product (or one with a long $t_{1/2}$)

    (3)  a value of $t_{1/2}$ long enough for preparation and use

    (4)  gamma-emitter only

    (5)  give a hot spot or a cold spot
27.

| | Isotope | Form | Purpose(s) |
|---|---|---|---|
| (a) | technetium-99$^m$ | $TcO_4^-$ | brain scanning |
| (b) | iodine-131 | $I^-$ | testing of thyroid function; treatment of thyroid cancer |
| (c) | iodine-123 | $I^-$ | same as for I-131 |
| (d) | cobalt-60 | Co | gamma ray source for cancer treatment |
| (e) | phosphorus-32 | $PO_4^{3-}$ | chronic leukemia treatment |

28.

| | Common Unit Used | SI Unit |
|---|---|---|
| (a) | electron volt | — — — |
| (b) | rem | — — — |
| (c) | rad | gray (Gy) |
| (d) | curie | becquerel (Bq) |

29. absorbed dose (or radiation absorbed dose), rad
30. 0.5 curie

31.  dose equivalent, rems
32.  10 microcuries
33.  the same

**Multiple-Choice**

1.  a
2.  b
3.  b
4.  d
5.  c
6.  b
7.  c
8.  d
9.  a, b, c, d
10.  a
11.  b
12.  b

ANSWERS TO

# PRACTICE AND REVIEW
# EXERCISES

IN

# Elements of General
# and
# Biological Chemistry

### Eighth Edition

## Practices Exercises, Chapter 1

1.  310 K
2.  (a)  $5.45 \times 10^8$          (b)  $5.67 \times 10^{12}$
    (c)  $6.454 \times 10^3$        (d)  $2.5 \times 10^1$
    (e)  $3.98 \times 10^{-5}$       (f)  $4.26 \times 10^{-3}$
    (g)  $1.68 \times 10^{-1}$       (h)  $9.87 \times 10^{-12}$
3.  (a)  $10^{-6}$
    (b)  $10^{-9}$
    (c)  $10^{-6}$
    (d)  $10^3$
4.  (a)  mL          (b)  μL          (c)  dL
    (d)  mm          (e)  cm          (f)  kg
    (g)  μg          (f)  mg
5.  (a)  kilogram
    (b)  centimeter
    (c)  deciliter
    (d)  microgram
    (e)  milliliter
    (f)  milligram
    (g)  millimeter
    (h)  microliter
6.  (a)  1.5 Mg     (b)  3.45 μL    (c)  3.6 mg
    (d)  6.2 mL     (e)  1.68 kg    (f)  5.4 dm
7.  (a)  275 kg     (b)  62.5 μL    (c)  82 nm or 0.082 μm
8.  (a)  95
    (b)  11.36
    (c)  0.0263
    (d)  1.3000
    (e)  16.1
    (f)  $3.8 \times 10^2$
    (g)  9.31
    (h)  $9.1 \times 10^2$
9.  (a)  $\dfrac{1\ g}{1000\ mg}$   or   $\dfrac{1000\ mg}{1\ g}$

    (b)  $\dfrac{1\ kg}{2.205\ lb}$   or   $\dfrac{2.205\ lb}{1\ kg}$
10. 0.324 g of aspirin

11.    40.0 °C
12.    59 °F (Quite cool.)
13.    20.7 mL
14.    32.1 g

## Review Exercises, Chapter 1

1.1    Both kittens and humans thrive on milk.
1.2    Anatomical, physiological, psychological, sociological, philosophical, religious.
1.3    It's a characteristic of something by means of which we can identify it and recognize it when we see it again.
1.4    A physical property can be studied and measured without changing the substance into some other substance.  A chemical property can be observed only as the substance changes chemically into another substance.
1.5    It is a number times a unit.
1.6    An object's inertia is its inherent resistance to any change in its motion or direction.  The mass is the quantitative measure of this inertia.
1.7    Volume, a derived unit, can be expressed as a particular product of a base unit; volume = length x length x length
1.8    Derived quantity.
1.9    Meter, for length.  Kilogram, for mass.  Second, for time.
       Kelvin, for temperature.  Mole, for quantity of substance.
1.10   (a)  second        (b)  kilogram
       (c)  kelvin        (d)  meter
       (e)  mole
1.11   A *reference standard* of measurement is the physical description or the physical embodiment of a *base unit*.  For example, the physical embodiment — the reference standard — for the base unit called the kilogram is a block of platinum-iridium alloy.
1.12   A reference standard should be free of risks of corrosion, fire, war, theft, and vandalism, and it should be accessible at any time to scientists in any country.
1.13   The kilogram mass
1.14   (a)  centimeter, cm
            millimeter, mm
       (b)  milliliter, mL
            microliter, µL
       (c)  gram, g
            milligram, mg

1.15    100 cm = 1 m
1.16    10 mm = 1 cm
1.17    1000 g = 1 kg
1.18    1000 mg = 1 g
1.19    Slightly shorter
1.20    Slightly shorter
1.21    It is the lowest degree of coldness possible when described in terms of Celsius degrees.
1.22    0 °C.  32 °F.  273 K  (before rounding, 273.15 K)
1.23    100 °C.  212 °F. 373 K (before rounding, 373.15 K)
1.24    (a)  212 degrees        (b)  100 degrees        (c)  100 degrees
1.25    They are identical
1.26    The °C is 9/5 times the °F.
1.27    273 K and 373 K
1.28    21 °C
1.29    410 °F
1.30    –40 °C
1.31    88 °F
1.32    No.  101 °F
1.33    42.8 °C.  No, regardless of the temperature *scale*, it's just as hot
1.34    34.6 °C
1.35    (a)  26 µg of vitamin E
         (b)  28 mm wide
         (c)  5.0 dL of solution
         (d)  55 km in distance
         (e)  46 µL of solution
         (f)  64 cm long
1.36    (a)  125 milligrams of water
         (b)  25.5 milliliters of coffee
         (c)  15 kilograms of salt
         (d)  12 deciliters of iced tea
         (e)  2.5 micrograms of ozone
         (f)  16 microliters of fluid
1.37    (a)  $1.3 \times 10^{-2}$ L          (b)  $6 \times 10^{-6}$ g
         (c)  $4.5 \times 10^{-3}$ m          (d)  $1.455 \times 10^3$ s
1.38    (a)  $2.4605 \times 10^4$ m          (b)  $6.54115 \times 10^5$ g
         (c)  $9.5 \times 10^{-9}$ L          (d)  $5.68 \times 10^{-6}$ s

1.39    (a)  1.3 cL           (b)  6 μg

        (c)  4.5 mm        (d)  1.455 ks

1.40    (a)  24.605 km     (b)  654.115 kg

        (c)  9.5 nL        (d)  5.68 μs

1.41    $3.94 \times 10^6$ people

1.42    (a)  Yes, the measurements correspond closely to the true value and the average of all measurements, 172.7 (correctly rounded), equals the true value.

        (b)  Yes, the data are precise to four significant figures, and successive measurements agree closely with each other.

1.43    (a)  F, H, and I

        (b)  B, C, D, and G

        (c)  A and E

1.44    Three

1.45    Two

1.46    (a)  144,549.1    (b)  $1.4455 \times 10^5$    (c)  $1.445 \times 10^5$

        (d)  $1.45 \times 10^5$    (e)  $1 \times 10^5$    (f)  $1.4 \times 10^5$

1.47    An infinite number

1.48    (a)  $\dfrac{5280 \text{ ft}}{1 \text{ mile}}$  or  $\dfrac{1 \text{ mile}}{5280 \text{ ft}}$

        (b)  $\dfrac{60 \text{ grains}}{1 \text{ dram}}$  or  $\dfrac{1 \text{ dram}}{60 \text{ grains}}$

        (c)  $\dfrac{453.6 \text{ g}}{1 \text{ lb}}$  or  $\dfrac{1 \text{ lb}}{453.6 \text{ g}}$

        (d)  $\dfrac{480 \text{ grains}}{1 \text{ ounce}}$  or  $\dfrac{1 \text{ ounce}}{480 \text{ grains}}$

        (e)  $\dfrac{39.37 \text{ in.}}{1 \text{ m}}$  or  $\dfrac{1 \text{ m}}{39.37 \text{ in.}}$

        (g)  $\dfrac{480 \text{ minims}}{1 \text{ liquid ounce}}$  or  $\dfrac{1 \text{ liquid ounce}}{480 \text{ minims}}$

1.49    (a)  0.520 liquid ounce

        (b)  $1.68 \times 10^3$ grain

        (c)  159 g

1.50    $\dfrac{\text{kg} \times \text{m}^2}{\text{sec}^2}$

1.51    (a)  192 cm

        (b)  111 lb

1.52    (a)  154 lb

        (b)  163 cm

1.53    16.9 liq. oz.

1.54    60.5 L

1.55    Do not cross; $2.1 \times 10^3$ kg > $1.5 \times 10^3$ kg (the limit)

1.56    $2.02 \times 10^3$ lb

1.57    3.5 g/pat

1.58    150 mg

1.59    $8.8477 \times 10^3$ m, 8.8477 km

1.60    $2.032 \times 10^4$ ft; 3.848 mi

1.61    There must be a difference in temperature between the two objects

1.62    All other forms of energy can be changed quantitatively into heat, so the other forms can be measured by letting them change into heat.

1.63    1 kcal

1.64    1.00 cal/g °C. Its high value means that the large water content of the body is able to take up or give off considerable heat without greatly changing the body temperature.

1.65    The density would not be affected because of the ratio of new mass to new volume, 0.5 g/ 0.5 mL or 1.0 g/mL, is the same as that of water.

1.66    The density increases.

1.67    23.7 lb of lead

1.68    $2.56 \times 10^3$ g, 2.56 kg, 5.63 lb of aluminum

1.69    28.3 mL of acetic acid

1.70    223 g of methyl alcohol

1.71    Density is always expressed as a ratio of mass to volume, typically in units of g/mL, whereas specific gravity has no units.

1.72    The density of water (to two significant figures) is 1.0 g/mL over the range of ordinary temperatures, so when the density of water is divided into the density of anything else, also expressed in g/mL, the units cancel and the numerical result is the same as dividing anything by 1 — there's no change.

1.73    The urinometer sinks to a lower level in a fluid of *low density*.

1.74    1.02 (100 mL of water has a mass of 100 g. The mass after adding 2 g of salt is 102 g, but the volume is still 100 mL, so the density of the solution is 102 g/100 mL = 1.02 g/mL. Divide this density by the density of water, 1.00 g/mL, and the result is 1.02; the units cancel.)

1.75    The specimen contains a relatively high amount of dissolved material.

## Practice Exercises, Chapter 2

1.    (a)  15        (b)  24        (c)  24
2.    (a)  5         (b)  18        (c)  20

3.     (a)  7 p, 7 n.  Electron shells:  2  5

       (b)  13 p, 14 n.  Electron shells:  2  8  3

       (c)  20 p, 20 n.  Electron shells:  2  8  8  2

4.     52.5

5.     (a)  Sn        (b)  Cl        (c)  Rb        (d)  Mg        (e)  Ar

6.     (a)  1         (b)  6         (c)  5         (d)  7

## Review Exercises, Chapter 2

2.1    Energy is not considered to have mass.  Matter both has mass and occupies space.

2.2    Solid, liquid, and gas

2.3    Solids have definite shapes and volumes, and neither depends on the specific container.  Gases have indefinite shapes and volumes; both depend on the specific container

2.4    In a chemical change a substance changes into one or more different substances.  Physical changes do not require this feature.

2.5    (a)  Physical

       (b)  Chemical

       (c)  Chemical

       (d)  Physical

       (e)  Physical

       (f)  Physical

       (g)  Physical

2.6    Chemical reaction

2.7    (a)  Reactants = hydrogen and oxygen

            Product = water

       (b)  Exothermic

2.8    Compounds are made from two or more elements that occur together in *definite proportions*.

2.9    The law of conservation of mass

2.10   Proton, 1+, 1 amu

       Neutron, 0 charge, 1 amu

       Electron, 1−

2.11   The electron and proton.

       (Nuclei, if considered as subatomic particles, also repel atomic nuclei and protons.)

2.13    3+

2.14    Change (a) because this change is the removal of an electron from a *neutral* particle, whereas change (b) is the removal of a negatively charged particle (the electron) from a particle that already is oppositely charged (1+).

2.15    (a)  Density = $3.2 \times 10^{15}$ g/cm$^3$

(b)  $3.2 \times 10^9$ metric ton/cm$^3$.  (3.2 billion metric tons per cubic centimeter)

2.16    They attract each other because $M$ has a charge of 3+ and $Y$ is oppositely charged with a charge of 2–.

2.17    $10^{-23}$ g

2.18    1.0000 g, almost identical with the atomic weight of hydrogen.

2.19    The electrons of an atom move about the hard, dense atomic nucleus in just a few allowed energy states.  Their movements resemble those of the planets about the sun.  As long as the electrons do not change their energy states, the atom does not emit or absorb energy.

2.20    An atom's electrons are confined to allowed energy states and the atom does not emit or absorb energy as long as the electrons do not change states.

2.21    (a)  8          (b)   18

2.22    Atomic number = 6.  Mass number = 13

Electron shells: 2  4

2.23    Atomic number = 18.  Mass number = 39

Electron shells: 2  8  8

2.24    Electron shells: 2  8  4

2.25    (a)  26          (b)   No

2.26    (a)  2  8  1

(b)  2  6

(c)  2  8  7

(d)  2  8  8  2

2.27    (a)  2  8  2

(b)  2  8  8  1

(c)  2  7

(d)  2  1

2.28    100

2.29    13

2.30    Metal

2.31    Nonmetal

2.32    Most elements consist of a mixture of a few isotopes that occur together in nature in definite proportions.  The isotopes of any given element all have identical numbers of protons in their nuclei but the numbers of neutrons vary from isotope to isotope.

2.33   *M* and *X* are isotopes.  *Q* and *Z* are isotopes.

2.34   Pair 1, because both have atomic number 10 so both are isotopes.

2.35   (a)  They have identical numbers of protons in their nuclei and identical numbers of electrons outside their nuclei.

(b)  The have different numbers of neutrons in their nuclei.

2.36   These two isotopes have identical chemical properties, so no specification that distinguishes each isotope has to be used.

2.37   The second letter in BN is not lower case, as it would have to be if it stood for an element.

2.38   (a)  I             (b)  Li            (c)  Zn
       (d)  Pb            (e)  N             (f)  Ba

2.39   (a)  C             (b)  Cl            (c)  Cu
       (d)  Ca            (e)  F             (f)  Fe

2.40   (a)  H             (b)  Al            (c)  Mn
       (d)  Mg            (e)  Hg            (f)  Na

2.41   (a)  O             (b)  Br            (c)  K
       (d)  Ag            (e)  P             (f)  Pt

2.42   (a)  Phosphorus    (b)  Platinum     (c)  Lead       (d)  Potassium
       (e)  Calcium       (f)  Carbon       (g)  Mercury    (h)  Hydrogen
       (i)  Bromine       (j)  Barium       (k)  Fluorine   (l)  Iron

2.43   (a)  Sulfur        (b)  Sodium       (c)  Nitrogen   (d)  Zinc
       (e)  Iodine        (f)  Copper       (g)  Oxygen     (h)  Lithium
       (i)  Manganese     (j)  Magnesium    (k)  Silver     (l)  Chlorine

2.44   Mass numbers are the sums of whole numbers—numbers of protons and neutrons per atom of an isotope.  The atomic weight of an element (in amu) is the average of the relative masses of the isotopes of the element taking into account the percentages that the various isotopes have in naturally occurring samples of the element.

2.45   127

2.46   80

2.47   64.8

2.48   64.7

2.49   (a)  Twice as heavy

(b)  The mass of the pile of magnesium atoms would be twice as much as that of the pile of carbon atoms.

(c)  A mass that is twice as much as that of the 2.0 g of carbon, 4.0 g of magnesium.

2.50   12 times heavier.

2.51   (a)  1.33 times as heavy.

        (b)  16.0 g of oxygen atoms.

2.52   The periodic law

2.53   Periods are horizontal rows of elements and groups are vertical columns of elements.

2.54   A representative element (one of the A-elements)

2.55   A representative element (because period 2 has no transition elements)

2.56   (a)  Group IA, alkali metal family

        (b)  Group VIIA, halogen family

        (c)  Group VIA, oxygen family

        (d)  Group IIA, alkaline earth metal family

2.57   (a)  Group VIIA, halogen family

        (b)  Group VA, nitrogen family

        (c)  Group IIA, alkaline earth metal family

        (d)  Group IA, alkali metal family

2.58   (a)  8. Yes (b) Nonmetal because its outside level has over 4 electrons, as nearly all nonmetals do. (c) Group VII, the halogen family.

2.59   (a)  Metal, because its outside level has fewer than 4 electrons.

        (b)  Group 2

2.60   (a)  9, 17, and 35

        (b)  15, 16, 17, and 18

        (c)  In order, from left to right:  VA, VIA, VIIA, 0

        (d)  Element 15:  5

             Element 16:  6

             Element 36:  8

        (e)  Element 36

        (f)  7

        (g)  Nonmetals

        (h)  In element 33, 5

             In element 34, 6

             In element 35, 7

2.61   Two sublevels.  One orbital in the $s$ sublevel and three orbitals in the $p$ sublevel.

2.62   The electrons must be spinning in opposite directions.

2.63   Alike in shape and energy.  Unlike in their orientations in space.

2.64   C: $1s^2 2s^2 2p_x^{1} 2p_y^{1}$  Na: $1s^2 2s^2 2p_x^{2} 2p_y^{2} 2p_z^{2} 3s^1$

## Practice Exercises, Chapter 3

1.  (a) AgBr      (b) $Na_2O$      (c) $Fe_2O_3$      (d) $CuCl_2$
2.  (a) Copper(II) sulfide; cupric sulfide
    (b) Sodium fluoride
    (c) Iron(II) iodide; ferrous iodide
    (d) Zinc bromide
    (e) Copper(I) oxide; cuprous oxide
3.  (a) 3+      (b) 2+      (c) 2+
4.  (a) 2 8 8 1.  1+
    (b) 2 8 6.  2–
    (c) 2 8 4.   No ion forms
5.  (a) 2 8 8
    (b) 2 8 8
6.  (a) $Cs^+$      (b) $F^-$      (c) $P^{3-}$      (d) $Sr^{2+}$
7.  (a)       $\begin{array}{c} Cl \\ | \\ Cl-C-Cl \\ | \\ Cl \end{array}$      (b) H–S–H      (c) $\begin{array}{c} Cl \\ | \\ Cl-N-Cl \end{array}$      (d) $\begin{array}{c} H \\ | \\ Br-C-Br \\ | \\ Br \end{array}$
8.  (a) $NaNO_3$      (b) KOH      (c) $Ca(OH)_2$
    (d) $MgCO_3$      (e) $Na_2SO_4$      (f) $(NH_4)_3PO_4$
9.  (a) Lithium carbonate      (b) Sodium bicarbonate
    (c) Potassium permanganate      (d) Sodium dihydrogen phosphate
    (e) Ammonium monohydrogen phosphate
10. (a)    $\delta+$ H          (b) $\delta+$  $\delta-$      (c) $\delta+$  $\delta-$  $\delta-$  $\delta+$
         |
    $\delta+$ H–N–H $\delta+$          H—F          H—O—O—H
         |
         $\delta-$

## Review Exercises, Chapter 3

3.1   A net force of electrical attraction between two atoms.
3.2   An atom is electrically neutral and an ion is always electrically charged.
3.3   The electron

3.4      2   1   +   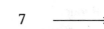   2   7   $\longrightarrow$

        Lithium atom                    Fluorine atom

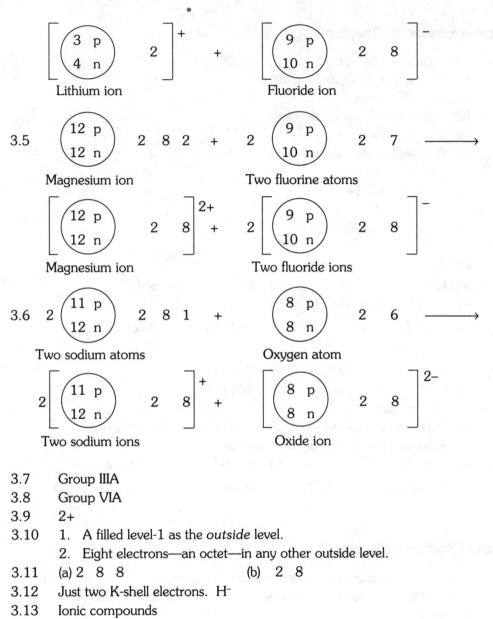

Lithium ion     Fluoride ion

Magnesium ion     Two fluorine atoms

Magnesium ion     Two fluoride ions

Two sodium atoms     Oxygen atom

Two sodium ions     Oxide ion

3.7   Group IIIA

3.8   Group VIA

3.9   2+

3.10   1.  A filled level-1 as the *outside* level.
       2.  Eight electrons—an octet—in any other outside level.

3.11   (a) 2  8  8          (b)  2  8

3.12   Just two K-shell electrons.  $H^-$

3.13   Ionic compounds

3.14   The sodium *ion*, $Na^+$

3.15   Reduced

3.16   A reduction

3.17   1+

3.18   4+

3.19  (a)  3+, $Bi^{3+}$     (b)  2+, $Cd^{2+}$     (c)  3+, $Cr^{3+}$
       (d)  3+, $Gd^{3+}$     (e)  2+, $Sn^{2+}$     (f)  3+, $Ti^{3+}$
3.20  (a)  7+              (b)  8+              (c)  5+              (d)  1+
3.21  Strontium ion
3.22  Telluride ion
3.23  (a)  $K^+$           (b)  $Al^{3+}$
       (c)  $I^-$           (d)  $Cu^+$
       (e)  $Ba^{2+}$       (f)  $S^{2-}$
       (g)  $Na^+$          (h)  $Fe^{3+}$
       (i)  $O^{2-}$        (j)  $Li^+$
       (k)  $Ag^+$          (l)  $Mg^{2+}$
       (m) $Cu^{2+}$        (n)  $Br^-$
       (o)  $Ca^{2+}$       (p)  $Fe^{2+}$
       (q)  $F^-$           (r)  $Cl^-$
       (s)  $Zn^{2+}$       (t)  $Ba^{2+}$
3.24  (a)  Sodium ion
       (b)  Iron (III) ion, ferric ion
       (c)  Lithium ion
       (d)  Oxide ion
       (e)  Sulfide ion
       (f)  Barium ion
       (g)  Copper(I) ion, cuprous ion
       (h)  Iodide ion
       (i)  Aluminum ion
       (j)  Potassium ion
       (k)  Zinc ion
       (l)  Fluoride ion
       (m) Iron(II) ion, ferrous ion
       (n)  Chloride ion
       (o)  Calcium ion
       (p)  Copper(II) ion, cupric ion
       (q)  Bromide ion
       (r)  Magnesium ion
       (s)  Silver ion
3.25  Mercurous ion
3.26  $Sn^{2+}$
3.27  Lead(II) ion

3.28    Gold(III) ion

3.29    None is violated.  It's just a convention to write the positive ion first.

3.30    (a)  LiCl          (b)  BaO
        (c)  $Al_2S_3$     (d)  NaBr
        (e)  CuO           (f)  $FeCl_3$

3.31    (a)  $Cu_2S$       (b)  KF
        (c)  $Na_2S$       (d)  $CaI_2$
        (e)  $MgCl_2$      (f)  $FeBr_2$

3.32    (a)  Iron(III) bromide, ferric bromide
        (b)  Magnesium chloride
        (c)  Sodium fluoride
        (d)  Zinc oxide
        (e)  Copper(II) bromide, cupric bromide
        (f)  Lithium oxide

3.33    (a)  Potassium iodide
        (b)  Calcium sulfide
        (c)  Barium chloride
        (d)  Aluminum oxide
        (e)  Iron(II) chloride, ferrous chloride
        (f)  Silver iodide

3.34    The noble gas family, because their atoms have outer octets (except for helium whose outer shell is filled with two electrons) and because they are chemically the least reactive elements.

3.35    The atoms of the reactive representative elements tend to undergo those chemical reactions that most directly give them electron configurations of the nearest noble gases.

3.36    They tend to gain electrons because this is the most direct way to acquire noble gas configuration.

3.37    A molecule is a small particle that consists of two or more atomic nuclei and enough electrons to ensure overall electrical neutrality.  An atom is a small, electrically neutral particle that has just *one* nucleus.  An ion is a small particle that is not electrically neutral.  It can have one nucleus or it can have two or more, but it always has too many or too few electrons for overall electrical neutrality.

3.38    An outer–level electron from one fluorine atom pairs with an outer-level electron from the other as the two are shared between the two fluorine nuclei in the fluorine molecule.  This may be symbolized as follows:

$$: \overset{..}{\underset{..}{F}} : \overset{..}{\underset{..}{F}} :$$

3.39    The level-1 (K-shell) electron of the hydrogen atom pairs with an outer-level (M-shell) electron of the chlorine atom as the two become shared between the two nuclei. This may be symbolized as follows:

$$H : \overset{..}{\underset{..}{Cl}} :$$

3.40    B

3.41    A

3.42    Electron shells:  2 8 4.  Number of covalent bonds = 4

3.43    (a)  4              (b)  IVA

3.44    H–P–H
          |
          H

3.45    S=C=S

3.46    H–Sb–H
           |
           H

3.47    H–O–Cl

3.48    (a)  Bicarbonate ion
        (b)  Sulfate ion
        (c)  Nitrate ion
        (d)  Hydroxide ion
        (e)  Ammonium ion
        (f)  Cyanide ion
        (g)  Monohydrogen phosphate ion
        (h)  Chromate ion
        (i)  Carbonate ion
        (j)  Permanganate ion
        (k)  Hydrogen sulfate ion
        (l)  Hydrogen sulfite ion
        (m)  Nitrite ion
        (n)  Phosphate ion
        (o)  Dichromate ion
        (p)  Dihydrogen phosphate ion
        (q)  Hydronium ion
        (r)  Acetate ion

3.49    (a)  $CO_3^{2-}$        (b)  $NO_3^-$
        (c)  $OH^-$            (d)  $NH_4^+$
        (e)  $PO_4^{3-}$       (f)  $CN^-$
        (g)  $H_3O^+$          (h)  $HPO_4^{2-}$

(i)  $NO_2^-$          (j)  $HCO_3^-$
(k)  $HSO_4^-$         (l)  $H_2PO_4^-$
(m)  $Cr_2O_7^{2-}$    (n)  $HSO_3^-$
(o)  $CrO_4^{2-}$      (p)  $SO_4^{2-}$
(q)  $SO_3^{2-}$       (r)  $C_2H_3O_2^-$

3.50   (a)  $(NH_4)_3PO_4$
       (b)  $K_2HPO_4$
       (c)  $MgSO_4$
       (d)  $CaCO_3$
       (e)  $LiHCO_3$
       (f)  $K_2Cr_2O_7$
       (g)  $NH_4Br$
       (h)  $Fe(NO_3)_3$

3.51   (a)  $NaH_2PO_4$
       (b)  $CuCO_3$
       (c)  $AgNO_3$
       (d)  $Zn(HCO_3)_2$
       (e)  $KHSO_4$
       (f)  $(NH_4)_2CrO_4$
       (g)  $Ca(C_2H_3O_2)_2$
       (h)  $Fe_2(SO_4)_3$

3.52   (a)  Sodium nitrate
       (b)  Calcium sulfate
       (c)  Potassium hydroxide
       (d)  Lithium carbonate
       (e)  Ammonium cyanide
       (f)  Sodium phosphate
       (g)  Potassium permanganate
       (h)  Magnesium dihydrogen phosphate

3.53   (a)  Potassium monohydrogen phosphate
       (b)  Sodium bicarbonate
       (c)  Ammonium nitrite
       (d)  Zinc chromate
       (e)  Lithium hydrogen sulfate
       (f)  Calcium acetate
       (g)  Potassium dichromate
       (h)  Sodium hydrogen sulfate

3.54    (a)  15          (b)  20          (c)  11
3.55    (a)  20          (b)  17          (c)  15
3.56    53
3.57    The electrons of the shared pair that make up the electron cloud have become located closer to the atom of Y.
3.58    Y, because a higher nuclear positive charge (hence, a higher number) is the cause of the higher electronegativity of Y compared to X, both X and Y being in the same period or row of the periodic table.
3.59    The chlorine atom's bonding shell (level 3) is screened from its atomic nucleus by both level 1 and level 2 electrons.  But the oxygen atom's bonding shell (level 2) is screened only by the level one electrons.
3.60    No, now one of the two atoms has fully accepted an electron from the other and both atoms are now ions.  Thus, the bond is an ionic bond.
3.61    Yes

δ–  δ+
X—Y

3.62    (a)  Not polar
        (b)  δ+  δ–
             H—F
        (c)  Not polar
        (d)  Not polar
        (e)  Not polar
        (f)  δ+  δ–
             H—I
3.63    X, because electronegativities increase as one moves upward within the same group or family in the periodic table.
3.64    Z, because electronegativities increase as one moves from left to right within the same period of the periodic table.
3.65    The overlapping of atomic orbitals, such as the overlapping of the 1s atomic orbitals of two hydrogen atoms to give the molecular orbital.
3.66    The pair is in the space, the molecular orbital, concentrated between two nuclei and so the electrons are attracted to both nuclei.  The high electron density in this space keeps the two nuclei attracted toward it and so held near each other; this attraction is the bond.
3.67    (a)  A 1s orbital of each H atom, one electron in each orbital.
        (b)  A $2p_z$ orbital of each F atom, one electron in each orbital.
        (c)  A 1s orbital of the H atom and a $2p_z$ orbital of the F atom, one electron in each orbital.

3.68 VS = valence shell; EP = electron pair; R = repulsion. The theory is used to explain the bond angles in a molecule.

3.69 The pairs of outer-level electrons are in four electron clouds that repel each other so that their axes are at angles of $109.5°$.

3.70 There are four outer-level electron clouds, like those in methane. (Two are not involved in bonding.) These repel each other to angles closer to $109.5°$.

## Practice Exercises, Chapter 4

1. $3O_2 \longrightarrow 2O_3$
2. $4Al + 3O_2 \longrightarrow 2Al_2O_3$
3. (a) $2Ca + O_2 \longrightarrow 2CaO$
   (b) $2KOH + H_2SO_4 \longrightarrow 2H_2O + K_2SO_4$
   (c) $Cu(NO_3)_2 + Na_2S \longrightarrow CuS + 2NaNO_3$
   (d) $2AgNO_3 + CaCl_2 \longrightarrow 2AgCl + Ca(NO_3)_2$
   (e) $2Al + 3H_2SO_4 \longrightarrow Al_2(SO_4)_3 + 3H_2$
   (f) $CH_4 + 2O_2 \longrightarrow 2H_2O + CO_2$
4. $8.68 \times 10^{22}$ atoms of gold per ounce
5. (a) 180        (b) 58.3        (c) 859
6. 0.500 mol of $H_2O$
7. 4.20 mol of $N_2$ and 4.20 mol of $O_2$
8. 450 mol of $H_2$ and 150 mole of $N_2$
9. 408 g of $NH_3$
10. 0.0380 mol of aspirin
11. 11.5 g of $O_2$
12. 18.4 g of Na and 46.8 g of NaCl
13. (a) 2.45 g of $H_2SO_4$
    (b) 9.00 g of $C_6H_{12}O_6$
14. 156 mL of 0.800 $M$ $Na_2CO_3$ solution
15. 9.82 mL of 0.112 $M$ $H_2SO_4$, when calculated step-by-step with rounding after each step.
    9.84 mL of 0.112 $M$ $H_2SO_4$, when found by a chain calculation.

## Review Exercises, Chapter 4

4.1 Carbon and oxygen react to give carbon monoxide in the proportion of 2 atoms of carbon used to 1 molecule of oxygen used and 2 molecules of carbon monoxide produced.

4.2    Nitrogen and hydrogen react to give ammonia in the proportion of 1 molecule of nitrogen used to 3 molecules of hydrogen used and 2 molecules of ammonia produced.

4.3    $H_3PO_4 + 2NaOH \longrightarrow NaHPO_4 + 2H_2O$

4.4    (a)  $2SO_2 + O_2 \longrightarrow 2SO_3$

      (b)  $CaO + 2HNO_3 \longrightarrow Ca(NO_3)_2 + H_2O$

      (c)  $2AgNO_3 + MgCl_2 \longrightarrow 2AgCl + Mg(NO_3)_2$

      (d)  $2HCl + Ca(OH)_2 \longrightarrow CaCl_2 + 2H_2O$

      (e)  $2C_2H_6 + 7O_2 \longrightarrow 4CO_2 + 6H_2O$

4.5    (a)  $2NaHCO_3 + H_2SO_4 \longrightarrow Na_2SO_4 + 2H_2O + 2CO_2$

      (b)  $Fe_2O_3 + 3H_2 \longrightarrow 2Fe + 3H_2O$

      (c)  $Ca(OH)_2 + 2HNO_3 \longrightarrow Ca(NO_3)_2 + 2H_2O$

      (d)  $2NO + O_2 \longrightarrow 2NO_2$

      (e)  $Al_2O_3 + 3H_2SO_4 \longrightarrow Al_2(SO_4)_3 + 3H_2O$

4.6    $6.02 \times 10^{23}$

4.7    Because this number equals the number of formula units of a substance whose total mass in grams is numerically equal to the formula weight of the substance.

4.8    $2.01 \times 10^{23}$ formula units of $H_2O$

4.9    $6.02 \times 10^{20}$ molecules of aspirin

4.10   0.150 g of medication

4.11   1.20 mg of impurity

4.12   The law of conservation of mass in chemical reactions.

4.13   (a)  40.0    (b)  100    (c)  98.1

      (d)  106    (e)  158    (f)  128

4.14   (a)  146    (b)  142    (c)  63.0

      (d)  98.1    (e)  132    (f)  310

4.15   Calculate the formula weight and write the unit *grams* after it.

4.16   Avogadro's number equals the number of formula units in 1 mol of the substance.

4.17   A different balance would be needed for *every* value of formula weight.

4.18   g $H_2O$/mol $H_2O$

4.19   $\dfrac{159.8 \text{ g } Br_2}{1 \text{ mol } Br_2}$  or  $\dfrac{1 \text{ mol } Br_2}{159.8 \text{ g } Br_2}$

4.20   (a)  100 g    (b)  250 g    (c)  245 g

      (d)  265 g    (e)  395 g    (f)  320 g

4.21   (a)  84.0 g    (b)  81.7 g    (c)  36.2 g

      (d)  56.4 g    (e)  75.9 g    (f)  178 g

4.22    (a)  1.88 mol    (b)  0.750 mol   (c)   0.765 mol
        (d)  0.708 mol   (e)  0.475 mol   (f)   0.586 mol

4.23    (a)  0.196 mol   (b)  0.201 mol   (c)   0.454 mol
        (d)  0.292 mol   (e)  0.217 mol   (f)   0.0923 mol

4.24    (a)  $\dfrac{4 \text{ mol Fe}}{3 \text{ mol O}_2}$    and    $\dfrac{3 \text{ mol O}_2}{4 \text{ mol Fe}}$

        (b)  $\dfrac{4 \text{ mol Fe}}{2 \text{ mol Fe}_2\text{O}_3}$    and    $\dfrac{2 \text{ mol Fe}_2\text{O}_3}{4 \text{ mol Fe}}$

        (c)  $\dfrac{3 \text{ mol O}_2}{2 \text{ mol Fe}_2\text{O}_3}$    and    $\dfrac{2 \text{ mol Fe}_2\text{O}_3}{3 \text{ mol O}_2}$

4.25    (a)  $\dfrac{2 \text{ mol C}_2\text{H}_6}{7 \text{ mol O}_2}$    and    $\dfrac{7 \text{ mol O}_2}{2 \text{ mol C}_2\text{H}_6}$

        (b)  $\dfrac{4 \text{ mol CO}_2}{2 \text{ mol C}_2\text{H}_6}$    and    $\dfrac{2 \text{ mol C}_2\text{H}_6}{4 \text{ mol CO}_2}$

        (c)  $\dfrac{6 \text{ mol H}_2\text{O}}{7 \text{ mol O}_2}$    and    $\dfrac{7 \text{ mol O}_2}{6 \text{ mol H}_2\text{O}}$

        (d)  $\dfrac{2 \text{ mol C}_2\text{H}_6}{6 \text{ mol H}_2\text{O}}$    and    $\dfrac{6 \text{ mol H}_2\text{O}}{2 \text{ mol C}_2\text{H}_6}$

4.26    (a)  26 mol of $O_2$
        (b)  50 mol of $H_2O$
        (c)  26 mol of $O_2$

4.27    (a)  525 mol of CO
        (b)  70 mol of Fe
        (c)  375 mol of CO

4.28    (a)  75.0 mol of $H_2$
        (b)  43 mol of $CH_4$

4.29    (a)  989 g of oxygen (when calculated in steps, rounding after each step)
             991 g of oxygen (when found by a chain calculation)
        (b)  171 g of $CO_2$ (step-wise calculation)
             170 g of $CO_2$ (chain calculation)
        (c)  52.0 g of $O_2$ (step-wise calculation)
             52.1 g of $O_2$ (chain calculation)

4.30    (a)  $1.28 \times 10^3$ g of NaCl
        (b)  872 g of NaOH (step-wise calculation with rounding after each step)
             873 g of NaOH (when found by a chain calculation)
        (c)  21.8 g of $H_2$ (using 2.00 g/mol for $H_2$)

4.31    (a)  $7.24 \times 10^2$ kg of $P_4O_{10}$
        (b)  $2.75 \times 10^2$ kg of $H_2O$ (step-wise calculation with rounding after each step)
             $2.76 \times 10^2$ kg of $H_2O$ (when found by a chain calculation)

4.32    48.7 g of $Na_2CO_3$ (step-wise calculation with rounding after each step)
        48.6 g of $Na_2CO_3$ (when found by a chain calculation)

4.33    The solution is unsaturated, but very nearly saturated since at 50 °C 37.0 g of NaCl can dissolve in 100 g of water.

4.34    Saturated, but very dilute

4.35    Start with some desired volume of water, warm it and then add solid KCl until no more dissolves (and some remains undissolved). Then cool this system to room temperature, and the solution above the undissolved solute is saturated.

4.36    To make it unsaturated, warm it.
        To try to make it supersaturated, carefully cool the solution and hope that excess solute does not precipitate out.

4.37    A *molecule* is a tiny particle, the smallest representative sample of a compound. A *mole* is Avogadro's number of molecules.
        *Molarity* is the ratio of the moles of a solute per liter of its solution.

4.38    Molar concentration

4.39    No, because when the solution is used, the *solution* is measured out, not the pure solvent that was used to prepare the solution.

4.40    (a)  5.85 g of NaCl
        (b)  5.63 g of $C_6H_{12}O_6$
        (c)  0.981 g of $H_2SO_4$
        (d)  11.2 g of KOH

4.41    (a)  3.18 g of $Na_2CO_3$
        (b)  2.00 g of NaOH
        (c)  7.50 g of $KHCO_3$
        (d)  8.55 g of $C_{12}H_{22}O_{11}$

4.42    $2.5 \times 10^2$ mL of 0.10 M HCl solution

4.43    2.5 mL of 1.0 M $H_2SO_4$ solution

4.44    $1.0 \times 10^2$ mL of 0.010 M $NaHCO_3$ solution

4.45    $7.5 \times 10^2$ mL of 0.10 M $H_2SO_4$ solution

4.46    250 mL of 1.00 M NaOH solution

4.47    $1.0 \times 10^2$ mL of 0.50 M $H_2SO_4$ solution

4.48    167 mL of 0.150 M $Na_2SO_4$ solution

4.49    80.2 mL of 0.100 M HCl solution (when calculated step-wise with rounding after each step). By a chain-calculation, 80.3 mL.

4.50    540 mL of 0.100 M HCl solution

4.51    21.2 g of $Na_2CO_3$ is needed. Since 40.0 g are taken, more than enough is supplied.

## Practice Exercises, Chapter 5

5.1    $1.31 \times 10^3$ mL of helium

5.2    547 mL

5.3    484 atm

## Review Exercises, Chapter 5

5.1    Pressure, volume, temperature, and moles

5.2    Pressure is the ratio of force to area.  Pressure $= \dfrac{\text{force}}{\text{area}}$ .

5.3    The gravitational attraction that the earth exerts on the gases in the atmosphere gives the atmosphere, at any given point, a weight per unit area, or a pressure.

5.4    The weight at sea level and 0 °C of a column of air uniformly 1 in.$^2$ in cross section and that extends to the end of the earth's atmosphere.  This is 14.7 lb/in.$^2$.  The standard atmosphere is also defined as the pressure that supports a column of mercury 760 mm high in a vacuum at 0 °C.

5.5    1 atm = 760 mm Hg

5.6    The pressure of the atmosphere supports the column of mercury.

5.7    350 mm Hg

5.8    0.197 atm

5.9    That the pressure is inversely proportional to the volume, and that this is true for all gases (all that Boyle studied, that is).

5.10    That the same mass of gas sample is used and that the temperature is the same.

5.11    $P_{\text{total}} = P_a + P_b + P_c + \ldots$

5.12    Nitrogen, oxygen, carbon dioxide, and water vapor

5.13    (a)  $n$, and $T$

(b)  $n$, and P

(c)  $T$, and V

(d)  $V$, $P$, and $T$

(e)  $n$, and $V$

5.14    The volume of a fixed mass of any gas is directly proportional to its Kelvin temperature, if the gas pressure is kept constant.

5.15    Equal volumes of gases have equal numbers of moles when compared at the same pressure and temperature.

5.16    829 mm Hg

5.17    532 mm Hg

5.18     73 mm Hg

5.19     31 mm Hg

5.20     1.84 L of $O_2$

5.21     −124 °C

5.22     1 atm (760 mm Hg) and 273 K

5.23     The volume occupied by one mole of a gas

5.24     At STP

5.25     $PV = nRT$

5.26     $P$ in mm Hg; $V$ in mL.; $T$ in kelvins; and $n$ in moles

5.27     $6.02 \times 10^{23}$ molecules of $H_2$

5.28     (a)   $3.54 \times 10^{-2}$ mol of gas

        (b)   32.1

        (c)   oxygen

5.29     (a)   $7.08 \times 10^{-2}$ mol of $H_2$

        (b)   $1.75 \times 10^3$ mL

5.30     (a)   2.00 mol of $H_2$

        (b)   200 L of $H_2$

        (c)   $1.55 \times 10^5$ mL of $H_2$

5.31     (a)   0.310 mol of $CO_2$

        (b)   $7.29 \times 10^3$ mL of $CO_2$

5.32     The postulates that describe an ideal gas; the postulates of the kinetic theory of gases.

5.33     It obeys all of the gas laws exactly.

5.34     Postulate 3.  The gas particles neither attract nor repel each other.

5.35     Gas pressure is the net result of the innumerable forces exerted on the fixed area of the container by the collisions made at the walls by the gas particles.

5.36     When the volume of the container is reduced, the area of its walls also decreases, so if *area* becomes smaller in the equation

$$\text{pressure} = \frac{\text{force}}{\text{area}}$$

the pressure must increase.

5.37     (a)   The Kelvin temperature

        (b)   The Kelvin temperature is proportional to the average kinetic energy.

5.38     A higher temperature means a higher average kinetic energy, and this means a higher average velocity.  In a fixed-volume container, this would mean a higher pressure, so if the pressure is to be kept constant (a condition of Charles' law), the volume must be allowed to expand with increasing temperature.

5.39    Since a higher temperature means a higher average velocity of the gas particles, they beat on the fixed walls of the container with greater average force. This means a higher pressure, since pressure = force/area.

5.40    The distances between particles in liquids and solids are virtually zero, so physical properties of liquids and solids are much more sensitive to the chemical natures of the particles.

5.41    (a)  Rapidly moving molecules of the liquid escape into the space above the liquid and thus create a partial pressure called the vapor pressure.

(b)  Extra vapor appears in the enclosed space to exert a higher pressure. This extra vapor appears because vaporization is endothermic. Hence the addition of heat (by raising the temperature) shifts the following equilibrium to the right (LeChatelier's principle):

$$\text{liquid} + \text{heat} \rightleftharpoons \text{vapor}$$

5.42    (a)  Water molecules in the vapor state can't return to the liquid state when they are blown away by the breeze.

(b)  Crushing the ice greatly increases the surface area from which water molecules can go into the liquid state, so the water molecules in ice can escape into their liquid state at many more places at the same time.

(c)  Water molecules can pass directly from the solid to the vapor state.

5.43    Electrical forces between the particles are weaker between nitrogen molecules and stronger between $Na^+$ and $Cl^-$ ions.

5.44    All of the chemical reactions that occur anywhere in the body

5.45    Insensible perspiration

5.46    Radiation, conduction, and convection

5.47    Radiation is the transfer of heat in the same manner as heat is lost from a warm iron. Conduction is the direct transfer of heat to a colder object.

5.48    Convection

5.49    Hypothermia is a decrease in body temperature, and reactions that sustain metabolism slow down dangerously.

5.50    Hyperthermia is an increase in body temperature, and the rates of the reactions that sustain metabolism increase. This increases the demand for oxygen, which increases the work load of the heart.

5.51    The electron clouds about each of the reacting particles must interpenetrate before the chemical change can happen, and this takes energy.

5.52    (a)  $CO(g)$ and $O_2(g)$

(b)  $CO_2 (g)$

(c)  C

(d)  B

(e) Exothermic, because the product is at a lower value of energy than the reactants, so this difference in energy must have been released to the environment.

(f) E

5.53

E
X
C + D
Y
A + B
Progress of reaction
X = Energy of activation
Y = Heat of reaction

5.54

A (3 kcal, energy of activation)
X + Y
E
B (10 kcal, heat of reaction)
Z
Progress of Reaction

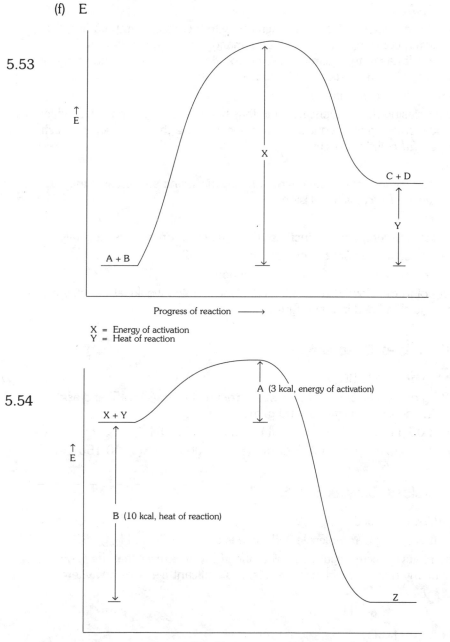

5.55    A rise in temperature increases the energies of the collisions of the reacting particles and raises the frequency with which their electron clouds successfully interpenetrate and give rise to a chemical reaction.

5.56    10 °C

5.57    As rates of metabolic reactions increase with increasing temperature, the rate of oxygen consumption rises. This demand for oxygen makes the heart pump harder. It has to try to circulate the blood through the lungs and then deliver the oxygen to the tissues that need it at a higher rate.

5.58    Increase their concentrations

5.59    By increasing the frequency of collisions between reactant particles, which we do by increasing their concentrations, we increase the frequency at which *successful* collisions occur.

5.60    c, d

5.61    The catalyst lowers (b), the energy of activation, and so increases (d), the frequency of successful collisions.

5.62    Enzymes

5.63    Increase the temperature, increase the concentration, and use a catalyst

5.64    A decrease in the core temperature of the body.

5.65    2–3 °F. Shivering and then loss of memory.

5.66    Alcohol in the blood dilates blood capillaries, which would let the cold blood near the skin flood into and further chill the body's core.

## Practice Exercises, Chapter 6

1.    10.2 g of 96% $H_2SO_4$

2.    1.25 g of glucose and 499 g of water (rounded from 498.75, and assuming that the density of water is 1.00 g/mL)

3.    (a) 2.00 g of $KMnO_4$        (b) 0.25 g of NaOH

4.    (a) 0.020 Osm  (b)  0.015 Osm  (c)  0.100 Osm  (d)  0.150 Osm

## Review Exercises, Chapter 6

6.1    (a) Covalent bond

        (b) It is a nonpolar molecule and there is no δ+ or δ– on H.

6.2    The relative electronegativities of C and H are so similar that the C–H bonds are nearly nonpolar. Hence there is no significant δ+ or δ– anywhere.

6.3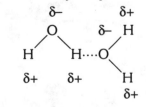

6.4

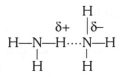

6.5  (a)  The boiling point of ammonia is much less than that of water. Since the sizes of their molecules are about the same, we can't explain this difference in boiling point by differences in transient forces between the molecules. Instead, there must be differences in the sizes of partial charges that exist more permanently, like those of hydrogen bonds.

(b)  Both $\delta+$ and $\delta-$ are larger in the water molecule than in the ammonia molecule.

(c)  Oxygen is more electronegative than nitrogen.

6.6  5 kcal/mol

6.7  (a)  A relatively large quantity of energy is needed to overcome the relatively strong forces of attraction between the molecules in these states.

(b)  The relatively large forces of attraction between water molecules in the liquid state make for a strong jamming together of these molecules at the surface.

6.8  On a waxy surface, wax molecules don't attract water molecules so the latter jam in on each other to form a bead. On a glass surface, the polar particles in the glass attract water molecules more strongly than the water molecules can attract each other, so this attraction by the glass for the water molecules makes the water spread out and form a film rather than beads.

6.9  Reduces it.

6.10  Detergents and soaps

6.11  Bile salts. Bile. The action of the surfactant helps to wash oils and fats from undigested food particles and it helps to break up the oil and fat globules into very fine globules that are more rapidly attacked by water in the process of digestion.

6.12  It is harder to digest fats and oils in the absence of the bile salts.

6.13  The $\delta-$ ends of polar water molecules.

6.14  It is surrounded by polar water molecules with their $\delta+$ ends pointing at it. (The drawing should be like that in the lower right corner of Figure 6.4).

6.15  Water molecules are strongly attracted to each other and they find nothing in molecules of $CCl_4$ with which to establish forces of attraction of comparable strength, so the water molecules stay together.

6.16  Make sure that the solution has undissolved $NaNO_3$ present.

6.17  $CaSO_4 \cdot 2H_2O(s) + heat \longrightarrow CaSO_4(s) + 2H_2O(g)$

6.18  They obey the law of definite proportions.

6.19  $CuSO_4(s) + 5H_2O \longrightarrow CuSO_4 \cdot 5H_2O(s)$

6.20   It can remove moisture from the air by forming a hydrate.
       Yes.

6.21   The pellets can draw enough moisture from the air to form a solution of sodium hydroxide that coats every exposed pellet.

6.22   $x = 2$. The formula is $M \cdot 2H_2O$.

6.23   $Z \cdot 10H_2O$

6.24   If a system is in equilibrium and something occurs to upset the equilibrium, the system will shift in whatever way helps to restore the equilibrium.

6.25   (a)  $NaCl(s) + heat \longrightarrow Na^+ (aq) + Cl^-(aq)$
       (b)  $Na^+(aq) + Cl^-(aq) \longrightarrow NaCl(s) + heat$
       (c)  Forward.  The forward reaction consumes energy as it occurs.
       (d)  Reverse
       (e)  Become equal
       (f)  Lower
       (g)  No
       (h)  The system is required to absorb more heat.
       (i)  Shift.  The equilibrium will shift to the right.

6.26   (a)  The system could be cooled so as to withdraw heat from it.
       (b)  The reverse change.  More ice forms as the equilibrium shifts to the left.
       (c)  The forward change must be slowed - by removing heat.  Lower the temperature.  (Salty water freezes below 0 °C.)

6.27   As equilibrium B

6.28   $NH_4Cl(s) + heat \rightleftharpoons NH_4^+(aq) + Cl^-(aq)$
       Cooling the solution would cause solid ammonium chloride to separate from the solution.

6.29   (a)  The total surface area of the powder is far greater than the total surface area of the larger pieces of solid, so the bombarding solvent molecules have far better access to the potential ions or molecules of the solute when it is a powder
       (b)  Stirring moves freshly dissolved solute particles (ions or molecules) away from the surface of the undissolved solid so the bombarding solvent molecules more readily get at undissolved solid.
       (c)  Heat increases the energy with which solvent molecules strike the surface of the undissolved solid, and this more readily moves the ions or molecules of the solid out of the solid state and into solution.

6.30   The solubility of a gas in a solution is directly proportionally to the partial pressure of the gas above the solution.  The high pressure in a hyperbaric chamber helps force additional oxygen into the patient's blood.

6.31    Using oxygen as an example, the equilibrium between its gaseous (undissolved) state and its solution is:

$$O_2(g) + \text{pressure} \rightleftharpoons O_2 (aq)$$

When the pressure is increased, the equilibrium shifts to the right in favor of more oxygen in solution.

6.32    The air in contact with this blood under a total pressure of 1 atm has a partial pressure of $CO_2$ equal to 30 mm Hg.

6.33    The first region (where the gas tension is 80 mm Hg).

6.34    A percentage of the carbon dioxide molecules in solution in water have reacted with the water to give carbonic acid.  This reaction increases the solubility of $CO_2$ in water over that of oxygen.

6.35    $\dfrac{1.2 \text{ g NaOH}}{100 \text{ g NaOH solution}}$  and  $\dfrac{100 \text{ g NaOH solution}}{1.2 \text{ g NaOH}}$

6.36    $\dfrac{1.5 \text{ g NaCl}}{100 \text{ mL NaCl solution}}$  and  $\dfrac{100 \text{ mL NaCl solution}}{1.5 \text{ g NaCl}}$

6.37    $\dfrac{12 \text{ volumes alcohol}}{100 \text{ volumes solution}}$  and  $\dfrac{100 \text{ volumes solution}}{12 \text{ volumes alcohol}}$

6.38    (a)   4.50 g of NaCl
          (b)   3.13 g of $NaC_2H_3O_2$
          (c)   5.00 g of $NH_4Cl$
          (d)   17.5 g of $Na_2CO_3$

6.39    (a)   0.500 g of NaI
          (b)   1.25 g of NaBr
          (c)   6.25 g of $C_6H_{12}O_6$
          (d)   15.0 g of $H_2SO_4$

6.40    (a)   12.5 g of NaCl
          (b)   5.00 g of KBr
          (c)   7.50 g of $CaCl_2$
          (d)   6.75 g of NaCl

6.41    (a)   12.5 g of $Mg(NO_3)_2$
          (b)   5.00 g of NaBr
          (c)   2.50 g of KI
          (d)   1.68 g of $Ca(NO_3)_2$

6.42    75.0 mL of methyl alcohol

6.43    25.0 mL of ethyl alcohol

6.44    (a)   70.0 mL of NaOH solution
          (b)   25.0 mL of $Na_2CO_3$ solution

(c)  20.0 mL of glucose solution

(d)  80.0 mL of NaOH solution

(e)  720 mL of glucose solution

6.45  (a)  11.0 mL of KOH solution

(b)  30.0 mL of HCl solution

(c)  650 mL of NaCl solution

(d)  281 mL of KOH solution

6.46  8.71 g of $NaC_2H_3O_2 \cdot 3H_2O$

6.47  24.7 g of $Na_2SO_4 \cdot 10H_2O$

6.48  A small sample taken in one place has the same composition and properties as a small sample from any other place.

6.49  The relative sizes of the dispersed or dissolved particles.

6.50  (a)  Suspensions

(b)  Colloidal dispersions

(c)  Colloidal dispersions

(d)  Solutions

(e)  Suspensions

6.51  Those that have the same kind of electrical charge, because they repel each other and can't coalesce into larger particles that would separate out.

6.52  Dissolved ions and small molecules make it a solution; dispersed protein molecules make it a colloidal dispersion; and various kinds of blood cells make blood a suspension.

6.53  The solute particles are too small to scatter light.

6.54  The uneven buffeting that randomly moving solvent molecules give to the particles in colloidal dispersion.

6.55  Test for the Tyndall effect.

6.56  It is a colloidal dispersion of one liquid in another.  Examples are cream, mayonnaise, and milk.

6.57  A colloidal dispersion of a solid in a liquid.  Examples are starch in water, jellies, and paints.

6.58  A sol that has adopted a semisolid, semirigid form.  Examples are gelatin desserts and fruit jellies.

6.59  $-1.86 \,^\circ C$

6.60  The solute in the second solution breaks up (ionizes) into two ions per formula unit when it dissolves, so the effective concentration is 2.00 mol/1000 g $H_2O$.

6.61  The rate at which water can move *into* the solution is greater than the rate at which it can return, because solute particles tend to block the return.  The figure should resemble part b of Figure 6.8.

6.62    The osmotic membrane doesn't let anything in the dissolved state go through, but the dialyzing membrane lets solutes through and blocks colloidally dispersed particles.

6.63    The blocking action of the solute particles is a function of their *presence* and *relative numbers*, not their chemical properties.

6.64    This solution is 2 $M$ in all solute particles, and osmolarity is a function of the molar concentrations of all osmotically active solute particles.

6.65    0.080 $M$ $Na_2SO_4$ solution, because it is 3 x 0.080 $M$ = 0.240 $M$ in its osmolarity.

6.66    10% NaCl, which has 1.7 mmol/100 g solution versus only 0.67 mmol/100 g solution for 10% NaI.

6.67    Solution A.  While the osmolarities of A and B are identical with respect to their dissolved salts and sugars, A has the higher concentration of starch.

6.68    They swell and split open.

6.69    (a)  Hypertonic

        (b)  (1)  Hemolysis              (2)   crenation

6.70    The loss of large molecules lowers the colloidal osmotic pressure of the blood to a value less than the osmotic pressure of the fluids just outside of the blood vessels.  As a result, there is a net flow of water from the blood to the outside of the blood vessels, and the blood volume decreases.

6.71    Oxygen and nitrogen.  The solubilities of gases increase with pressure (Henry's law).

6.72    The dissolved gases quickly come out of solution and their microbubbles block blood flow at capillaries.

6.73    It gives time for excess dissolved oxygen to be used up by normal metabolism and for excess dissolved nitrogen to come out much more slowly and so generate much fewer troublesome microbubbles.

6.74    Allow 20 minutes of decompression time for each extra atmosphere of pressure.

6.75    Dissolve 12 mL of 17 $M$ $HC_2H_3O_2$ in water and make the final volume equal to 100 L.

6.76    Dissolve 31 mL of 16 $M$ $HNO_3$ in water and make the final volume equal to 500 mL.

6.77    Add concentrated sulfuric acid *to* the water, slowly.  In this way the large amount of heat generated as the acid dissolves has a chance to dissipate into a maximum volume of water.   Otherwise, the solution could boil and spatter hot acid.

6.78    125 g of stock solution

6.79    12.5 g of stock solution, or 11.9 mL of stock solution

6.80   (a)   35.9% (w/w) HCl

   (b)   118 mL of concentrated solution

6.81   (a)   71.0% (w/w) $HNO_3$

   (b)   24.8 mL of concentrated solution

6.82   It is a procedure by which blood is diverted from the body and passed through a tube made of a dialyzing membrane.  The purpose is to remove unwanted solutes in blood by dialysis.

6.83   It is the solution on the other side of the dialyzing membrane from the blood. It is free of the substances whose removal from the blood is sought, but it has the other desirable solutes in concentrations equal to those present in blood.

6.84   (a)  Same        (b)  Same        (c)  Less

## Practice Exercises, Chapter 7

1.   $HNO_3(aq) + KOH(aq) \longrightarrow H_2O + KNO_3(aq)$

   $H^+(aq) + NO_3^-(aq) + K^+(aq) + OH^-(aq) \longrightarrow H_2O + K^+(aq) + NO_3^-(aq)$

   $H^+ (aq) + OH^-(aq) \longrightarrow H_2O$

2.   $2NaHCO_3(aq) + H_2SO_4(aq) \longrightarrow 2CO_2(g) + 2H_2O + Na_2SO_4(aq)$

   $2Na^+(aq) + 2HCO_3^-(aq) + 2H^+(aq) + SO_4^{2-}(aq) \longrightarrow$

   $$2CO_2(g) + 2H_2O + 2Na^+(aq) + SO_4^{2-}(aq)$$

   $HCO_3^-(aq) + H^+(aq) \longrightarrow CO_2(g) + H_2O$

3.   $K_2CO_3(aq) + H_2SO_4(aq) \longrightarrow CO_2(g) + H_2O + K_2SO_4(aq)$

   $2K^+(aq) + CO_3^{2-}(aq) + 2H^+(aq) + SO_4^{-2}(aq) \longrightarrow$

   $$CO_2(g) + H_2O + 2K^+(aq) + SO_4^{2-}(aq)$$

   $CO_3^{2-}(aq) + 2H^+(aq) \longrightarrow CO_2(g) + H_2O$

4.   $MgCO_3(s) + 2HNO_3(aq) \longrightarrow CO_2(g) + H_2O + Mg(NO_3)_2(aq)$

   $MgCO_3(s) + 2H^+(aq) + 2NO_3^-(aq) \longrightarrow CO_2(g) + H_2O + Mg^{2+}(aq) + 2NO_3^-(aq)$

   $MgCO_3(s) + 2H^+(aq) \longrightarrow CO_2(g) + H_2O + Mg^{2+}(aq)$

5.   $Mg(OH)_2(s) + 2HCl(aq) \longrightarrow 2H_2O + MgCl_2(aq)$

   $Mg(OH)_2(s) + 2H^+(aq) \longrightarrow 2H_2O + Mg^{2+}(aq)$

6.   (a)  $NH_3(aq) + HBr(aq) \longrightarrow NH_4Br(aq)$

      $NH_3(aq) + H^+(aq) \longrightarrow NH_4^+(aq)$

(b)  $2NH_3(aq) + H_2SO_4(aq) \longrightarrow (NH_4)_2SO_4(aq)$

   $NH_3(aq) + H^+(aq) \longrightarrow NH_4^+(aq)$

7.  $Mg(s) + 2HCl(aq) \longrightarrow H_2(g) + MgCl_2(aq)$

   $Mg(s) + 2H^+(aq) \longrightarrow H_2(g) + Mg^{2+}(aq)$

8.  All are weak Brønsted bases.

9.  (a)  Weak    (b)  Weak    (c)  Strong    (d)  Strong

10.  $Cu(NO_3)_2(aq) + Na_2S(aq) \longrightarrow CuS(s) + 2NaNO_3(aq)$

   $Cu^{2+}(aq) + S^{2-}(aq) \longrightarrow CuS(s)$

11.  The acetate ion, $C_2H_3O_2^-(aq)$, binds $H^+$ ions from $HCl(aq)$ because $C_2H_3O_2^-$ a relatively strong Brønsted base:

   $C_2H_3O_2^-(aq) + H^+(aq) \rightleftharpoons HC_2H_3O_2(aq)$

12.  (a)  $Ag^+(aq) + Cl^-(aq) \longrightarrow AgCl(s)$

   (b)  $CaCO_3(s) + 2H^+(aq) \longrightarrow Ca^{2+}(aq) + H_2O + CO_2(g)$

   (c)  No reaction

## Review Exercises, Chapter 7

7.1  Acids, bases, and salts

7.2  (a)  No

   (b)  Yes, any polyatomic ion such as $SO_4^{2-}$

   (c)  No

   (d)  No

7.3  $2H_2O \rightleftharpoons H_3O^+ + OH^-$

   Hydronium      Hydroxide
     ion            ion

7.4  They furnish hydronium ions in water.

7.5  Hydrogen ion and hydronium ion

7.6  (a)  $[H_3O^+] > [OH^-]$

   (b)  $[H_3O^+] < [OH^-]$

   (c)  $[H_3O^+] = [OH^-]$

7.7  Salts do not supply a *common* ion in solution, whereas aqueous acids furnish hydrogen ions and aqueous bases supply hydroxide ions.

7.8  (a)  Acidic    (b)  Neutral    (c)  Basic

7.9  They consist of oppositely charged ions, and between these are strong forces of attraction that set up rigid solids.

7.10    (1) An aqueous solution that conducts electricity, such as a salt solution.

(2) An ionic compound in its pure form which, if either melted or put into solution in water, conducts electricity.

Examples are NaCl, KBr, NaOH, and $Mg(NO_3)_2$.

7.11    Cathode

7.12    Positive

7.13    Cations

7.14    Cations take electrons from the cathode simultaneously as anions deliver electrons to the anode, so the net effect is just as if electrons transferred directly from the cathode to the anode—an electron flow.

7.15    Methyl alcohol does not furnish any ions of any kind in water.

7.16    Strong electrolyte

7.17    It is a molecular compound. It does not consist of ions in the liquid state.

7.18    Hydrochloric acid is the aqueous solution of the gas, hydrogen chloride.

7.19    In $H_3O^+$, because the weaker bond in H–Cl breaks to form the stronger bond in the hydronium ion by the following reaction.

$$HCl(g) + H_2O \longrightarrow H_3O^+(aq) + Cl^-(aq)$$

7.20    Weak acid

7.21    Monoprotic. Acetic acid

7.22    Hydrofluoric acid, HF(aq)

Hydrochloric acid, HCl(aq)

Hydrobromic acid, HBr(aq)

Hydriodic acid, HI(aq)

7.23    $H_2A(aq) + H_2O \rightleftharpoons HA^-(aq) + H_3O^+(aq)$

or, $H_2A(aq) \rightleftharpoons HA^-(aq) + H^+(aq)$

$HA^-(aq) + H_2O \rightleftharpoons A^{2-}(aq) + H_3O^+(aq)$

or, $HA^-(aq) \rightleftharpoons A^{2-}(aq) + H^+(aq)$

7.24    The second $H^+$ ion has to pull away from a particle that has more negative charge than the first $H^+$ ion. Unlike charges attract, so the more unlike they are the more energy is needed.

7.25    The extra oxygen atom provides more electron-withdrawing action, and this weakens the H–O bond more.

7.26    Sulfuric acid is the stronger acid. Its molecules have one more oxygen to exert electron-withdrawing action on the electron pairs of the H–O bonds.

7.27    $CO_2(aq) + H_2O \rightleftharpoons H_2CO_3(aq)$

7.28    Letting $H^+$ represent the hydronium ion:

$H_2CO_3(aq) \rightleftharpoons H^+(aq) + HCO_3^-(aq)$

$HCO_3^-(aq) \rightleftharpoons H^+(aq) + CO_3^{2-}(aq)$

7.29    What does dissolve in water ionizes 100 percent.

7.30    Only a very low percentage of all ammonia molecules in solution react with water to generate hydroxide ions in the equilibrium:

$$NH_3(aq) + H_2O \rightleftharpoons NH_4^+(aq) + OH^-(aq)$$

7.31    Sodium hydroxide, NaOH, and potassium hydroxide, KOH

7.32    $CO_2(aq) + NaOH(aq) \longrightarrow NaHCO_3(aq)$

7.33    A solution of ammonia in water. Since it contains only a very low percentage of ammonium and hydroxide ions, it's inappropriate to call it ammonium hydroxide (although some chemical suppliers do).

7.34    Hydrochloric acid, $HCl(aq)$

Hydrobromic acid, $HBr\ (aq)$

Hydriodic acid, $HI(aq)$

Nitric acid, $HNO_3(aq)$

Sulfuric acid. $H_2SO_4(aq)$

7.35    Sodium hydroxide, NaOH (very soluble)

Potassium hydroxide, KOH (very soluble)

Magnesium hydroxide, $Mg(OH)_2$ (slightly soluble)

Calcium hydroxide, $Ca(OH)_2$ (slightly soluble)

7.36    (a)  Yes          (b)  No

7.37    (a)  $HNO_3(aq) + NaOH(aq) \longrightarrow NaNO_3(aq) + H_2O$

$H^+(aq) + OH^-(aq) \longrightarrow H_2O$

(b)  $2HCl(aq) + K_2CO_3(aq) \longrightarrow 2KCl(aq) + H_2O + CO_2(g)$

$2H^+(aq) + CO_3^{2-}(aq) \longrightarrow H_2O + CO_2(g)$

(c)  $2HBr(aq) + CaCO_3(s) \longrightarrow CaBr_2(aq) + H_2O + CO_2(g)$

$2H^+(aq) + CaCO_3(s) \longrightarrow Ca^{2+}(aq) + H_2O + CO_2(g)$

(d)  $HNO_3(aq) + NaHCO_3(aq) \longrightarrow NaNO_3(aq) + H_2O + CO_2(g)$

$H^+(aq) + HCO_3^-(aq) \longrightarrow H_2O + CO_2(g)$

(e)  $HI(aq) + NH_3(aq) \longrightarrow NH_4I(aq)$

$H^+(aq) + NH_3(aq) \longrightarrow NH_4^+(aq)$

(f)  $2HNO_3(aq) + Mg(OH)_2(s) \longrightarrow Mg(NO_3)_2(aq) + 2H_2O$

$2H^+(aq) + Mg(OH)_2(s) \longrightarrow Mg^{2+}(aq) + 2H_2O$

(g)  $2HBr(aq) + Zn(s) \longrightarrow ZnBr_2(aq) + H_2(g)$

$2H^+(aq) + Zn(s) \longrightarrow Zn^{2+}(aq) + H_2(g)$

7.38    (a)  $2KOH(aq) + H_2SO_4(aq) \longrightarrow K_2SO_4(aq) + 2H_2O$

$OH^-(aq) + H^+(aq) \longrightarrow H_2O$

(b)  $Na_2CO_3(aq) + 2HNO_3(aq) \longrightarrow 2NaNO_3(aq) + CO_2(g) + H_2O$

$CO_3^{2-}(aq) + 2H^+(aq) \longrightarrow CO_2(g) + H_2O$

(c)  $KHCO_3(aq) + HCl(aq) \longrightarrow KCl(aq) + CO_2(g) + H_2O$

$HCO_3^-(aq) + H^+(aq) \longrightarrow CO_2(g) + H_2O$

(d)  $MgCO_3(s) + 2HI(aq) \longrightarrow MgI_2(aq) + CO_2(g) + H_2O$

$MgCO_3(s) + 2H^+(aq) \longrightarrow Mg^{2+}(aq) + CO_2(g) + H_2O$

(e)  $NH_3(aq) + HBr(aq) \longrightarrow NH_4Br(aq)$

$NH_3(aq) + H^+(aq) \longrightarrow NH_4^+(aq)$

(f)  $Ca(OH)_2(s) + 2HCl(aq) \longrightarrow CaCl_2(aq) + 2H_2O$

$Ca(OH)_2(s) + 2H^+(aq) \longrightarrow Ca^{2+}(aq) + 2H_2O$

(g)  $2Al(s) + 6HCl(aq) \longrightarrow 2AlCl_3(aq) + 3H_2(g)$

$2Al(s) + 6H^+(aq) \longrightarrow 2Al^{3+}(aq) + 3H_2(g)$

7.39    (a)  $OH^-(aq) + H^+(aq) \longrightarrow H_2O$

(b)  $HCO_3^-(aq) + H^+(aq) \longrightarrow CO_2(g) + H_2O$

(c)  $CO_3^{2-}(aq) + 2H^+(aq) \longrightarrow CO_2(g) + H_2O$

(d)  $NH_3(aq) + H^+(aq) \longrightarrow NH_4^+(aq)$

7.40    $MCO_3(s) + 2H^+(aq) \longrightarrow M^{2+}(aq) + CO_2(g) + H_2O$

7.41    $M(OH)_2(s) + 2H^+(aq) \longrightarrow M^{2+}(aq) + 2H_2O$

7.42    $M(s) + 2H^+(aq) \longrightarrow M^{2+}(aq) + H_2(g)$

7.43    (a)  To be higher in the activity series means, in a qualitative sense, that the atoms of the element have a greater tendency to become ions that those of elements lower in the series.

(b)  The lower ionization energies of sodium and potassium also indicated this greater tendency to become ions.

7.44    Zinc reacts more rapidly with 1 $M$ $HNO_3$. The reaction is with the hydronium ion, and the molar concentration of this ion is much higher in 1 $M$ $HNO_3$ (which is 100% ionized) than in 1 $M$ $HC_2H_3O_2$, which is a weak acid and weakly ionized in water.

7.45    0.250 mol of $NaHCO_3$

7.46    0.800 mol of $KOH$

7.47    6.68 g of $Na_2CO_3$

7.48   5.46 g of $CaCO_3$

7.49   4.91 g of $NaHCO_3$

7.50   1.40 g of $K_2CO_3$

7.51   29.5 mL of NaOH solution

7.52   35.2 mL of KOH solution

7.53   $CaCO_3(s) + 2HCl(aq) \longrightarrow CO_2(g) + H_2O + CaCl_2(aq)$
       $CaCO_3(s) + 2H^+(aq) \longrightarrow CO_2(g) + H_2O + Ca^{2+}(aq)$
       47.8 g of $CaCO_3$
       191 mL of 5.00 $M$ HCl

7.54   $Na_2CO_3(s) + 2HCl(aq) \longrightarrow CO_2(g) + H_2O + 2NaCl(aq)$
       $Na_2CO_3(s) + 2H^+(aq) \longrightarrow CO_2(g) + H_2O + 2Na^+(aq)$
       18.3 L of $CO_2$
       79.5 g of $Na_2CO_3$

7.55   (a)  $HC_2H_3O_2$ and $H_3O^+$
       (b)  $HC_2H_3O_2$

7.56   $NH_3$

7.57   (a)  $NH_2^-$        (b)  $OH^-$        (c)  $S^{2-}$

7.58   (a)  $H_2PO_4^-$      (b)  $H_2SO_3$      (c)  $NH_4^+$

7.59   (a)  HCl            (b)  $H_2O$         (c)  $HSO_4^-$

7.60   Potassium hydroxide, KOH
       $KOH(aq) + HCl(aq) \longrightarrow KCl(aq) + H_2O$
       Potassium bicarbonate, $KHCO_3$
       $KHCO_3(aq) + HCl(aq) \longrightarrow KCl(aq) + CO_2(g) + H_2O$
       Potassium carbonate, $K_2CO_3$
       $K_2CO_3(aq) + 2HCl(aq) \longrightarrow 2KCl(aq) + CO_2(g) + H_2O$

7.61   Lithium hydroxide, LiOH
       $LiOH(aq) + HBr(aq) \longrightarrow LiBr(aq) + H_2O$
       Lithium bicarbonate, $LiHCO_3$
       $LiHCO_3(aq) + HBr(aq) \longrightarrow LiBr(aq) + CO_2(g) + H_2O$
       Lithium carbonate, $Li_2CO_3$
       $Li_2CO_3(aq) + 2HBr(aq) \longrightarrow 2LiBr(aq) + CO_2(g) + H_2O$

7.62   (c), (d)

7.63   (d), (e)

7.64   (b), (c)

7.65   (d), (f)

7.66   (a)  $Ag^+(aq) + Cl^-(aq) \longrightarrow AgCl(s)$
       (b)  No reaction

(c)  $H^+(aq) + OH^-(aq) \longrightarrow H_2O$

(d)  $Pb^{2+}(aq) + 2Cl^-(aq) \longrightarrow PbCl_2(s)$

(e)  No reaction

(f)  $Cu^{2+}(aq) + S^{2-}(aq) \longrightarrow CuS(s)$

(g)  $Ba^{2+}(aq) + SO_4^{2-}(aq) \longrightarrow BaSO_4(s)$

(h)  $H^+(aq) + OH^-(aq) \longrightarrow H_2O$

(i)  $Ni^{2+}(aq) + S^{2-}(aq) \longrightarrow NiS(s)$

(j)  $Ag^+(aq) + Cl^-(aq) \longrightarrow AgCl(s)$

(k)  $H^+(aq) + HCO_3^-(aq) \longrightarrow CO_2(g) + H_2O$

(l)  $Ca^{2+}(aq) + 2OH^-(aq) \longrightarrow Ca(OH)_2(s)$

7.67  (a)  $Cd^{2+}(aq) + S^{2-}(aq) \longrightarrow CdS(s)$

(b)  $H^+(aq) + OH^-(aq) \longrightarrow H_2O$

(c)  $Ba^{2+}(aq) + SO_4^-(aq) \longrightarrow BaSO_4(s)$

(d)  $Pb^{2+}(aq) + SO_4^{2-}(aq) \longrightarrow PbSO_4(s)$

(e)  No reaction

(f)  $H^+(aq) + HCO_3^-(aq) \longrightarrow CO_2(g) + H_2O$

(g)  $Ni^{2+}(aq) + S^{2-}(aq) \longrightarrow NiS(s)$

(h)  $H^+(aq) + OH^-(aq) \longrightarrow H_2O$

(i)  No Reaction

(j)  $H^+(aq) + HCO_3^-(aq) \longrightarrow CO_2(g) + H_2O$

(k)  No reaction

(l)  $Pb^{2+}(aq) + CrO_4^{2-}(aq) \longrightarrow PbCrO_4(s)$

7.68  3.5 – 5.0 meq $K^+$/L

7.69  4.2 – 5.2 meq $Ca^{2+}$/L

7.70  3.76 g or 3.76 x $10^3$ mg of $Cl^-$

7.71  3.11 g or 3.11 x $10^3$ mg of $Na^+$

7.72  5.01 meq of $K^+$

7.73  2.00 meq $Mg^{2+}$

7.74  Both electrons in the bond came from one atom, from the oxygen atom in this case

7.75  No difference.

7.76    Because of the high electronegativity of the oxygen atom in C=O.

7.77    Because $H^+$ has to transfer away from something, $HCO_3^-$, already negatively charged, and unlike charges attract.

7.78    $CO_3^{2-}$. It has twice the ability to attract an oppositely charged particle ($H^+$).

7.79    Bicarbonate ion, $HCO_3^-$. Citric acid.

7.80    The reactants that combine to give $CO_2$ have no mobility in the solid state, so they cannot react.

7.81    Moisture from humid air slowly enters the tablets and gives localized mobility to the reactants so that $CO_2$ is slowly lost over a period of time.

7.82    It was hard to use soap in it.

7.83    It was hard water only until it was boiled.

7.84    $Ca^2$, $Mg^{2+}$, and $Fe^{2+}$ (or $Fe^{3+}$). Yes.

7.85    $HCO_3^-$. When heated in water it decomposes to $CO_2$ and $CO_3^{2-}$.

7.86    First, $2HCO_3^-(aq) \xrightarrow{\text{heat}} CO_3^{2-}(aq) + CO_2(g) + H_2O$

Then, $CO_3^{2-}(aq) + Mg^{2+} \longrightarrow MgCO_3(s)$ A similar reaction would occur with any of the other hardness cations.

7.87    First, $NH_3(aq) + H_2O \rightleftharpoons NH_4^+(aq) + OH^-(aq)$

Then, $2OH^-(aq) + Mg^{2+}(aq) \longrightarrow Mg(OH)_2(s)$

7.88    The hardness ions become lodged in the zeolite and their places in the trickling water are taken by sodium ions from the zeolite.

7.89    Anion gap = 10 meq/L. This is in the normal range of 5 - 14 meq/L, so no serious disturbance in metabolism is indicated.

7.90    Anion gas = 19 meq/L. This is above the normal range of 5 - 14 meq/L, so this anion gap suggests a disturbance in metabolism.

## Practice Exercises, Chapter 8

1.    (a)  $2.5 \times 10^{-6}$ mol $OH^-$/L.  Basic

(b)  $9.1 \times 10^{-8}$ mol $OH^-$/L.  Acidic

(c)  $1.1 \times 10^{-7}$ mol $OH^-$/L.  Basic

2.    (a)  $1 \times 10^{-7}$ to $1 \times 10^{-8}$ mol $H^+$/L

(b)  Slightly basic

(c)  Acidosis

(d)  6.90

4.    (a)  basic        (b)  basic

(c)  basic        (d)  acidic

(e)  basic        (f)  basic

5.    Basic. The acetate ion, a Brønsted base, hydrolyzes.

6.    Acidic. The hydrated $Cu^{2+}$ ion hydrolyzes.
7.    Yes. Decrease the pH. The $NH_4^+$ is a weak Brønsted acid.
8.    0.105 $M$ NaOH
9.    0.125 $M$ $H_2SO_4$

## Review Exercises, Chapter 8

8.1    $K_w = [H^+][OH^-]$
8.2    $1.00 \times 10^{-14}$
8.3    The forward reaction in the equilibrium: $H_2O \rightleftharpoons H^+ + OH^-$ is endothermic, so the addition of heat shifts the equilibrium to the right in accordance with Le Chatelier's principle.
8.4    $2.43 \times 10^{-14}$
8.5    $9.0 \times 10^{-16}$
8.6    $[H^+] = 1 \times 10^{-pH}$
8.7    Acidic
8.8    Acidic
8.9    2.0
8.10   pH = 12
8.11   Because in water at 25 °C, $[H^+] = 1.00 \times 10^{-7}$.
8.12   More alkaline. Alkalosis
8.13   More acidic. Acidosis
8.14   $1 \times 10^{-5}$ mol/L. Slightly acidic
8.15   $1 \times 10^{-4}$ to $1 \times 10^{-5}$ mol/L
8.16   Strong. At a pH of 1, the value of $[H^+]$ is $1 \times 10^{-1}$ mol/L, which is numerically identical to the initial concentration of the acid, so all of the acid is ionized.
8.17   Weak. At a pH of 4.56, the value of $[H^+]$ is $1 \times 10^{-4}$ to $1 \times 10^{-5}$ mol/L, but the initial concentration of the acid is much higher ($1 \times 10^{-2}$ mol/L), over 100 times higher than $[H^+]$. Therefore, only a small percent of the acid molecules can be ionized.
8.18   Basic (pH = 7.9)
8.19   pH = 7.20. This means acidosis since 7.20 is less than 7.35, the normal pH of blood.
8.20   The acetate ions react slightly with water as follows, which produces a slight excess of hydroxide ions in the solution.

$$C_2H_3O_2^-(aq) + H_2O \rightleftharpoons HC_2H_3O_2(aq) + OH^-(aq)$$

8.21   (a) Neutral    (b) Acidic    (c) Basic    (d) Acidic    (e) Basic
8.22   (a) Neutral    (b) Basic    (c) Basic    (d) Acidic    (e) Basic

8.23    It maintains a steady pH even as slight amounts of acids or bases are added.

8.24    $HCO_3^-$ and $H_2CO_3$

8.25    Acid is neutralized by:  $H^+(aq) + HCO_3^-(aq) \longrightarrow H_2CO_3(aq)$

Base is neutralized by;  $OH^-(aq) + H_2CO_3(aq) \longrightarrow H_2O + HCO_3^-(aq)$

8.26    The permanent removal of $CO_2$ at the lungs means that the following equilibria all shift to the right so that one $H^+$ disappears for each $CO_2$ that is lost.

$$H^+(aq) + HCO_3^-(aq) \rightleftharpoons H_2CO_3(aq) \rightleftharpoons H_2O + CO_2(g)$$

8.27    (a)  Alkalosis

(b)  For each $CO_2$ lost at the lungs, one $H^+$ ion is permanently neutralized (as described in the answer to Review Exercise 8.26).  The loss of $H^+$ ion, of course, means a rise in the value of the pH of the blood.

8.28    (a)  Acidosis

(b)  Hypoventilation helps the body to retain $CO_2$, which helps it retain $H_2CO_3$ that is made from $CO_2$ and $H_2O$.  Retaining $H_2CO_3$ means retaining the $H^+$ ion that results from the slight ionization of $H_2CO_3$ as an acid.

8.29    $HPO_4^{2-}$ and $H_2PO_4^-$

8.30    Slightly acidic

8.31    It means that the concentration of the solution is accurately known.

8.32    When the color change of the indicator occurs.

8.33    The indicator should undergo its color change at a pH that matches the pH of the solution that would result if it were prepared from the salt that is made by the titration.

8.34    Titration of any strong acid (such as HCl) with any strong base (such as NaOH).

8.35    Titration of acetic acid (or any weak acid) with sodium hydroxide (or any strong base).

8.36    Titration of ammonia solution with a strong acid such as HCl.

8.37    (a)  0.4000 $M$ HCl                  (b)  0.4000 $M$ HBr

8.38    (a)  0.4000 $M$ HI                   (b)  0.2000 $M$ $H_2SO_4$

8.39    400.0

8.40    128.0

8.41    (a)  9.115 g of HCl                 (b)  5.435 g of $HNO_3$

(c)  0.3678 g of $H_2SO_4$

8.42    (a)  0.8091 g of HBr                (b)  2.455 g of $H_2SO_4$

(c)  66.25 g of $Na_2CO_3$

8.43    (a)  0.04457 $M$ $Na_2CO_3$           (b)  4.723 g of $Na_2CO_3$/L

8.44    (a)  0.1267 $M$ NaOH                 (b)  5.068 g of NaOH/L

8.45    5.6. Because of dissolved $CO_2$, some of which exists as carbonic acid.

8.46    5.6

8.47    They rest on limestone-rich rock or soil, which contains carbonates able to neutralize $H^+$.

8.48    Sulfur dioxide, $SO_2$, and sulfur trioxide, $SO_3$

8.49    The combustion of sulfur-containing fuels gives $SO_2$; some of this is changed by atmospheric reactions to $SO_3$.

8.50    By passing the gases over beds of wet calcium hydroxide, the gas's $SO_2$ becomes solid $CaSO_3$.

$$Ca(OH)_2(s) + SO_2 \longrightarrow CaSO_3(s) + H_2O$$

8.51    $SO_3$. It reacts with water to give a *strong* acid, sulfuric acid.

8.52    Nitrogen monoxide (nitric oxide), NO, and nitrogen dioxide, $NO_2$. $NO_2$ causes the color of smog.

8.53    In engines: $N_2(g) + O_2(g) \longrightarrow 2NO(g)$

In air: $2NO(g) + O_2(g) \longrightarrow 2NO_2(g)$

8.54    $NO_2$ reacts with water:

$$2NO_2(g) + H_2O \longrightarrow HNO_3(aq) + HNO_2(aq)$$

8.55    Acidic water leaches toxic metal ions from the ground.

## Practice Exercises, Chapter 9

1.    (a)  $CH_3–CH_2–CH_3$

(b) 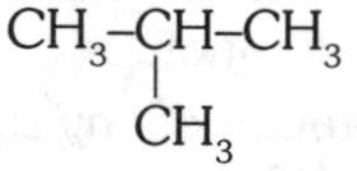  $CH_3–CH–CH_3$
   $\phantom{CH_3–CH}|$
   $\phantom{CH_3–C}CH_3$

(c)

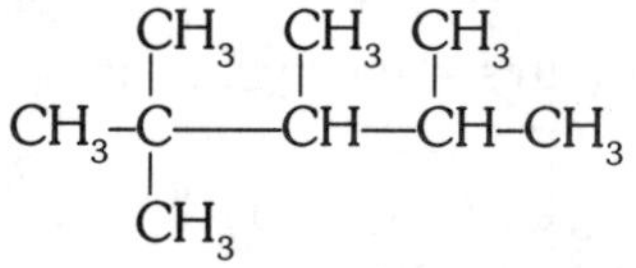

2.    (a)  $CH_3CH_2CH_3$

(b) 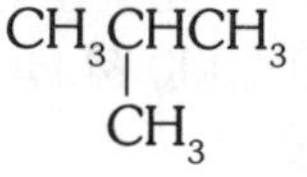  $CH_3CHCH_3$
   $\phantom{CH_3}|$
   $\phantom{CH_3}CH_3$

(c)

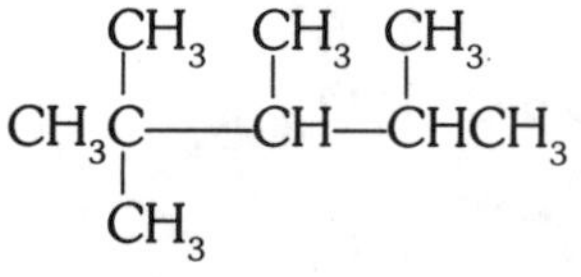

Note that *vertical* bonds are always shown and that sometimes horizontal bonds are left in so that there will be room for other groups.

3.    (a)

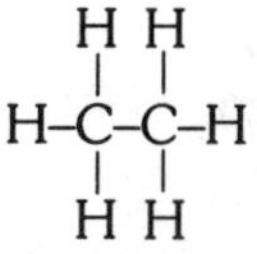

(b)

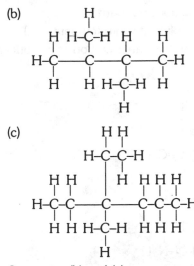

(c)

4.     Structures (b) and (c) cannot represent real compounds.

5.     (a)  Identical
       (b)  Isomers
       (c)  Identical
       (d)  Isomers
       (e)  Different in another way.

## Review Exercises, Chapter 9

9.1    They occur widely in living organisms.

9.2    A vital force that only living things possess.

9.3    Frederick Wohler

9.4    He heated a solution of ammonium cyanate, which scientists at the time regarded as an inorganic compound, and when the water had boiled away the residue was crystalline urea, an organic compound.

9.5    Covalent bond.

9.6    (b), (d), and (e)

9.7    Ionic. Inorganic

9.8    Very few organic compounds consist of ions or furnish ions in water.

9.9    (a)  Molecular (All ionic compounds are *solids* at room temperature.)
       (b)  Ionic (The compound is a carbonate or a bicarbonate.)
       (c)  Molecular (Most ionic compounds melt above this temperature, and few burn in air.)

(d)  Ionic (It has a high melting point and does not burn.)

(e)  Molecular (All ionic compounds are solids at room temperature.)

9.10    Straight-chain.  No carbon has *more* than two other carbon atoms attached to it.

9.11    None is possible.

9.12    (a) 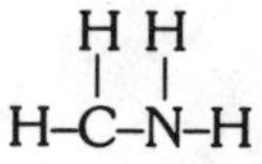
```
    H H
    | |
  H-C-N-H
```

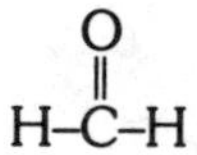

(g) 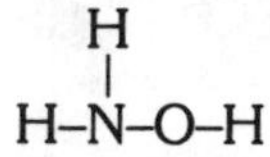
```
   H
   |
 H-N-O-H
```

(b)
```
   H
   |
 Br-C-Br
   |
   H
```

(h)  H–C≡C–H

(c)
```
   H
   |
 Cl-C-Cl
   |
   Cl
```

(i) 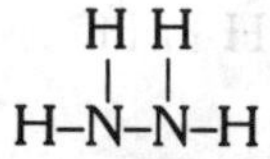
```
   H H
   | |
 H-N-N-H
```

(d)
```
   H H
   | |
 H-C-C-H
   | |
   H H
```

(j)  H–C≡N

(e)
```
     O
     ‖
   H-C-O-H
```

(k) 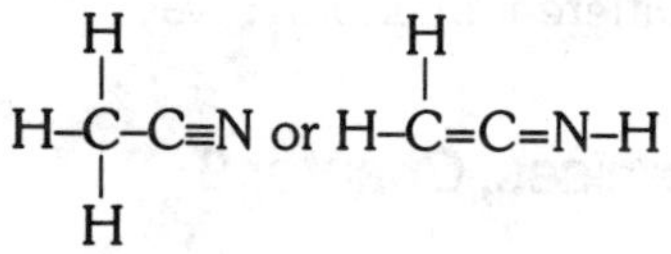
```
   H                    H
   |                    |
 H-C-C≡N  or  H-C=C=N-H
   |
   H
```

(f)
```
     O
     ‖
   H-C-H
```

(l) 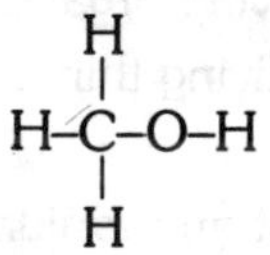
```
   H
   |
 H-C-O-H
   |
   H
```

9.13    (a)  $CH_3CH_2CH_2CH_3$

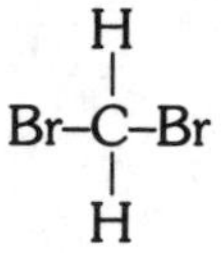

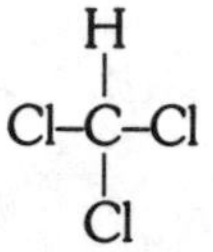

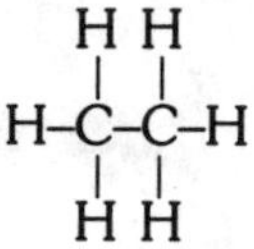

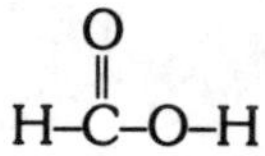

(b) 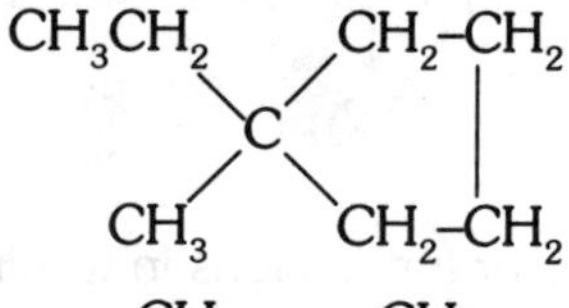
```
 CH₃CH₂        CH₂-CH₂
        \      /
         C
        /      \
     CH₃      CH₂-CH₂
```

(c)
```
        CH₂      CH
       /    \   //  \
   CH₂      C      CH₂
    |       ‖       |
   CH₂      C      CH₂
       \   //  \   /
        CH      CH₂
```

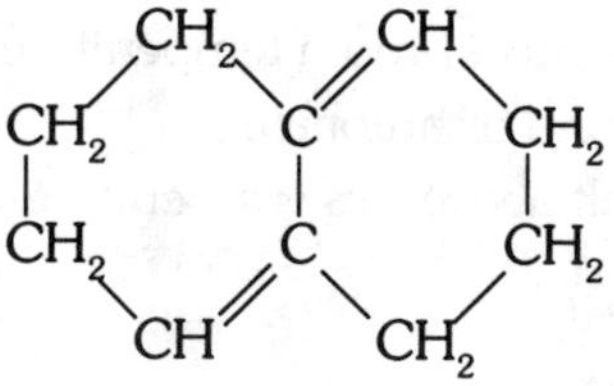

9.14    (a)  Identical            (h)  Identical
        (b)  Identical            (i)  Isomers
        (c)  Unrelated            (j)  Isomers
        (d)  Isomers              (k)  Identical
        (e)  Identical            (l)  Identical
        (f)  Identical            (m)  Isomers
        (g)  Isomers              (n)  Unrelated

9.15    (a)  Alkene               (b)  Alcohol
        (c)  Thioalcohol          (d)  Alkyne
        (e)  Ester                (f)  Aldehyde
        (g)  Carboxylic acid      (h)  Ketone
        (i)  Amine                (j)  Ether

9.16    (a)  Alkane               (b)  Alkane
        (c)  Alcohols             (d)  Alkene and cycloalkane
        (e)  Aldehyde             (f)  Alkane
        (g)  Amines               (h)  Carboxylic acid
        (i)  Ester and carboxylic acid
        (j)  Ester and alcohol — carboxylic acid
        (k)  Ketone
        (l)  Alkane
        (m)  Amides
        (n)  These are inorganic compounds.

## Practice Exercises, Chapter 10

1.      The second structure
2.      (a)  3-Metylhexane
        (b)  4-$t$-Butyl-2, 3-dimethylheptane
        (c)  5-$sec$-Butyl-2, 4-dimethylnonane
        (d)  2-Bromo-1-chloro-3-iodopropane
3.      (a)  Ethyl chloride
        (b)  Butyl bromide
        (c)  Isobutyl chloride
        (d)  $t$-Butyl bromide
4.      (a)  2-Methylpropene         (b)  4-Isobutyl-3, 6-dimethyl-3-heptene
        (c)  1-Chloropropene         (d)  3-Bromopropene

5. (a)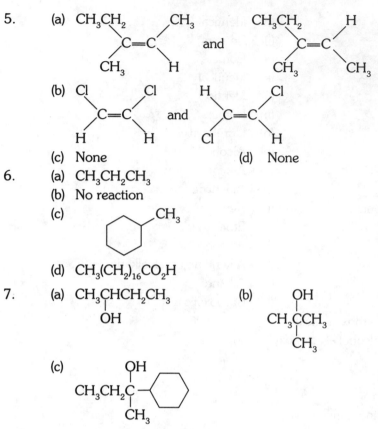

(c) None  (d) None

6. (a) $CH_3CH_2CH_3$
   (b) No reaction
   (c) 

   (d) $CH_3(CH_2)_{16}CO_2H$

7. (a) $CH_3CHCH_2CH_3$  (b) $\underset{\underset{CH_3}{|}}{CH_3\overset{\overset{OH}{|}}{C}CH_3}$
         $\underset{OH}{|}$

   (c)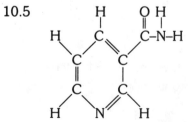

## Review Exercises, Chapter 10

10.1 (a) alkane (or cycloalkane)  (b) alkene
     (c) alkyne  (d) alkane
10.2 (a) alkene (or cycloalkene)  (b) alkane (or cycloalkane)
     (c) alkene  (d) alkyne
10.3 (a) and (c)
10.4 none
10.5

10.6

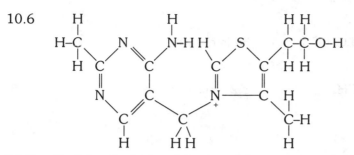

10.7 B. It has the higher formula weight.

10.8 B. It has more polar OH groups, and gasoline is a nonpolar solvent.

10.9 When water is added to pentane, *two* liquid layers appear because pentane won't dissolve in water. When water is added to hydrochloric acid, only one layer forms because the two mix in all proportions.

10.10 Methyl alcohol dissolves in water but hexane doesn't.

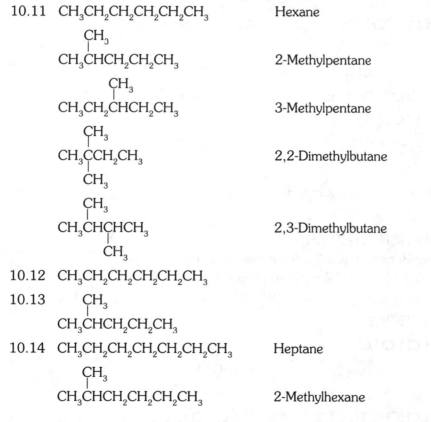

10.11 $CH_3CH_2CH_2CH_2CH_2CH_3$ Hexane

$$\begin{array}{l} \;\;\;CH_3 \\ \;\;\;\;| \\ CH_3CHCH_2CH_2CH_3 \end{array}$$ 2-Methylpentane

$$\begin{array}{l} \;\;\;\;\;\;\;\;\;\;CH_3 \\ \;\;\;\;\;\;\;\;\;\;\;| \\ CH_3CH_2CHCH_2CH_3 \end{array}$$ 3-Methylpentane

$$\begin{array}{l} \;\;\;CH_3 \\ \;\;\;\;| \\ CH_3CCH_2CH_3 \\ \;\;\;\;| \\ \;\;\;CH_3 \end{array}$$ 2,2-Dimethylbutane

$$\begin{array}{l} \;\;\;CH_3 \\ \;\;\;\;| \\ CH_3CHCHCH_3 \\ \;\;\;\;\;\;\;\;| \\ \;\;\;\;\;\;\;CH_3 \end{array}$$ 2,3-Dimethylbutane

10.12 $CH_3CH_2CH_2CH_2CH_2CH_3$

10.13
$$\begin{array}{l} \;\;\;CH_3 \\ \;\;\;\;| \\ CH_3CHCH_2CH_2CH_3 \end{array}$$

10.14 $CH_3CH_2CH_2CH_2CH_2CH_2CH_3$ Heptane

$$\begin{array}{l} \;\;\;CH_3 \\ \;\;\;\;| \\ CH_3CHCH_2CH_2CH_2CH_3 \end{array}$$ 2-Methylhexane

$$\underset{\underset{CH_3}{|}}{CH_3CH_2CHCH_2CH_2CH_3} \qquad \text{3-Methylhexane}$$

$$\underset{\underset{CH_3}{|}}{\overset{\overset{CH_3}{|}}{CH_3CCH_2CH_2CH_3}} \qquad \text{2,2-Dimethylpentane}$$

$$\underset{\underset{CH_3}{|}}{\overset{\overset{CH_3}{|}}{CH_3CHCHCH_2CH_3}} \qquad \text{2,3-Dimethylpentane}$$

$$\overset{\overset{CH_3 \quad CH_3}{|\qquad |}}{CH_3CHCH_2CHCH_3} \qquad \text{2,4-Dimethylpentane}$$

$$\underset{\underset{CH_3}{|}}{\overset{\overset{CH_3}{|}}{CH_3CH_2CCH_2CH_3}} \qquad \text{3,3-Dimethylpentane}$$

$$\underset{\underset{}{}}{\overset{\overset{CH_2CH_3}{|}}{CH_3CH_2CHCH_2CH_3}} \qquad \text{3-Ethylpentane}$$

$$\underset{\underset{CH_3}{|}}{\overset{\overset{CH_3 \quad CH_3}{|\qquad |}}{CH_3C\!\!-\!\!-\!\!CHCH_3}} \qquad \text{2,2,3-Trimethylbutane}$$

10.15 (a) $C_3CH_2CH_2CH_2CH_2CH_2CH_3$

(b) $\underset{\underset{}{}}{\overset{\overset{CH_3}{|}}{CH_3CHCH_2CH_2CH_2CH_3}}$

10.16 (a) 5-sec-Butyl-5-ethyl-2,3,3,9-tetramethyldecane

(b) 7-t-Butyl-5-isobutyl-2-methyl-6-propyldecane

10.17 (a) $\underset{\underset{}{}}{\overset{\overset{CH_3}{|}}{CH_3CHCH_2Br}}$      (b) $CH_3CH_2I$

(c) $CH_3CH_2CH_2Cl$      (d) 

$$\overset{\overset{CH_3}{|}}{\underset{\underset{CH_3}{|}}{CH_3\!-\!C\!-\!\bigcirc}}$$

10.18 (a) $\underset{\underset{Cl}{|}}{CH_3CHCH_2CH_3}$      (b) $CH_3CH_2CH_2CH_2I$

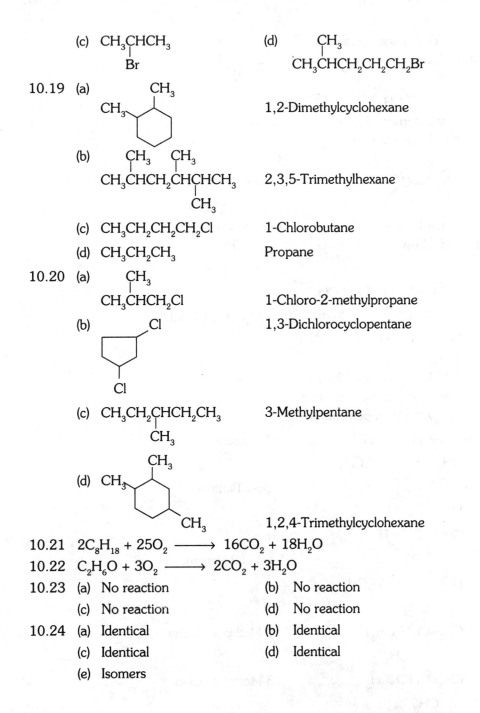

(c) $CH_3CHCH_3$
    $\overset{|}{Br}$

(d) $\overset{CH_3}{\overset{|}{CH_3CHCH_2CH_2CH_2Br}}$

10.19 (a) 1,2-Dimethylcyclohexane

(b) $\overset{CH_3 \quad CH_3}{\overset{| \quad \quad |}{CH_3CHCH_2CHCHCH_3}}$ 2,3,5-Trimethylhexane
    $\overset{|}{CH_3}$

(c) $CH_3CH_2CH_2CH_2Cl$    1-Chlorobutane

(d) $CH_3CH_2CH_3$    Propane

10.20 (a) $\overset{CH_3}{\overset{|}{CH_3CHCH_2Cl}}$    1-Chloro-2-methylpropane

(b) 1,3-Dichlorocyclopentane

(c) $\overset{}{CH_3CH_2CHCH_2CH_3}$    3-Methylpentane
    $\overset{|}{CH_3}$

(d) 1,2,4-Trimethylcyclohexane

10.21 $2C_8H_{18} + 25O_2 \longrightarrow 16CO_2 + 18H_2O$

10.22 $C_2H_6O + 3O_2 \longrightarrow 2CO_2 + 3H_2O$

10.23 (a) No reaction    (b) No reaction

(c) No reaction    (d) No reaction

10.24 (a) Identical    (b) Identical

(c) Identical    (d) Identical

(e) Isomers

10.25 (a) No cis-trans isomers

(b)

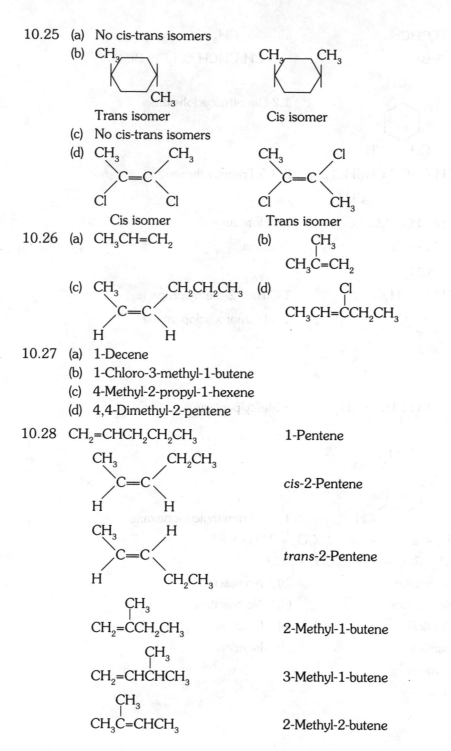

Trans isomer          Cis isomer

(c) No cis-trans isomers

(d)

Cis isomer          Trans isomer

10.26 (a) $CH_3CH=CH_2$

(b)
$$CH_3C=CH_2$$ with $CH_3$ above the C

(c)
$CH_3$ and $CH_2CH_2CH_3$ on $C=C$ with H and H below

(d)
$$CH_3CH=CCH_2CH_3$$ with Cl above the second C

10.27 (a) 1-Decene

(b) 1-Chloro-3-methyl-1-butene

(c) 4-Methyl-2-propyl-1-hexene

(d) 4,4-Dimethyl-2-pentene

10.28 $CH_2=CHCH_2CH_2CH_3$          1-Pentene

$CH_3$ and $CH_2CH_3$ on top, H and H below on $C=C$          cis-2-Pentene

$CH_3$ and H on top, H and $CH_2CH_3$ below on $C=C$          trans-2-Pentene

$$CH_2=CCH_2CH_3$$ with $CH_3$ above          2-Methyl-1-butene

$$CH_2=CHCHCH_3$$ with $CH_3$ above          3-Methyl-1-butene

$$CH_3C=CHCH_3$$ with $CH_3$ above          2-Methyl-2-butene

10.29  The products of the reactions are as follows.

(a)          $CH_3$
          |
     $CH_3CHCH_3$

(b)          $CH_3$
          |
     $CH_3CCH_3$
          |
          OH

10.30  The products of the reactions are as follows.

(a) 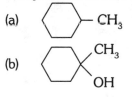 —$CH_3$

(b)  (cyclopentane with $CH_3$ and OH)

10.31  The products of the reactions are as follows.
(a)  $(CH_3)_2CHCH_2CH_3$
(b)  $(CH_3)_2CCH_2CH_3$
              |
             OH

10.32  The products of the reactions are as follows.

(a) (cyclohexane)— $CH_3$

(b) (cyclohexane with) $CH_3$ / OH

10.33  51.4 g of $KMnO_4$
10.34  12.7 g of $K_2C_6H_8O_4$
10.35  (a)

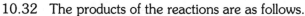

etc.$-C-CH_2-C-CH_2-C-CH_2-C-CH_2-$etc. with $CH_3$ substituents

(b)          $CH_3$
          |
     $+C-CH_2+_n$
          |
          $CH_3$

10.36

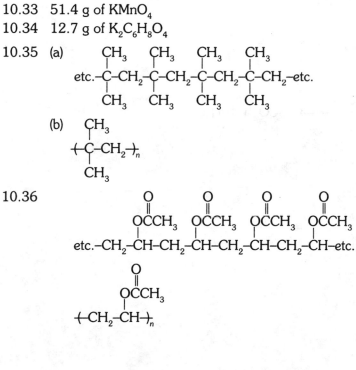

etc.$-CH_2-CH-CH_2-CH-CH_2-CH-CH_2-CH-$etc. with $OCCH_3$ groups

$+CH_2-CH+_n$ with $OCCH_3$

10.37   Its ring is not a benzene ring.

10.38   It has a benzene ring.

10.39   (a)  $C_6H_6 + SO_3 \xrightarrow{H_2SO_4} C_6H_5SO_3H$

(b)  $C_6H_6 + HNO_3 \xrightarrow{H_2SO_4} C_6H_5NO_2 + H_2O$

(c)  No reaction

(d)  No reaction

(e)  No reaction without an iron catalyst

10.40   Addition reactions would permanently destroy the cyclic electron system that gives the ring so much stability.

10.41   (a)  $CH_3CH_2CH_2CH_3$

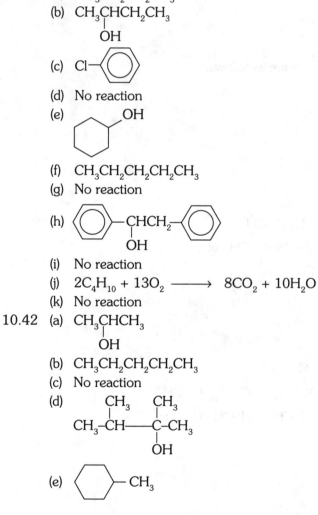

(b)  $CH_3CHCH_2CH_3$
             |
            $OH$

(c)  Cl—⬡

(d)  No reaction

(e)  [cyclohexane ring]—$OH$

(f)  $CH_3CH_2CH_2CH_2CH_3$

(g)  No reaction

(h)  ⬡—$CHCH_2$—⬡
              |
             $OH$

(i)  No reaction

(j)  $2C_4H_{10} + 13O_2 \longrightarrow 8CO_2 + 10H_2O$

(k)  No reaction

10.42   (a)  $CH_3CHCH_3$
                  |
                 $OH$

(b)  $CH_3CH_2CH_2CH_2CH_3$

(c)  No reaction

(d)         $CH_3$      $CH_3$
              |           |
    $CH_3–CH——C–CH_3$
                          |
                         $OH$

(e)  ⬡— $CH_3$

(f)  No reaction

(g)  $C_6H_5Br$

(h)  No reaction

(i)  $C_6H_5CH_3 + 9O_2 \longrightarrow 7CO_2 + 4H_2O$

(j)  No reaction

10.43  Hydrocarbons

10.44  (a)  5–10       (b)  12–18       (c)  20 and higher

10.45  Smaller molecules form and unsaturated hydrocarbons are also made.

10.46  $CH_3CH_2{}^+$, the ethyl carbocation

10.47  $CH_3\overset{+}{C}HCH_2CH_3$

10.48  $CH_3CH_2\overset{+}{O}H_2$.  A proton-transfer

10.49  $CH_3CHCH_2CH_3 \qquad CH_3CHCH_2CH_3$

$\qquad\quad {}^+OH_2 \qquad\qquad\qquad OH$

## Practice Exercises, Chapter 11

1.   (a)  Alcohol               (b)  Phenol
     (c)  Carboxylic acid       (d)  Alcohol
     (e)  Alcohol               (f)  Alcohol

2.   (a)  Monohydric, secondary
     (b)  Monohydric, secondary
     (c)  Dihydric, unstable
     (d)  Dihydric
     (e)  Monohydric, primary
     (f)  Monohydric, primary
     (g)  Monohydric, tertiary
     (h)  Monohydric, secondary
     (i)  Trihydric, unstable

3.   (a)  $CH_3CH=CH_2$          (b)  $CH_3CH=CH_2$

     (c)      $CH_3$            (d)
          $CH_2=CCH_3$

4.   (a)       $CH_3$                          $CH_3$
          $CH_3CHCH=O$  and   $CH_3CHCO_2H$

     (b)  $C_6H_5-CH=O$   and   $C_6H_5-CO_2H$

     (c)  $CH_2=O$   and   $HCO_2H$

5.    (a)
$$CH_3\overset{\displaystyle O}{\overset{\|}{C}}CH_2CH_3$$

(b)    $$C_6H_5-\overset{\displaystyle O}{\overset{\|}{C}}CH_3$$

(c) 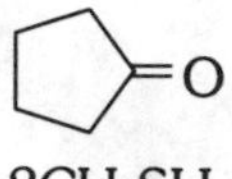    [cyclopentanone ring]=O

6.    (a)    $2CH_3SH$

(b)    $(CH_3)_2CH-S-S-CH(CH_3)_2$

(c)    $HSCH_2CH_2CH_2CH_2SH$

(d) 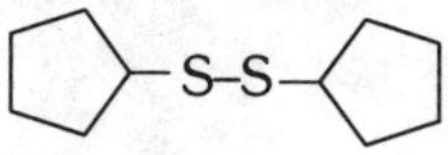    [cyclopentyl]–S–S–[cyclopentyl]

7.    (a)    $CH_3-O-CH_3$

(b)    $CH_3CH_2CH_2-O-CH_2CH_2CH_3$

(c) 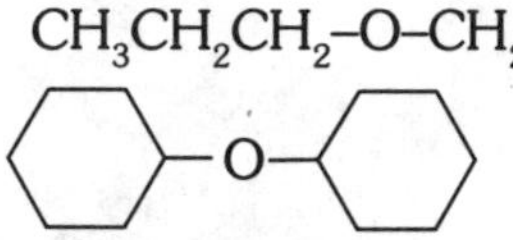    [cyclohexyl]–O–[cyclohexyl]

8.    (a)    $C_6H_5-NH_3^+$

(b)    $CH_3\overset{+}{N}HCH_3$
      $\phantom{CH_3}|$
      $\phantom{CH_3N}CH_3$

(c)    $^+NH_3CH_2CH_2NH_3^+$

9.    (a) 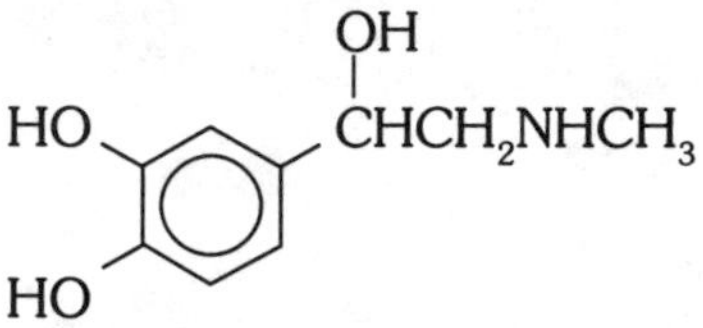
$$HO-[benzene ring]-\overset{\displaystyle OH}{\overset{|}{C}}HCH_2NHCH_3$$
$$HO-$$

(b) 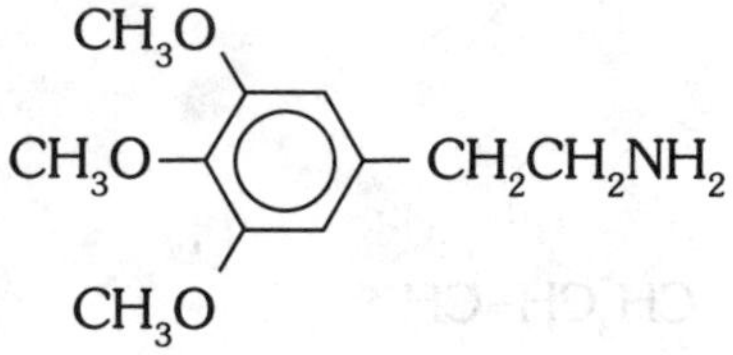
$$CH_3O-[benzene ring]-CH_2CH_2NH_2$$
$$CH_3O-$$
$$CH_3O-$$

## Review Exercises, Chapter 11

11.1    1, phenol.  2, ether.  3, alkene

11.2    1, alkene.  2, alkene.  3, aldehyde

11.3    1, 1° alcohol.  2, ketone.  3, 2° alcohol.  4, ketone.  5, alkene

11.4    1, ketone.  2, carboxylic acid.  3, 2° alcohol.  4, 2° alcohol

11.5 (a) 1 = aliphatic and heterocyclic amine
(b) 1 = aliphatic amine
2 = ester
3 = aromatic amine
(c) 1 = heterocyclic amine. (This ring is also an aromatic system.)
2 = aliphatic amine; also heterocyclic
(d) 1 = 2° alcohol
2 = aliphatic amine

11.6 (a) 1 = alkene
2 = ester
3 = aliphatic and heterocyclic amine
(b) 1 = aliphatic and heterocyclic amine
2 = 1° alcohol
3 = ester
(c) 1 = alkene
2 = 2° alcohol
3 = ether
4 = heterocyclic amine
(d) 1 = amide
2 = aliphatic and heterocyclic amine
3 = alkene
4 = aliphatic and heterocyclic amine

11.7 (a) 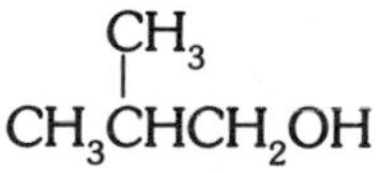 $CH_3$
   $CH_3CHCH_2OH$

(b) 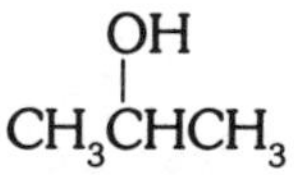 $OH$
   $CH_3CHCH_3$

(c) $CH_3CH_2CH_2OH$

(d) 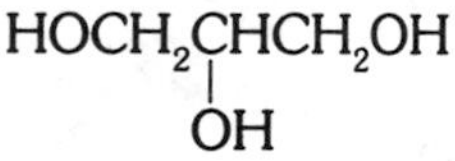 $HOCH_2CHCH_2OH$
   $OH$

11.8 (a) $CH_3OH$

(b) 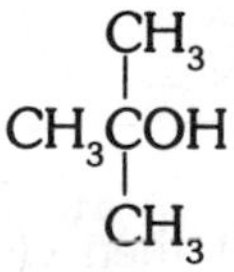 $CH_3$
   $CH_3COH$
   $CH_3$

(c) $CH_3CH_2OH$

(d) $CH_3CH_2CH_2CH_2OH$

11.9 (a) Isopropyl alcohol
(b) Ethylene glycol
(c) t-Butyl alcohol
(d) Glycerol

11.10 (a) Ethyl alcohol
(b) Propyl alcohol
(c) Isobutyl alcohol
(d) Butyl alcohol

**11.11** $HOCH_2CH_2OH$, ethylene glycol

**11.12** $HOCH_2CHCH_2OH$, glycerol
$\quad\quad\quad\;\;\; |$
$\quad\quad\quad\;\; OH$

**11.13**

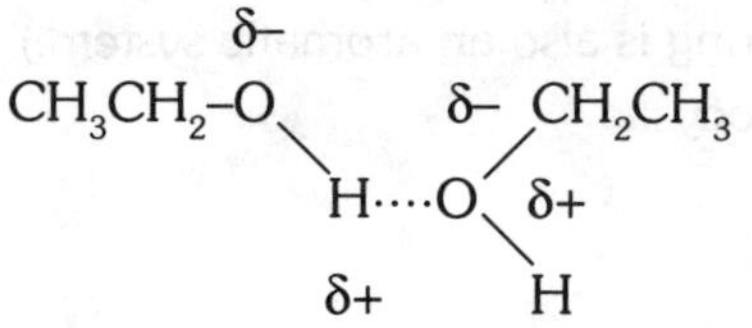

**11.14**

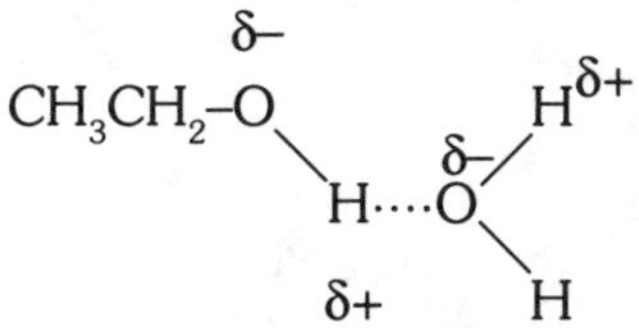

**11.15**  D < B < A < C

**11.16**  C < B < A

**11.17**  (a)  $CH_3CH=CHCH_3$ (plus some $CH_2=CHCH_2CH_3$)

(b)
$\quad\quad\quad CH_3$
$\quad\quad\quad\; |$
$\quad\quad CH_3C=CHCH_2CH_3$ (plus some $CH_2=CCH_2CH_2CH_3$)
$\quad\quad\quad\quad\quad\quad\quad\quad\quad\quad\quad\quad\quad\quad\quad\quad |$
$\quad\quad\quad\quad\quad\quad\quad\quad\quad\quad\quad\quad\quad\quad\quad CH_3$

(c)

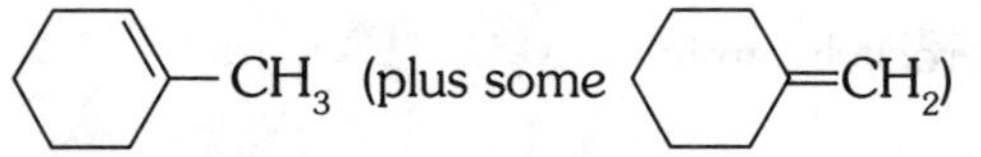

(d)  $C_6H_5-CH=CH_2$

(e)  $C_6H_5-CH=CH-C_6H_5$

(f)

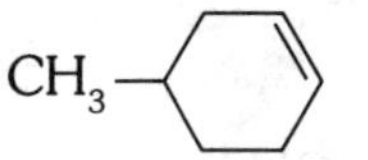

**11.18**  (a)
$\quad\quad\quad\;\; O$
$\quad\quad\quad\;\; ||$
$\quad\quad CH_3CCH_2CH_3$

(b)  No reaction

(c)  No reaction

(d)

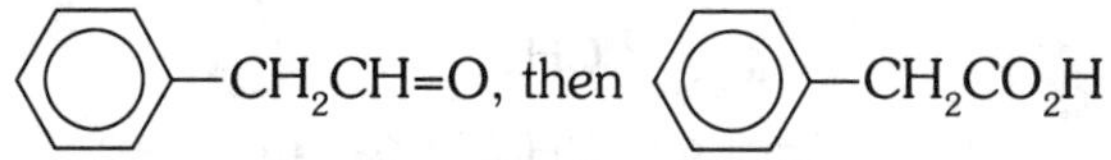

(e)

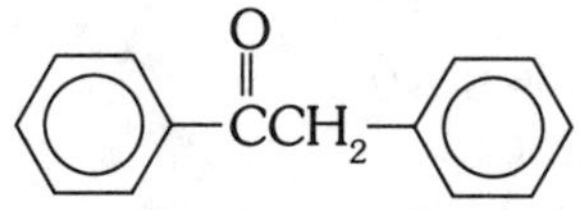

(f)

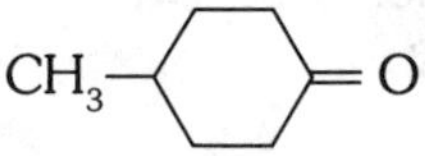

11.19 (a) $CH_3CH_2OH$

(b) $CH_3CH_2CH_2OH$ or $CH_3\underset{\underset{OH}{|}}{C}HCH_3$

(c)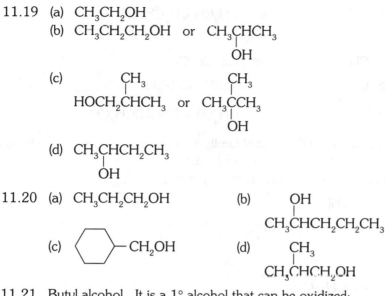

(d) $CH_3\underset{\underset{OH}{|}}{C}HCH_2CH_3$

11.20 (a) $CH_3CH_2CH_2OH$

(b) $\underset{\underset{CH_3\underset{\underset{OH}{|}}{C}HCH_2CH_2CH_3}{}}{\overset{OH}{|}}$

(c) a cyclohexane ring — $CH_2OH$

(d) $\overset{CH_3}{\underset{|}{}}$ $CH_3CHCH_2OH$

11.21 Butyl alcohol. It is a 1° alcohol that can be oxidized:

$$CH_3CH_2CH_2CH_2OH + (O) \longrightarrow CH_3CH_2CH_2CO_2H$$

The 3° alcohol could not be oxidized in this manner.

11.22 *sec*-butyl alcohol. It, but not the alkane (pentane), can be oxidized by the oxidizing agent:

$$CH_3CH_2\overset{\overset{OH}{|}}{C}HCH_3 + (O) \longrightarrow CH_3CH_2\overset{\overset{O}{\|}}{C}CH_3$$

11.23 Phenol is a strong enough acid to be able to neutralize sodium or potassium hydroxide:

$$\text{C}_6\text{H}_5\text{—O–H} + \text{OH} \longrightarrow \text{C}_6\text{H}_5\text{—O}^- + H_2O$$

11.24 It is compound A, the phenol, because it reacts with sodium hydroxide to form an ion that, like the carboxylate ion, is soluble in water.

$$CH_3CH_2CH_2CH_2\text{—C}_6\text{H}_4\text{—OH} + OH^- \longrightarrow$$

Insoluble in water

$$CH_3CH_2CH_2CH_2\text{—C}_6\text{H}_4\text{—O}^- + H_2O$$

Soluble in water

11.25 The unknown is I. II is a phenol that would react with aqueous KOH to form a water-soluble salt, but I is just a 3° alcohol and does not react with hydroxide ion.

11.26  (a)  CH$_3$CHCH$_3$
                    |
                    SH
        (b)  CH$_3$CH$_2$CH$_2$CH$_2$SH

       (c)  CH$_3$–S–S–CH$_3$
        (d)  HS–CH$_2$CH$_2$–SH

11.27  (a)  CH$_3$–S–S–CH$_3$
        (b)  2 (CH$_3$)$_2$CH–SH

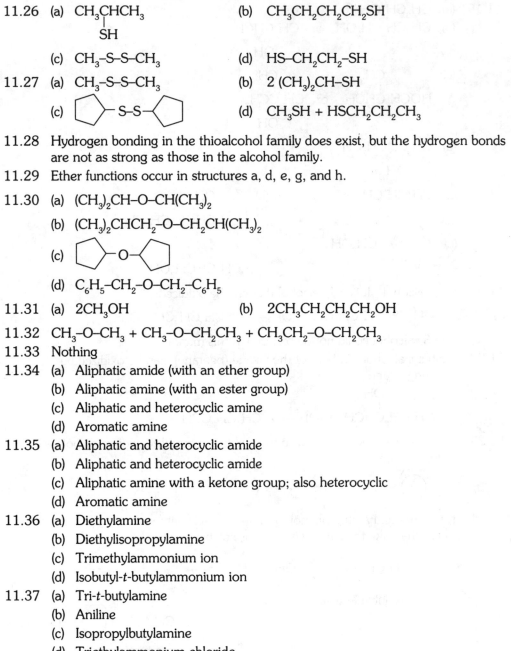

       (c)  ⬠–S–S–⬠
        (d)  CH$_3$SH + HSCH$_2$CH$_2$CH$_3$

11.28  Hydrogen bonding in the thioalcohol family does exist, but the hydrogen bonds are not as strong as those in the alcohol family.

11.29  Ether functions occur in structures a, d, e, g, and h.

11.30  (a)  (CH$_3$)$_2$CH–O–CH(CH$_3$)$_2$

        (b)  (CH$_3$)$_2$CHCH$_2$–O–CH$_2$CH(CH$_3$)$_2$

        (c)  ⬠–O–⬠

        (d)  C$_6$H$_5$–CH$_2$–O–CH$_2$–C$_6$H$_5$

11.31  (a)  2CH$_3$OH
        (b)  2CH$_3$CH$_2$CH$_2$CH$_2$OH

11.32  CH$_3$–O–CH$_3$ + CH$_3$–O–CH$_2$CH$_3$ + CH$_3$CH$_2$–O–CH$_2$CH$_3$

11.33  Nothing

11.34  (a)  Aliphatic amide (with an ether group)
        (b)  Aliphatic amine (with an ester group)
        (c)  Aliphatic and heterocyclic amine
        (d)  Aromatic amine

11.35  (a)  Aliphatic and heterocyclic amide
        (b)  Aliphatic and heterocyclic amide
        (c)  Aliphatic amine with a ketone group; also heterocyclic
        (d)  Aromatic amine

11.36  (a)  Diethylamine
        (b)  Diethylisopropylamine
        (c)  Trimethylammonium ion
        (d)  Isobutyl-*t*-butylammonium ion

11.37  (a)  Tri-*t*-butylamine
        (b)  Aniline
        (c)  Isopropylbutylamine
        (d)  Triethylammonium chloride

11.38 (a) No reaction

(b) 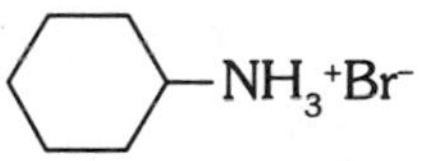 (Also acceptable: 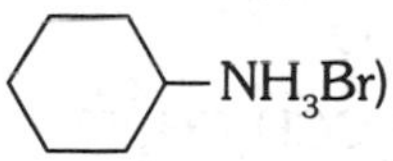)

(c) $NH_3$ (+ $H_2O$ + NaCl)

(d) 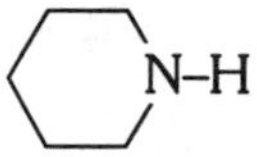

(e) $CH_3NH_2$

(f) $NH_2CH_2CO_2^-$

11.39 (a) $C_6H_5NH_3^+Cl^-$ (Also acceptable: $C_6H_5NH_3Cl$)

(b) 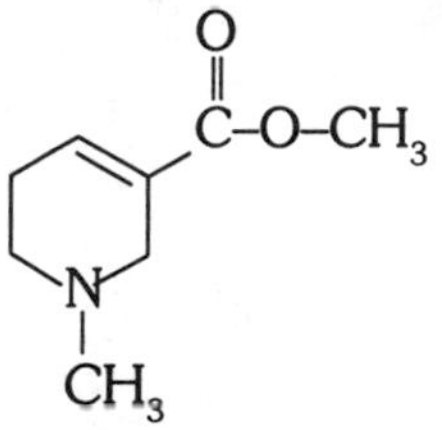

(c) 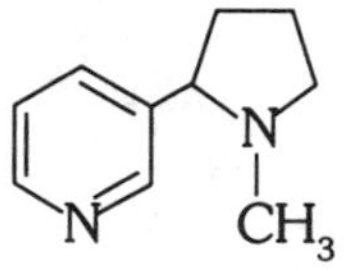

(d) 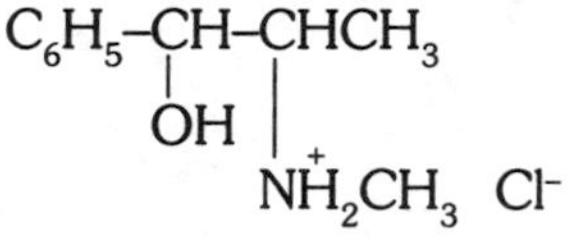

(e) 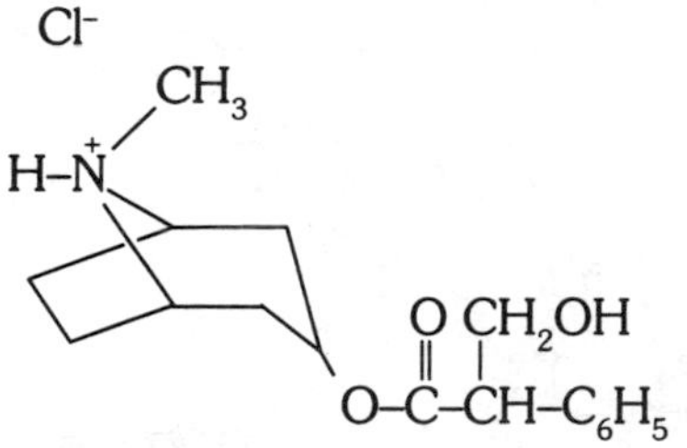

11.40 B. A is an amide, but B is an amine (as well as a ketone).

11.41 A. A is an amine, but B has no unshared pair of electrons on N for accepting a proton.

11.42 (a) $(CH_3)_2C{=}O$

(b) No reaction

(c) No reaction

(d) No reaction

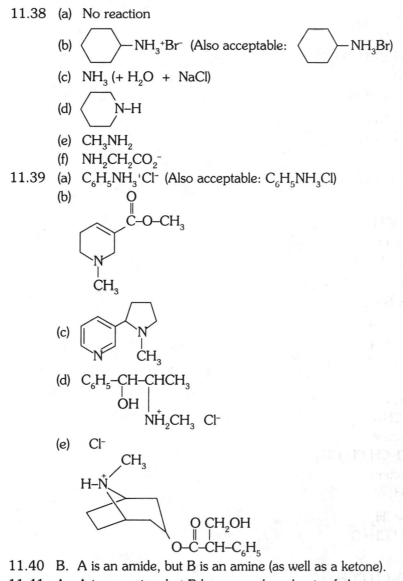

(e) $CH_3-O-CH_3$

(f) 
$$CH_3CH_2\overset{\displaystyle O}{\overset{\displaystyle \|}{C}}CH_2CH_3$$

(g) No reaction

(h) 
$$CH_3CH_2\underset{\displaystyle \underset{\displaystyle OH}{|}}{C}HCH_3$$

(i) $CH_3CH=CH_2$

(j) 
$$CH_3\overset{\displaystyle O}{\overset{\displaystyle \|}{C}}CH_2CH_3$$

(k) 
$$CH_3\underset{\displaystyle \underset{\displaystyle OH}{|}}{C}HCH_3$$

(l) No reaction

(m) $C_6H_5CO_2H$

(n) $CH_3CH_2CH_2NH_3{}^+Cl^-$

(o) $CH_3-S-S-CH_3$

11.43 (a) $CH_3CH_2CH_3$

(b) No reaction

(c)

(d) No reaction

(e) $C_6H_5-CH=CH_2$

(f) No reaction

(g) $CH_3-O-CH_2CH_2CH_3$

(h) No reaction

(i) 
$$CH_3\underset{\displaystyle \underset{\displaystyle CH_3}{|}}{C}=CH_2$$

(j) $CH_3CH_2CH=O$

(k) 
$$CH_3-O-CH_2\overset{\displaystyle O}{\overset{\displaystyle \|}{C}}CH_3$$

(l) No reaction

(m) $Cl^-\overset{+}{N}H_3-CH_2CH_2CH_2-NH_3{}^+Cl^-$

(n) $CH_3-O-CH_2CH_2CO_2H$

(o) 
$$CH_3CH_2\underset{\displaystyle \underset{\displaystyle OH}{|}}{C}HCH_2CH_3$$

(p) $2CH_3SH$

## Practice Exercises, Chapter 12

1.   (a)

$$CH_3CH_2\overset{\overset{\displaystyle O}{\|}}{C}\underset{\underset{\displaystyle CH_3}{|}}{C}HCH_3$$

    (b)

$$CH_3-\overset{\overset{\displaystyle O}{\|}}{C}-C_6H_5$$

  (c)

$$CH_3CH_2CH_2\overset{\overset{\displaystyle O}{\|}}{C}CH_2CH_2CH_3$$

    (d)

$$CH_3\overset{\overset{\displaystyle CH_3}{|}}{\underset{\underset{\displaystyle CH_3}{|}}{C}}\underset{}{\overset{\overset{\displaystyle O}{\|}}{\phantom{C}}}\overset{\overset{\displaystyle CH_3}{|}}{\underset{\underset{\displaystyle CH_3}{|}}{C}}CCH_3$$

2.   (a)

$$CH_3\overset{\overset{\displaystyle OH}{|}}{C}HCH_2CH_3$$

    (b)

$$CH_3\overset{\overset{\displaystyle CH_3}{|}}{C}HCH_2CH_2OH$$

  (c) $\langle\!\!\bigcirc\!\!\rangle$—OH

3.   (a) $CH_3\overset{\overset{\displaystyle }{}}{\underset{\underset{\displaystyle OH}{|}}{C}}HOCH_3$

    (b) $CH_3CH_2CH_2\overset{}{\underset{\underset{\displaystyle OH}{|}}{C}}H-O-CH_2CH_3$

  (c) $C_6H_5-\overset{\overset{\displaystyle OH}{|}}{C}H-O-CH_2CH_2CH_3$

    (d) $HO-CH_2-O-CH_3$

4.   (a) $CH_3CH_2CH=O \ + \ HOCH_3$

  (b) $CH_3CH_2OH \ + \ O=CHCH_2CH_3$

5.   (a) $CH_3CH_2\overset{\overset{\displaystyle OCH_3}{|}}{\underset{\underset{\displaystyle OCH_3}{|}}{C}}H$

    (b) $CH_3\overset{\overset{\displaystyle OCH_2CH_2CH_3}{|}}{\underset{\underset{\displaystyle OCH_2CH_2CH_3}{|}}{C}}CH_3$

6.   (a) $2CH_3OH + CH_2=O$

  (b) No reaction

  (c) $CH_3\overset{\overset{\displaystyle CH_3}{|}}{\underset{\underset{\displaystyle CH_3}{|}}{C}}HC=O \ + \ 2\ HOCH_3$

## Review Exercises, Chapter 12

12.1   Two carbons are always directly attached to the carbon atom of the carbonyl group in a ketone, whereas in aldehydes only one carbon is attached to this carbon (together with the aldehydic H atom).

12.2   (a)  $CH_3CH_2CH_2CH=O$          (b)  $CH_3CH=O$

      (c)  $CH_3CH_2CO_2H$          (d)  $HCO_2H$

12.3   (a)  $CH_3CO_2H$          (b)  $CH_3CH_2CH=O$

      (c)  $HCH=O$          (d)  $CH_3CH_2CH_2CO_2H$

12.4   (a)  Benzaldehyde          (b)  Acetaldehyde

      (c)  Butyric acid          (d)  Propionaldehyde

12.5   (a)  Formic acid          (b)  Benzaldehyde

      (c)  Propionic acid          (d)  Formaldehyde

12.6   Valeraldehyde

12.7   Anisic acid

12.8   A > D > B > C

12.9   C > D > A > B

12.10  A > D > B > C

12.11  C > D > A > B

12.12
$$
\begin{array}{c}
CH_3 \quad \delta - \quad \delta + \\
\diagdown \\
\phantom{xx}C=O \cdots H{-}O \\
\diagup \phantom{xxxxxxxx} \diagdown \\
CH_3 \phantom{xxxxxxx} H
\end{array}
$$

12.13
$$
\begin{array}{c}
CH_3 \quad \delta - \quad \delta + \\
\diagdown \\
\phantom{xx}C=O \cdots H{-}O \\
\diagup \phantom{xxxxxxxx} \diagdown \\
H \phantom{xxxxxxx} CH_3
\end{array}
$$

12.14  (a)

$$
\begin{array}{c}
O \\
\parallel \\
CH_3CCH_2CH_3
\end{array}
$$

      (b)  O= (cyclopentanone)

      (c)  $C_6H_5CH=O$          (d)  $(CH_3)_3CCH=O$

12.15  (a)  $CH_3CH=O$

      (b)  This compound cannot exist in pure form because it has two OH groups on the same carbon.

      (c)  This, a 3° alcohol, cannot be oxidized to an aldehyde or a ketone.

      (d)

$$
\begin{array}{c}
O \\
\parallel \\
CH_3CCH_2CO_2H
\end{array}
$$

12.16  $C_3H_6O$ is $CH_3CH_2CH=O$

      $C_3H_6O_2$ is $CH_3CH_2CO_2H$

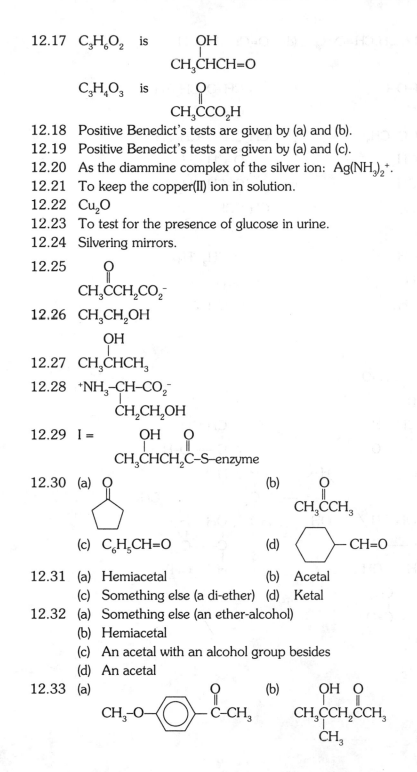

12.17 $C_3H_6O_2$ is

OH
|
$CH_3CHCH{=}O$

$C_3H_4O_3$ is

O
||
$CH_3CCO_2H$

12.18 Positive Benedict's tests are given by (a) and (b).
12.19 Positive Benedict's tests are given by (a) and (c).
12.20 As the diammine complex of the silver ion: $Ag(NH_3)_2^+$.
12.21 To keep the copper(II) ion in solution.
12.22 $Cu_2O$
12.23 To test for the presence of glucose in urine.
12.24 Silvering mirrors.

12.25

O
||
$CH_3CCH_2CO_2^-$

12.26 $CH_3CH_2OH$

OH
|
12.27 $CH_3CHCH_3$

12.28 $^+NH_3{-}CH{-}CO_2^-$
|
$CH_2CH_2OH$

12.29 I =

OH O
| ||
$CH_3CHCH_2C{-}S{-}$enzyme

12.30 (a)

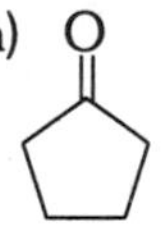

(b)

O
||
$CH_3CCH_3$

(c) $C_6H_5CH{=}O$

(d)

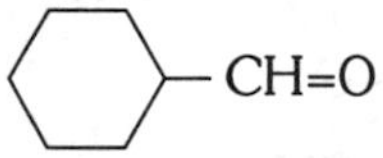

12.31 (a) Hemiacetal (b) Acetal
(c) Something else (a di-ether) (d) Ketal

12.32 (a) Something else (an ether-alcohol)
(b) Hemiacetal
(c) An acetal with an alcohol group besides
(d) An acetal

12.33 (a)

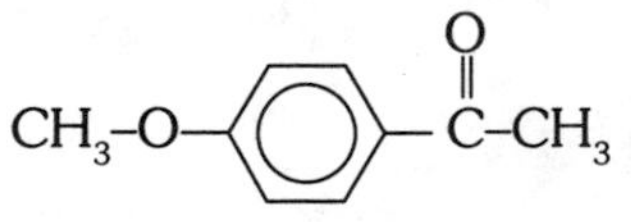

(b)

OH O
| ||
$CH_3CCH_2CCH_3$
|
$CH_3$

(c)  $CH_3CH_2OCH_2CH=O$

(d)  $O=CHCHOCH_3$
       $|$
       $CH_3$

12.34  (a)  $CH_3CH-O-CH_3$
                    $|$
                    $OH$

            $CH_3CH-O-CH_3$
                    $|$
                    $O-CH_3$

(b)  $CH_3CH-O-CH_2CH_3$
                $|$
                $OH$

      $CH_3CH-O-CH_2CH_3$
                $|$
                $O-CH_2CH_3$

12.35  (a)         $O-CH_3$
                       $|$
            $CH_3CCH_3$
                       $|$
                     $OH$

                   $O-CH_3$
                       $|$
            $CH_3CCH_3$
                       $|$
                   $O-CH_3$

(b)         $O-CH_2CH_3$
                $|$
      $CH_3CCH_3$
                $|$
              $OH$

            $O-CH_2CH_3$
                $|$
      $CH_3CCH_3$
                $|$
            $O-CH_2CH_3$

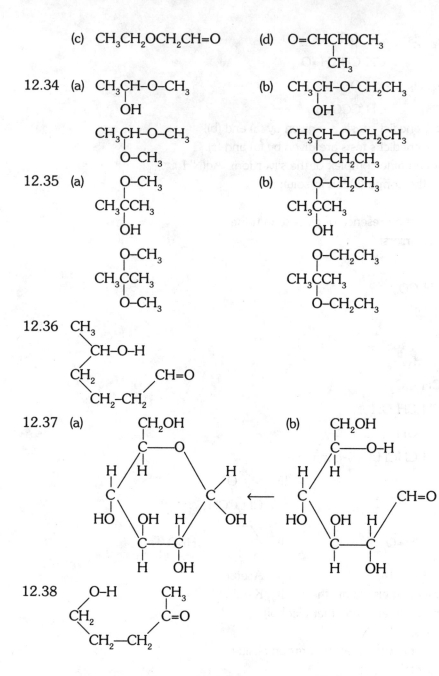

12.36  $CH_3$
           $|$
           $CH-O-H$
           $|$
           $CH_2$        $CH=O$
           $|$               $|$
           $CH_2-CH_2$

12.37  (a)                    $CH_2OH$
                                   $|$
                                   $C----O$
                           $H$ / $H$          $H$
                              $C$                $C$
                         $HO$  $OH$  $H$   $OH$
                                   $C----C$
                                 $H$     $OH$

(b)            $CH_2OH$
                    $|$
                    $C----O-H$
            $H$ / $H$
               $C$                $CH=O$
          $HO$  $OH$  $H$
                    $C----C$
                  $H$     $OH$

$\leftarrow$

12.38    $O-H$          $CH_3$
             $|$               $|$
             $CH_2$          $C=O$
             $|$               $|$
             $CH_2-CH_2$

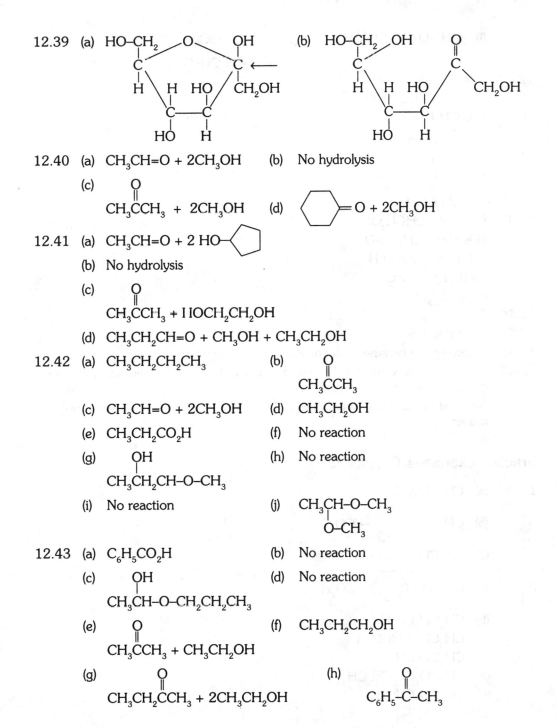

12.39 (a) [structure] (b) [structure]

12.40 (a) $CH_3CH=O + 2CH_3OH$    (b) No hydrolysis

(c) [structure] $CH_3CCH_3 + 2CH_3OH$    (d) [structure] $=O + 2CH_3OH$

12.41 (a) $CH_3CH=O + 2$ HO—[structure]

(b) No hydrolysis

(c) [structure] $CH_3CCH_3 + HOCH_2CH_2OH$

(d) $CH_3CH_2CH=O + CH_3OH + CH_3CH_2OH$

12.42 (a) $CH_3CH_2CH_2CH_3$    (b) [structure] $CH_3CCH_3$

(c) $CH_3CH=O + 2CH_3OH$    (d) $CH_3CH_2OH$

(e) $CH_3CH_2CO_2H$    (f) No reaction

(g) [structure] $CH_3CH_2CH-O-CH_3$ with OH    (h) No reaction

(i) No reaction    (j) $CH_3CH-O-CH_3$ with $O-CH_3$

12.43 (a) $C_6H_5CO_2H$    (b) No reaction

(c) [structure] $CH_3CH-O-CH_2CH_2CH_3$ with OH    (d) No reaction

(e) [structure] $CH_3CCH_3 + CH_3CH_2OH$    (f) $CH_3CH_2CH_2OH$

(g) [structure] $CH_3CH_2CCH_3 + 2CH_3CH_2OH$    (h) [structure] $C_6H_5-C-CH_3$

(i)   $CH_3-O-CH_2CH_2CH_3$

(j)   $OCH_3$
$CH_3CH_2CH-O-CH_3$

12.44   A is $CH_3CH_2CH=O$
B is $CH_3CH_2CH_2OH$
C is $CH_3CH=CH_2$
D is $CH_3CHCH_3$
        $OH$

E is    $O$
        $\|$
        $CH_3CCH_3$

12.45   F is $(CH_3)_2CHCH_2OH$
G is $(CH_3)_2CHCH=O$
H is $(CH_3)_2CHCO_2H$
I is $(CH_3)_2C=CH_2$
J is $(CH_3)_3COH$

12.46   Acetone

12.47   Formaldehyde

12.48   A keto group because of the *one* in the name estrone.

12.49   Compound A. Compound B would be derived from a 1,1–diol and these are generally too unstable to exist.

12.50   Acetone. It dissolves "super glue". At a notions counter as nail polish remover.

## Practice Exercises, Chapter 13

1.      (a)  $CH_3CH_2CO_2^-$

(b) 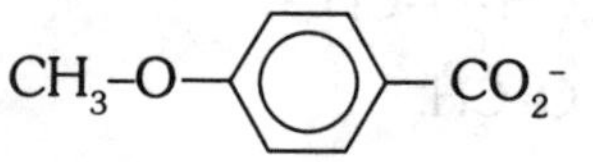  $CH_3-O-\langle\bigcirc\rangle-CO_2^-$

(c)  $CH_3CH=CHCO_2^-$

2.      (a) 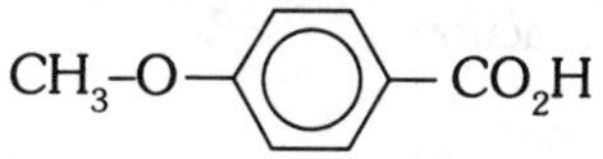  $CH_3-O-\langle\bigcirc\rangle-CO_2H$

(b)  $CH_3CH_2CO_2H$
(c)  $CH_3CH=CHCO_2H$

3.      (a)  $CH_3CO_2CH_3$
(b)  $CH_3CO_2CH_2CH_2CH_3$
(c)  $CH_3CO_2CHCH_3$
            $CH_3$

4.  (a)  $HCO_2CH_2CH_3$
    (b)  $CH_3CH_2CO_2CH_2CH_3$
    (c)  $C_6H_5-CO_2CH_2CH_3$
5.  (a)  $CH_3CO_2H + CH_3OH$
    (b)  $CH_3CH_2CO_2H + CH_3CHCH_3$
                                         |
                                        OH

    (c)       $CH_3$
               |
         $CH_3CHCO_2H + CH_3CH_2CH_2OH$
6.  (a)  $C_6H_5-OH + CH_3CO_2^-$

    (b)  $CH_3OH + CH_3-O-$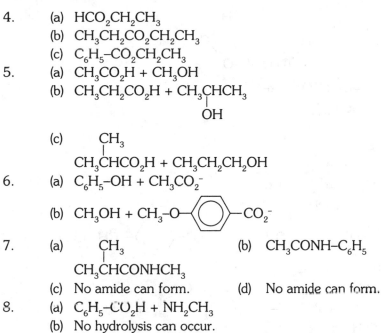$-CO_2^-$

7.  (a)       $CH_3$                    (b)  $CH_3CONH-C_6H_5$
               |
         $CH_3CHCONHCH_3$
    (c)  No amide can form.          (d)  No amide can form.
8.  (a)  $C_6H_5-CO_2H + NH_2CH_3$
    (b)  No hydrolysis can occur.
    (c)  $CH_3CO_2H = C_6H_5-NH_2$
    (d)  $2CH_3CO_2H + NH_2CH_2CH_2NH_2$

## Review Exercises, Chapter 13

13.1    O
        ‖
        $-C-O-H$ (which is often written as $-CO_2H$ and sometimes as $-COOH$)  For a
        compound to be an alcohol its molecules have the $-OH$ group attached to a
        carbon that has only *single* bonds.  A compound is a ketone if its molecules
        have a carbonyl group attached on both sides to carbon atoms.
13.2    From fats and oils in the diet
13.3    Acetic acid.
13.4    (a)  $CH_3CH_2CH_2CO_2H$          (b)  $CH_3CO_2H$
        (c)  $HCO_2H$                     (d)  $C_6H_5CO_2H$
13.5    (a)  $CH_3CO_2Na$                 (b)  $CH_3CH_2CO_2H$
        (c)  $C_6H_5CO_2Na$              (d)  $CH_3CH_2CH_2CO_2K$
13.6                δ−  δ+

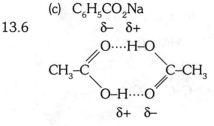

13.7

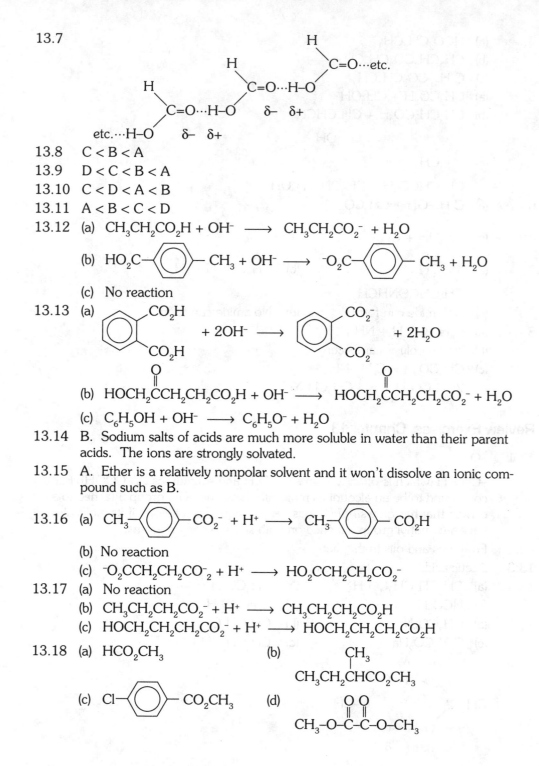

13.8   C < B < A

13.9   D < C < B < A

13.10  C < D < A < B

13.11  A < B < C < D

13.12  (a) $CH_3CH_2CO_2H + OH^- \longrightarrow CH_3CH_2CO_2^- + H_2O$

(b) $HO_2C-\bigcirc-CH_3 + OH^- \longrightarrow {}^-O_2C-\bigcirc-CH_3 + H_2O$

(c) No reaction

13.13  (a)

(b) $HOCH_2\overset{O}{\overset{||}{C}}CH_2CH_2CO_2H + OH^- \longrightarrow HOCH_2\overset{O}{\overset{||}{C}}CH_2CH_2CO_2^- + H_2O$

(c) $C_6H_5OH + OH^- \longrightarrow C_6H_5O^- + H_2O$

13.14  B. Sodium salts of acids are much more soluble in water than their parent acids. The ions are strongly solvated.

13.15  A. Ether is a relatively nonpolar solvent and it won't dissolve an ionic compound such as B.

13.16  (a) $CH_3-\bigcirc-CO_2^- + H^+ \longrightarrow CH_3-\bigcirc-CO_2H$

(b) No reaction

(c) $^-O_2CCH_2CH_2CO_2^- + H^+ \longrightarrow HO_2CCH_2CH_2CO_2^-$

13.17  (a) No reaction

(b) $CH_3CH_2CH_2CO_2^- + H^+ \longrightarrow CH_3CH_2CH_2CO_2H$

(c) $HOCH_2CH_2CH_2CO_2^- + H^+ \longrightarrow HOCH_2CH_2CH_2CO_2H$

13.18  (a) $HCO_2CH_3$

(b) $CH_3CH_2\overset{CH_3}{\overset{|}{C}H}CO_2CH_3$

(c) $Cl-\bigcirc-CO_2CH_3$

(d) $CH_3-O-\overset{O}{\overset{||}{C}}-\overset{O}{\overset{||}{C}}-O-CH_3$

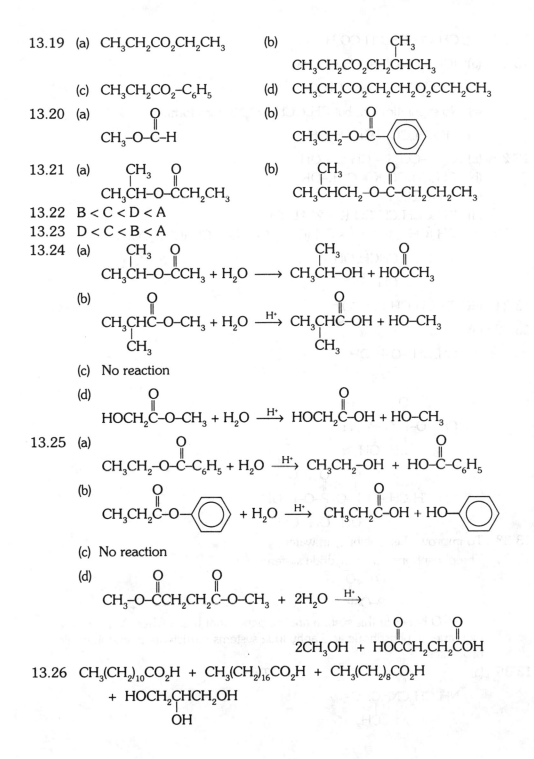

13.19  (a)  $CH_3CH_2CO_2CH_2CH_3$    (b)

CH_3CH_2CO_2CH_2CHCH_3 with CH_3 substituent

(c)  $CH_3CH_2CO_2-C_6H_5$    (d)  $CH_3CH_2CO_2CH_2CH_2O_2CCH_2CH_3$

13.20  (a)

$$CH_3-O-\overset{\overset{O}{\|}}{C}-H$$

(b)

$$CH_3CH_2-O-\overset{\overset{O}{\|}}{C}-\text{(phenyl)}$$

13.21  (a)

$$\underset{CH_3}{CH_3CH}-O-\overset{\overset{O}{\|}}{C}CH_2CH_3$$

(b)

$$\underset{CH_3}{CH_3CHCH_2}-O-\overset{\overset{O}{\|}}{C}-CH_2CH_2CH_3$$

13.22  B < C < D < A

13.23  D < C < B < A

13.24  (a)

$$\underset{CH_3}{CH_3CH}-O-\overset{\overset{O}{\|}}{C}CH_3 + H_2O \longrightarrow \underset{CH_3}{CH_3CH}-OH + HO\overset{\overset{O}{\|}}{C}CH_3$$

(b)

$$\underset{CH_3}{CH_3CH}\overset{\overset{O}{\|}}{C}-O-CH_3 + H_2O \xrightarrow{H^+} \underset{CH_3}{CH_3CH}\overset{\overset{O}{\|}}{C}-OH + HO-CH_3$$

(c)  No reaction

(d)

$$HOCH_2\overset{\overset{O}{\|}}{C}-O-CH_3 + H_2O \xrightarrow{H^+} HOCH_2\overset{\overset{O}{\|}}{C}-OH + HO-CH_3$$

13.25  (a)

$$CH_3CH_2-O-\overset{\overset{O}{\|}}{C}-C_6H_5 + H_2O \xrightarrow{H^+} CH_3CH_2-OH + HO-\overset{\overset{O}{\|}}{C}-C_6H_5$$

(b)

$$CH_3CH_2\overset{\overset{O}{\|}}{C}-O-\text{(phenyl)} + H_2O \xrightarrow{H^+} CH_3CH_2\overset{\overset{O}{\|}}{C}-OH + HO-\text{(phenyl)}$$

(c)  No reaction

(d)

$$CH_3-O-\overset{\overset{O}{\|}}{C}CH_2CH_2\overset{\overset{O}{\|}}{C}-O-CH_3 + 2H_2O \xrightarrow{H^+}$$

$$2CH_3OH + HO\overset{\overset{O}{\|}}{C}CH_2CH_2\overset{\overset{O}{\|}}{C}OH$$

13.26  $CH_3(CH_2)_{10}CO_2H + CH_3(CH_2)_{16}CO_2H + CH_3(CH_2)_8CO_2H$

$+ HOCH_2\underset{OH}{CH}CH_2OH$

13.27  $HOCH_2CH_2CH_2CH_2CO_2H$

13.28  (a)  $(CH_3)_2CHOH + CH_3CO_2Na$

(b)  $(CH_3)_2CHCO_2Na + CH_3OH$

(c)  No saponification, but $CH_3OCH_2CO_2Na$ does form.

(d)  $HOCH_2CO_2Na + CH_3OH$

13.29  (a)  $C_6H_5-CO_2K + CH_3CH_2OH$

(b)  $CH_3CH_2CO_2K + C_6H_5-OK$

(c)  No reaction

(d)  $KO_2CCH_2CH_2CO_2K + 2CH_3OH$

13.30  (a)  $CH_3(CH_2)_{10}CO_2Na + CH_3(CH_2)_{16}CO_2Na + CH_3(CH_2)_8CO_2Na$

$+ HOCH_2\overset{\displaystyle |}{\underset{\displaystyle OH}{C}}HCH_2OH$

13.31  $HOCH_2CH_2CH_2CH_2CO_2^-$

13.32  (a)

$$CH_3CH_2-O-\overset{\displaystyle O}{\overset{\displaystyle \|}{\underset{\displaystyle |}{\underset{\displaystyle OH}{P}}}}-OH$$

(b)

$$CH_3-O-\overset{\displaystyle O}{\overset{\displaystyle \|}{\underset{\displaystyle |}{\underset{\displaystyle OH}{P}}}}-O-\overset{\displaystyle O}{\overset{\displaystyle \|}{\underset{\displaystyle |}{\underset{\displaystyle OH}{P}}}}-OH$$

(c)

$$CH_3CH_2CH_2-O-\overset{\displaystyle O}{\overset{\displaystyle \|}{\underset{\displaystyle |}{\underset{\displaystyle OH}{P}}}}-O-\overset{\displaystyle O}{\overset{\displaystyle \|}{\underset{\displaystyle |}{\underset{\displaystyle OH}{P}}}}-O-\overset{\displaystyle O}{\overset{\displaystyle \|}{\underset{\displaystyle |}{\underset{\displaystyle OH}{P}}}}-OH$$

13.33  To improve their solubility in water

13.34  The phosphoric acid anhydride system:

$$-\overset{\displaystyle O}{\overset{\displaystyle \|}{P}}-O-\overset{\displaystyle O}{\overset{\displaystyle \|}{P}}-$$

The P–O bonds in this system are the bonds that break when ATP or other high energy phosphoric acid anhydride systems participate in endothermic processes.

13.35  (a)                                        (b)  Two

$$NH_2CH_2\overset{\displaystyle O}{\overset{\displaystyle \|}{C}}NH\overset{\displaystyle |}{\underset{\displaystyle CH_3}{C}}H\overset{\displaystyle O}{\overset{\displaystyle \|}{C}}-$$

13.36 (a) $CH_3CONH_2$  (b) $CH_3CH_2CONHCH_3$
(c) $C_6H_5CON(CH_3)_2$  (d) $HCONHCH_2CH_3$

13.37 (a) $CH_3CH_2CH_2CONHCH_3$  (b) $CH_3CON(CH_3)_2$
(c) $HCONH_2$  (d) $CH_3CH_2CONH_2$

13.38 (a) $CH_3NH_2 + CH_3CH_2CO_2H$
(b) No reaction
(c) $CH_3CO_2H + (CH_3)_2NH$
(d) $(CH_3)_2CHCO_2H + CH_3NH_2$

13.39 (a) $2NH_2CH_2CO_2H$
(b) $(CH_3)_2NCH_2CO_2H + NH_2CH_2CH_2CO_2H + CH_3NH_2$
(c) $NH_2CH_2CH_2CH_2CH_2CO_2H$
(d) $(CH_3)_2NCH_2CH_2NH_2 + HO_2CCH_2CH_2CO_2H + CH_3NH_2$

13.40 (a) The addition of water to a double bond to give an alcohol:

$$\mathord{>}C=C\mathord{<} + H_2O \xrightarrow{H^+} -\underset{H}{\overset{|}{C}}-\underset{OH}{\overset{|}{C}}-$$

The hydrolysis of an acetal or a ketal to give an aldehyde or a ketone plus two molecules of an alcohol:

$$-\underset{O-R}{\overset{O-R}{C}}- + H_2O \xrightarrow{H^+} -\overset{O}{\overset{\|}{C}}- + 2HOR$$

The hydrolysis of an ester to give an acid and an alcohol:

$$R-\overset{O}{\overset{\|}{C}}-O-R' + H_2O \longrightarrow R-CO_2H + HOR'$$

The hydrolysis of an amide to give an acid and an amine (or ammonia):

$$R-\overset{O}{\overset{\|}{C}}-\overset{|}{N}- + H_2O \longrightarrow R-CO_2H + H-\overset{|}{N}-$$

(b) The reduction of a disulfide to two molecules of mercaptan:

$$R-S-S-R + 2(H) \longrightarrow 2RSH$$

The reduction of an alkene to give an alkane:

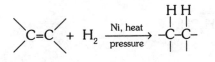

The reduction of an aldehyde or a ketone to give an alcohol:

$$\underset{\displaystyle -\overset{\textstyle O}{\overset{\|}{C}}-}{} \ + \ H_2 \ \xrightarrow[\text{pressure}]{\text{Ni, heat}} \ \underset{\displaystyle -\overset{\textstyle OH}{\overset{|}{C}H}-}{}$$

(c) The oxidation of a 1° alcohol or a 2° alcohol to an aldehyde or a ketone:

$$\underset{\displaystyle -\overset{\textstyle OH}{\overset{|}{C}H}-}{} \ + \ (O) \ \longrightarrow \ \underset{\displaystyle -\overset{\textstyle O}{\overset{\|}{C}}-}{} \ + \ H_2O$$

The oxidation of a mercaptan to a disulfide:

$$2RSH + (O) \ \longrightarrow \ R\text{-}S\text{-}S\text{-}R + H_2O$$

The oxidation of an aldehyde to a carboxylic acid:

$$RCH{=}O + (O) \ \longrightarrow \ RCO_2H$$

(Note. Strong oxidizing agents also attack sidechains on benzene rings and double bonds.)

13.41 The acetal (or ketal) group in carbohydrates; the ester group in fats and oils; and the amide group in proteins.

13.42 (a) $CH_3CO_2H$

(b) No reaction

(c) $CH_3CO_2CH_3$

(d)
$$CH_3CH_2\overset{\textstyle O}{\overset{\|}{C}}CH_3$$

(e) $CH_3CH{=}CH_2$

(f) $CH_3OH + HO_2CCH(CH_3)_2$

(g) $CH_3CH_2\underset{\displaystyle OH}{\overset{|}{C}}HCH_3$

(h) $CH_3CHO + 2CH_3OH$

(i) $(CH_3)_2CHCO_2Na + HOCH_3$

(j) No reaction

(k) $CH_3CH_2CO_2Na$

(l) $CH_3\underset{\displaystyle OH}{\overset{|}{C}}HCH_3$

(m) $CH_3OH + CH_3CO_2H$

(n) No reaction

(o) $CH_3CH_2CH{=}O + 2HOCH_2CH_3$

(p) $C_6H_5\text{-}CO_2H$

(q)  $CH_3CH_2CO_2Na + CH_3CH_2OH$

(r)        $OCH_3$
           |
     $CH_3CH-OCH_3$

(s)  No reaction

(t)  No reaction

(u)  No reaction

(v)  $CH_3CH_2CO_2CH_3$

(w)  $CH_3CH_2CH_2NH_3{}^+Cl^-$

(x)  $HCO_2H + NH_3$

(y)  $CH_3-S-S-CH_3$

(z)  $2CH_3NH_2 + HO_2CCH_2C_2CO_2H$

13.43 (a)  $CH_3CH_2CO_2H$

(b)  No reaction

(c)  No reaction

(d)  $CH_3CH_2CH_2OH + HO_2CCH_2CH_2OH + HO_2CCH_3$

(e)  $C_6H_5CH=O + 2CH_3CH_2OH$

(f)  No reaction

(g)  $CH_3C_2CO_2CH(CH_3)_2$

(h)  No reaction

(i)  $CH_3OH + NaO_2CCH_2CH_2OH + NaO_2CCH_2CH_3$

(j)  $CH_3(CH_2)_6CO_2Na$

(k)  $HO_2CCH_2CH_2CH_2CO_2H$

(l)  $CH_3O_2CCH_2CH_2CH_2CO_2CH_3$

(m) $C_6H_5CO_2Na + CH_3CH_2OH$

(n)        $OCH_3$
           |
     $C_6H_5-CH-OCH_3$

(o)  No reaction

(p)  No reaction

(q)  $NaO_2CCH_2CH_2CH_3$

(r)  $CH_3CH_2\overset{\displaystyle O}{\overset{\displaystyle \|}{C}}CH_3 + 2HOCH_2CH_3$

(s)  No reaction

(t)  $2CH_3OH + HO_2CCH_2CH_2CO_2H$

(u)  $Cl^-\ NH_3{}^+-CH_2CH_2CH_2-NH_3{}^+Cl^-$

(v)  $CH_3-O-CH_2CH_2CO_2H$

(w)  $CH_3(CH_2)_4CO_2CH_2CH_3$

(x)  $2CH_3CH_2OH + CH_3CH=O$

(y)          OH
             |
    $CH_3CH_2CHCH_2CH_3$

(z)  $2CH_3SH$

13.44  Acetate ion

13.45  Molds, yeasts, and bacteria

13.46  *para*-Hydroxybenzoic acid.  Additives for cosmetics, pharmaceuticals and food to inhibit molds and yeasts.

13.47  Suppress pain.  Reduce fever

13.48  Aspirin is less irritating to the stomach.

13.49  Oil of wintergreen

13.50  A polymer made from two (sometimes more) monomers, not just one.

13.51  They are the same.  Mylar is the film form.  Dacron is the fiber form.

13.52  Ethylene glycol

13.53  Amide

13.54  Hydrogen bond

13.55  It allows the long molecules to line up better side by side.

13.56  The two monomers each have 6 carbons per molecule.

13.57  Amide

13.58  Depresses it.  Rates of breathing and heart beat.

13.59  Ethyl alcohol

## Review Exercises, Chapter 14

14.1    Materials, energy, and information.

14.2    Carbohydrates, lipids, and proteins.

14.3    Enzymes and nucleic acids, the latter carrying genetic bluprints.

14.4    Organization

14.5    (a)  C          (b)  D          (c)  A          (d)  B and D

14.6    (a)  $CH_2CH-CH-CH-CH=O$
                |    |     |     |
               OH  OH   OH   OH

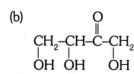

        (b)              O
                         ‖
             $CH_2-CH-C-CH_2$
               |     |       |
              OH   OH     OH

14.7    Polysaccharide.

14.8    Nonreducing carbohydrate.

14.9    (a) Glucose.    (b)  Glucose.

14.10  Fructose.

14.11

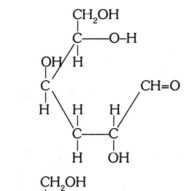

14.12

14.13

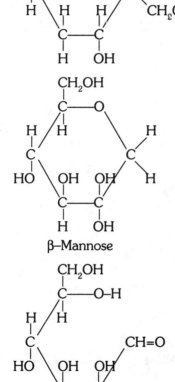

β–Mannose

Open-form of mannose

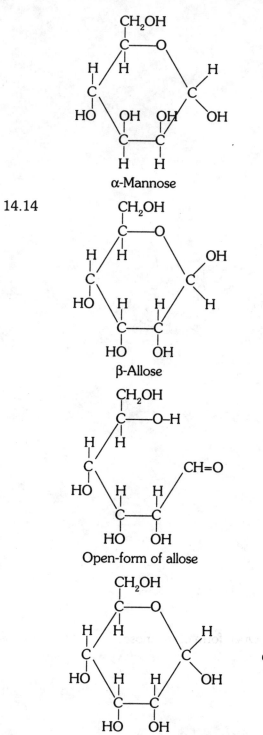

α-Mannose

14.14

β-Allose

Open-form of allose

α-Allose

14.15   As molecules of the open-chain form are oxidized, the equilibrium continuously shifts to make more from the cyclic forms.

14.16   As the beta form is used, molecules of the other forms continuously change into it as the equilibrium shifts.

14.17

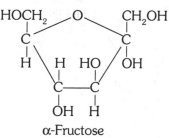

α-Fructose

14.18   The structure is that of α-fructose tipped over.

14.19

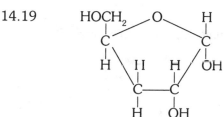

14.20   No, it has no –OH on carbon-4 that would be necessary to the formation of the cyclic hemiacetal with a five-membered ring.

14.21   Its molecules cannot be made to superimpose onto molecules that are mirror images.

14.22   Yes, it has the hemiacetal system from which the aldehyde group is released as the ring opens up.

14.23   When the structures are drawn with the six-membered ring on its side and the ring oxygen atom in the upper right hand corner, the $CH_2OH$ projects above the plane of the ring and is to the left of the ring oxygen atom.  In the L-family, the ring oxygen would be in the upper left corner with the $CH_2OH$ group to its right but still projecting upward.

14.24   (a)  Dextrorotatory

(b)  It will rotate the plane of the plane polarized light 26° to the left.

14.25   Their molecules are related as an object to its mirror image.

14.26   The enzymes involved have their own chiralities, and they do not accept molecules of L-glucose.

14.27   Maltose, lactose, and sucrose.

14.28   A 50:50 mixture of glucose and fructose.

14.29   Its molecules have no hemiacetal or hemiketal group at which ring-opening and ring-closing can occur.

14.30   (a) and (b)

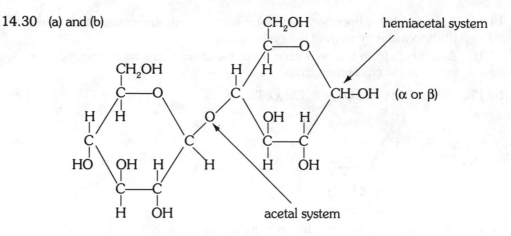

(c)  Yes, it has the hemiacetal system.

(d)  The bridge between the two glucose units is β(1 → 4) instead of α(1 → 4).

(e)  Two molecules of glucose:

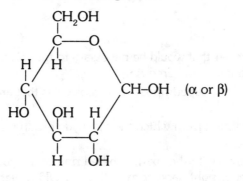

14.31   (a)  No, it has no hemiacetal or hemiketal system.

(c)  Two molecules of glucose

14.32

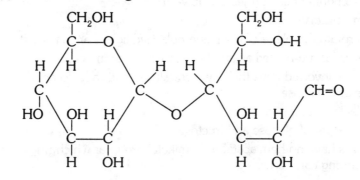

14.33

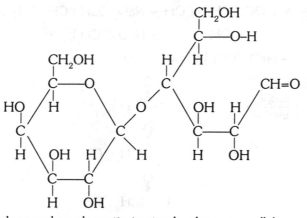

14.34  Amylose and amylopectin in starch; glycogen; cellulose

14.35  The oxygen bridges in amylose are α(1 → 4) and in cellulose they are β(1 → 4).

14.36  They are polymers of α-glucose.

14.37  Amylose is an entirely linear polymer of α-glucose in which all of the oxygen bridges are α(1 → 4), and amylopectin has branching.

14.38  Humans lack the enzyme for catalyzing this reaction.

14.39  A test for starch.  The reagent is a solution of iodine in aqueous potassium iodide.  It is used to test for starch, and a positive test is the immediate appearance of a blue-black color.

14.40  They are very similar except that glycogen is more branched.

14.41  To store glucose units.

14.42  The sun.

14.43  Carbon dioxide and water are made into glucose units and oxygen.

14.44  Chlorophyll

14.45  Marine phytoplankton and algae.

## Practice Exercises, Chapter 15

1.

$$CH_3(CH_2)_{26}\overset{\displaystyle O}{\overset{\displaystyle \|}{C}}O(CH_2)_{25}CH_3$$

2.

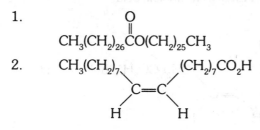

3.   $1 + 3NaOH \longrightarrow HOCH_2CHCH_2OH + NaO_2C(CH_2)_7CH=CH(CH_2)_7CH_3$

$$\overset{\displaystyle |}{OH} \qquad + NaO_2C(CH_2)_{14}CH_3$$

$$+ NaO_2C(CH_2)_7CH=CHCH_2CH=CH(CH_2)_4CH_3$$

4.   $1 + 3H_2 \xrightarrow[\text{heat, pressure}]{\text{catalyst}}$

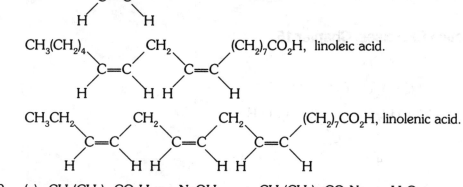

## Review Exercises, Chapter 15

15.1   It is not obtainable from living plants or animals.

15.2   It is extractable from animal and plant sources by relatively nonpolar solvents.

15.3   It is soluble in water, and it isn't present in plant or animal sources.

15.4   It is present in undecomposed plant or animal materials and is extractable by relatively nonpolar solvents.

15.5   $CH_3(CH_2)_{14}CO_2(CH_2)_{17}CH_3$

15.6   C.  A is ruled out because *both* the acid and alcohol portions of the wax molecule are usually long-chain.  B is ruled out because both of these portions are likely to have an even number of carbons.

15.7   Palmitic acid;  $CH_3(CH_2)_{14}CO_2H$
Stearic acid;  $CH_3(CH_2)_{16}CO_2H$

15.8   $CH_3(CH_2)_7$ \qquad $(CH_2)_7CO_2H$, oleic acid.

$$\underset{H \qquad\quad H}{\overset{\quad}{C=C}}$$

$CH_3(CH_2)_4$ \qquad $CH_2$ \qquad $(CH_2)_7CO_2H$,  linoleic acid.

$$\underset{H \qquad H\ H \qquad H}{C=C \qquad C=C}$$

$CH_3CH_2$ \qquad $CH_2$ \qquad $CH_2$ \qquad $(CH_2)_7CO_2H$, linolenic acid.

$$\underset{H \qquad H\ H \qquad H\ H \qquad H}{C=C \qquad C=C \qquad C=C}$$

15.9   (a)  $CH_3(CH_2)_{14}CO_2H + NaOH \longrightarrow CH_3(CH_2)_{14}CO_2Na + H_2O$

(b)  $CH_3(CH_2)_{14}CO_2H + CH_3OH \xrightarrow{H^+} CH_3(CH_2)_{14}CO_2CH_3 + H_2O$

15.10  (a)  $CH_3(CH_2)_7CH=CH(CH_2)_7CO_2H + Br_2 \longrightarrow$

$$CH_3(CH_2)_7\underset{\underset{Br}{|}}{CH}-\underset{\underset{Br}{|}}{CH}(CH_2)_7CO_2H$$

(b)  $CH_3(CH_2)_7CH=CH(CH_2)_7CO_2H + KOH \longrightarrow$

$$CH_3(CH_2)_7CH=CH(CH_2)_7CO_2K + H_2O$$

(c)  $CH_3(CH_2)_7CH=CH(CH_2)_7CO_2H + H_2 \xrightarrow[\text{heat, pressure}]{Ni} CH_3(CH_2)_{16}CO_2H$

(d)  $CH_3(CH_2)_7CH=CH(CH_2)_7CO_2H + CH_3CH_2OH \xrightarrow[\text{heat}]{H^+}$

$$CH_3(CH_2)_7CH=CH(CH_2)_7CO_2CH_2CH_3 + H_2O$$

15.11  B. Molecules of fatty acids rarely have branched chains or an uneven number of carbon atoms.

15.12  Fatty acids whose molecules have five-membered rings to which long side chains are joined.

15.13

$$CH_2-O-\overset{\overset{\displaystyle O}{\|}}{C}(CH_2)_7CH=CHCH_2CH=CHCH_2CH=CHCH_2CH_3$$
$$|$$
$$CH-O-\overset{\overset{\displaystyle O}{\|}}{C}(CH_2)_7CH=CH(CH_2)_7CH_3$$
$$|$$
$$CH_2-O-\overset{\overset{\displaystyle O}{\|}}{C}(CH_2)_{12}CH_3$$

15.14

$$CH_2-O-\overset{\overset{\displaystyle O}{\|}}{C}(CH_2)_{16}CH_3$$
$$|$$
$$CH-O-\overset{\overset{\displaystyle O}{\|}}{C}(CH_2)_7CH=CH(CH_2)_7CH_3$$
$$|$$
$$CH_2-O-\overset{\overset{\displaystyle O}{\|}}{C}(CH_2)_{14}CH_3$$

15.15  $CH_2OH + 2\ HO_2C(CH_2)_7CH=CH(CH_2)_7CH_3$
$$|$$
$$CH-OH + HO_2C(CH_2)_{12}CH_3$$
$$|$$
$$CH_2OH$$

15.16    $CH_2OH$ + 2 $NaO_2C(CH_2)_7CH=CH(CH_2)_7CH_3$

$CH-OH$ + $NaO_2C(CH_2)_{12}CH_3$

$CH_2OH$

15.17

$$CH_2-O-\overset{\overset{\displaystyle O}{\|}}{C}(CH_2)_{10}CH_3$$

$$CH-O-\overset{\overset{\displaystyle O}{\|}}{C}(CH_2)_7CH=CHCH_2CH=CH(CH_2)_4CH_3$$

$$CH_2-O-\overset{\overset{\displaystyle O}{\|}}{C}(CH_2)_7CH=CH(CH_2)_7CH_3$$

This structure shows just one possibility. The fatty acyl units can be joined in different orders to the glycerol unit.

15.18

$$CH_2-O-\overset{\overset{\displaystyle O}{\|}}{C}(CH_2)_{10}CH_3$$

$$CH-O-\overset{\overset{\displaystyle O}{\|}}{C}(CH_2)_{10}CH_3$$

$$CH_2-O-\overset{\overset{\displaystyle O}{\|}}{C}(CH_2)_7CH=CH(CH_2)_7CH_3$$

The order in which the acyl groups are joined to the glycerol can, of course, be different than shown here.

15.19    There are more alkene double bonds per molecule in vegetable oils than in animal fats.

15.20    The triacylglycerol molecules that are present have several alkene units per molecule, so the substances are more "polyunsaturated" than the animal fats.

15.21    Hydrogenation.

15.22    Butter melts on the tongue; lard and tallow do not.

15.23    Glycerol-based (phosphoglycerides and plasmalogens) and sphingosine-based(sphingolipids).

15.24    Phosphoglycerides molecules have a fatty acyl unit joined to the glycerol unit where plasmalogens have an unsaturated ether link.

15.25    The cerebrosides are glycolipids in which a sugar unit is joined to the sphingosine unit where, in the sphingomyelins, a phosphate diester is joined.

15.26    A sugar unit such as galactose or glucose.

15.27   Their molecules have electrically charged sites.

15.28   Cell membranes.

15.29   Sphingomyelins and cerebrosides.

15.30   By glycosidic links.  These involve an acetal system, not an ordinary ether group.  The glycosidic link is more easily hydrolyzed.

15.31   (a)

$$CH_2-O-\overset{\overset{\displaystyle O}{\|}}{C}(CH_2)_7CH=CHCH_2CH=CHCH_2CH=CHCH_2CH_3$$

$$CH-\overset{\overset{\displaystyle O}{\|}}{C}-\overset{\overset{\displaystyle O}{\|}}{C}(CH_2)_7CH=CH(CH_2)_7CH_3$$

$$CH_2-O-\overset{\overset{\displaystyle O}{\|}}{\underset{\underset{\displaystyle O^-}{|}}{P}}-O-CH_2CH_2\overset{+}{N}(CH_3)_3$$

(b)   Phosphoglyceride, because glycerol is one of the products of hydrolysis.

(c)   Lecithin, because it gives choline when hydrolyzed.

15.32   (a)

$$CH_2-O-\overset{\overset{\displaystyle O}{\|}}{C}(CH_2)_{10}CH_3$$

$$CH-O-\overset{\overset{\displaystyle O}{\|}}{C}(CH_2)_7CH=CH(CH_2)_7CH_3$$

$$CH_2-O-\overset{\overset{\displaystyle O}{\|}}{\underset{\underset{\displaystyle O^-}{|}}{P}}-O-CH_2CH_2NH_3{}^+$$

(b)   Phosphoglyceride, because glycerol is one of the products of hydrolysis.

(c)   Cephalin, because aminoethanol is a hydrolysis product.

15.33   The anion of cholic acid.

15.34   Vitamin $D_3$.

15.35   Estradiol, progesterone, testosterone, and androsterone.

15.36   Cholesterol.

15.37   The membrane consists mostly of two layers of phospholipid molecules whose hydrophobic parts intermingle with each other between the layers and whose hydrophilic parts face toward the aqueous solutions whether they are inside the cell or outside.

15.38   The hydrophobic tails intermesh with each other between the two layers of the bilayer.

15.39   Proteins

15.40  The water-avoiding properties of the hydrophobic units and the water-attracting properties of the hydrophilic units.

15.41  Yes, there is a concentration gradient because the sugar concentration is not uniform throughout. In time, the process of diffusion makes the concentration uniform, and there is no chance that this uniformity will *spontaneously* change to reestablish a gradient.

15.42  Plasma.

15.43  Cell fluid.

15.44  Cell fluid.

15.45  Energy-consuming reactions carry various chemical species through the membrane.

15.46  It is a membrane-bound mechanism for carrying sodium and potassium ions through a membrane against their individual concentration gradients.

15.47  Their molecules act as receptor sites for molecules that must either enter the cell or cause the cell to do something. They also serve as channels and pumps.

15.48  Arachidonic acid

15.49  Aspirin inhibits the synthesis of prostaglandins. Since prostaglandins enhance a fever, aspirin reduces a fever by this action.

15.50  It has a double bond at the third carbon counting from the $\omega$-carbon, the one most remote from the carboxyl group.

15.51  Marine oils.

15.52  Some evidence suggests that they protect one against the heart disease associated with elevated cholesterol levels.

15.53  Detergent. Soap is just one example of a detergent.

15.54  A mixture of the sodium or potassium salts of long-chain fatty acids.

15.55  A synthetic detergent, because it works better in hard water.

15.56  The detergent properties are caused by an anion.

15.57  The hydrophobic tails of the detergent ions become embedded in the grease layer and the hydrophobic heads stick out into the wash solution. As the grease breaks up, its tiny globules become pincushioned with detergent ions, and the effect of the charges is to help bring the globules into a colloidal dispersion.

## Practice Exercises, Chapter 16

1.    Glycine:       $^+NH_3CH_2CO_2^-$

      Alanine:       $^+NH_3CHCO_2^-$
      $\qquad\qquad\qquad\quad |$
      $\qquad\qquad\qquad\ CH_3$

Lysine:         $^+NH_3CHCO_2^-$
                 |
                 $(CH_2)_4$
                 |
                 $NH_2$

Glutamic acid:   $^+NH_3CHCO_2^-$
                 |
                 $(CH_2)_2$
                 |
                 $CO_2H$

2.    (a)  $^+NH_3CHCO_2^-$
           |
           $CH_2CO_2^-$

      (b)  $^+NH_3CHCO_2^-$
           |
           $CH_2CONH_2$

3.    $^+NH_3CHCO_2^-$          $NH_2^+$
      |                         ‖
      $CH_2CH_2CH_2NHCNH_2$

4.    Hydrophilic; neutral side chain.

5.    $^+NH_3CHCONHCHCO_2^-$          $^+NH_3CHCONHCHCO_2^-$
      |             |                 |             |
      $CH_3$     $CH_2CH_2CO_2H$      $CH_2$       $CH_3$
                                      |
                                      $CH_2CO_2H$

## Review Exercises, Chapter 16

16.1   $^+NH_3CH_2CO_2H$

16.2   $NH_2CHCO_2^-$
       |
       $CH_3$

16.3   $NH_2CHCO_2CH_2CH_3$
       |
       $CH_3$

The polarity of this molecule is much, much less than the polarity of the dipolar ionic structure of alanine, so the ester molecules stick together with weaker forces and the compound has a lower melting point.

16.4   The ester has an $-NH_2$ as the proton acceptor, whereas alanine itself has the $-CO_2^-$ group as the acceptor.  And this group is made a weak acceptor by the electron withdrawal of the adjacent $-NH_3^+$ group in the dipolar ion.  (The withdrawal of electron density from a proton–accepting site renders this site less able to take and hold a proton.)

16.5    B, because of the polar sites at the end of the side chain.

16.6    A, because the side chain is purely hydrocarbonlike.

16.7    (a)  pH 1, because at a relatively high concentration of $H^+$ the molecule tends to be fully protonated at all of its proton accepting sites.

      (b)  Cathode.

16.8    (a)  At a pH of 10, because at a relatively high concentration of $OH^-$ few proton-accepting sites can retain protons.

      (b)  Anode.

16.9    1+

16.10   L-family

16.11   $^+NH_3CH_2CONHCHCO_2^-$     and     $^+NH_3CHCONHCH_2CO_2^-$

$$\begin{matrix} & | & & & | \\ & (CH_2)_4 & & & (CH_2)_4 \\ & | & & & | \\ & NH_2 & & & NH_2 \end{matrix}$$

16.12   $^+NH_3CHCONHCHCO_2^-$     and     $^+NH_3CHCONHCHCO_2^-$

$$\begin{matrix} | & & | & & & | & & | \\ CH_2SH & & CH_2 & & & CH_2 & & CH_2SH \\ & & | & & & | & & \\ & & CH_2CO_2H & & & CH_2CO_2H & & \end{matrix}$$

16.13   Lys·Glu·Ala      Glu·Ala·Lys      Ala·Lys·Glu

        Lys·Ala·Glu      Glu·Lys·Ala      Ala·Glu·Lys

16.14   Cys·Gly·Ala      Gly·Cys·Ala      Ala·Gly·Cys

        Cys·Ala·Gly      Gly·Ala·Cys      Ala·Cys·Gly

16.15   $^+NH_3CHCONHCHCONHCHCO_2^-$

$$\begin{matrix} | & & | & & | \\ CHCH_3 & & CHCH_3 & & CH_2 \\ | & & | & & | \\ CH_3 & & CH_2 & & C_6H_5 \\ & & | & & \\ & & CH_3 & & \end{matrix}$$

16.16   $^+NH_3CHCONHCHCO_2CH_3$

$$\begin{matrix} | & & | \\ CH_2 & & CH_2 \\ | & & | \\ CO_2H & & C_6H_5 \end{matrix}$$

16.17   $^+NH_3CHCONHCHCONHCHCONHCH_2CONHCHCO_2^-$

$$\begin{matrix} | & & | & & | & & & | \\ CH_3 & & CHCH_3 & & CH_2 & & & CH_2 \\ & & | & & | & & & | \\ & & CH_3 & & C_6H_5 & & & CHCH_3 \\ & & & & & & & | \\ & & & & & & & CH_3 \end{matrix}$$

16.18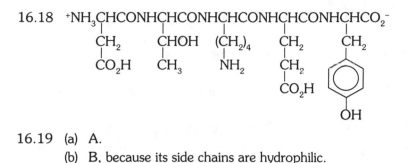

16.19  (a)  A.

(b)  B, because its side chains are hydrophilic.

16.20  D, because its side chains are all hydrophobic.

16.21  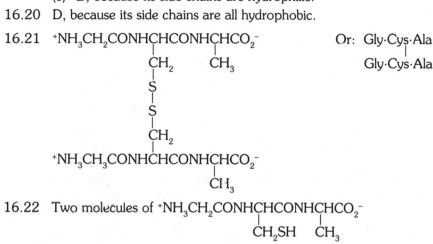  Or:  Gly·Cys·Ala

Gly·Cys·Ala

16.22  Two molecules of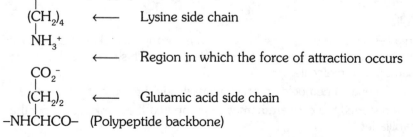

16.23  The hydrogen bond:  C=O···H–N

16.24  The hydrogen bond.

16.25  Three polypeptide strands are wrapped into a triple helix, and these helices wrap into a right-handed super helix.

16.26  It is needed to make collagen.

16.27  The water-avoiding and the water-seeking characteristics of hydrophobic and hydrophilic side chains.  Enzymes are not needed.

16.28  A force of attraction between a (+) and a (–) charge on different side chains as in:

$$-NHCHCO-$$  (Polypeptide backbone)

$(CH_2)_4$  ⟵  Lysine side chain

$NH_3^+$

⟵  Region in which the force of attraction occurs

$CO_2^-$

$(CH_2)_2$  ⟵  Glutamic acid side chain

$-NHCHCO-$  (Polypeptide backbone)

16.29  It involves the association of four polypeptide molecules (each bearing a heme unit).

16.30  (a)  Hemoglobin has four associated polypeptide units.
          Myoglobin has just one.
       (b)  Hemoglobin is in the bloodstream.  Myoglobin is in heart muscle.
       (c)  The same.
       (d)  Hemoglobin transports oxygen.  Myoglobin temporarily stores oxygen.

16.31  $^+NH_3CHCO_2^-$ + $^+NH_3CHCO_2^-$ + $^+NH_3CHCO_2^-$ + $^+NH_3CHCO_2^-$
             |                  |                  |                  |
            $CH_3$            $CH_2OH$            $CH_2$             $CH_2$
                                                   |                  |
                                                 $C_6H_5$             S
                                                                      |
          + $^+NH_3CH_2CO_2^-$                                        S
                                                                      |
                                                                    $CH_2$
                                                                      |
                                                              $^+NH_3CHCO_2^-$

16.32  (a), (c), and (d) could form, but (b) could not; (b) has an amide bond that is not joined to the *alpha* amino group of lysine but is joined to its side chain amino group, instead..

16.33  At this pH, the protein molecules are not like-charged, so they cannot repel each other.  Yet they are very polar, so they can attract each other and collect into a much larger particle that precipitates from the solution.

16.34  Digestion is the hydrolysis of a protein into its amino acid units.  Denaturation is the disorganization of the secondary, tertiary, or quaternary structure of a protein so that it loses its ability to function biologically.

16.35  Fibrous proteins are insoluble in water but globular proteins do dissolve.

16.36  Collagen changes to gelatin when boiled in water.

16.37  They both have strengthening functions in tissue; both are fibrous proteins.  The action of hot water on collagen turns it to gelatin, but elastin is unaffected in this way.

16.38  The globulins are less soluble in water than the albumins and they need the presence of dissolved salts to dissolve.

16.39  The protein that forms a blood clot.  Fibrinogen is changed to fibrin by the clotting mechanism.

16.40  A lipoprotein

16.41  In a β-subunit, a valine residue with an isopropyl side chain has replaced a glutamic acid residue with a $CO_2H$ side chain.  This changes the shape of the hemoglobin molecule.

16.42  Deoxygenated hemoglobin precipitates inside the red cell.

16.43  The distorted red cells are harder to pump and they can clump together to plug capillaries.

# Review Exercises, Chapter 17

17.1   (a)  A catalyst.

       (b)  It consists of a protein.

17.2        Each enzyme catalyzes a reaction for a specific substrate or a specific kind of reaction.

17.3   (a)  An apoenzyme is the wholly polypeptide part of the enzyme.

       (b)  A cofactor is a non-polypeptide molecule or ion needed to make the complete enzyme.

       (c)  A coenzyme is one kind of cofactor, an organic molecule

17.4        Nicotinamide.

17.5        Riboflavin

17.6        $NADH + H^+$

17.7        $H^+ + NADH + FMN \longrightarrow NAD^+ + FMNH_2$

17.8   (a)  Sucrose    (b)  Glucose    (c)  A protein    (d)  An ester

17.9   (a)  An oxidation (or a dehydrogenation).

       (b)  The transfer of a methyl group from one molecule to another.

       (c)  The hydrolysis of some bond.

       (d)  An electron-transfer event.

17.10  Lactose is a disaccharide and the substrate for the enzyme, lactase.

17.11  Hydrolysis is a kind of reaction catalyzed by the enzyme hydrolase.

17.12  Enzymes of identical function but slightly different in structure.

17.13  CK(MM) in skeletal muscle but also in heart muscle; CK(BB) in the brain; and CK(MB) in heart muscle.

17.14  Active site.

17.15  Specificity arises from the necessity of the fitting of the substrate molecule to the surface of the enzyme much as a key must fit to a particular lock.  Rate enhancement stems from the ability of the enzyme to bring the reactants very close together.

17.16  The substrate molecule includes a change in the shape of the enzyme that makes the fit between the two possible.

17.17  A molecule of a product of the work of the enzyme combines with the enzyme to inactivate it.

17.18  The product molecule is the stimulus that shuts off a mechanism (the enzyme) for some process until a reaction removes the product and lets the process run again.

17.19  When a substrate molecule manages to force its way onto one active site, a configurational change occurs in the enzyme to make all other active sites easily accessible.

17.20  (a)  By the loss of a small polypeptide fragment a zymogen is changed into the active enzyme form.
       (b)  Plasmin is made from plasminogen, and plasmin catalyzes the hydrolysis of the fibrin of a blood clot.

17.21  (a)  An effector is a nonsubstrate molecule that activates an enzyme.
       (b)  Calcium ion
       (c)  Nerve signals open calcium channels to let calcium ions into the cell, and active transport processes pump them back out.
       (d)  The calcium ions activating it are removed from the cell.

17.22  (a)  $CN^-$ inactivates an enzyme involved in cellular respiration.
       (b)  $Hg^{2+}$ reacts with HS– groups on cysteine side chains and precipitates the enzyme.
       (c)  These inactivate cholinesterase, an enzyme needed in the signal-sending activities in the nervous system.

17.23  Antimetabolites are compounds that interfere with the metabolism of disease-causing bacteria.  Antibiotics are those antimetabolites that are made by microorganisms.

17.24  Feedback inhibition.

17.25  The levels of these enzymes rise in blood as the result of a disease or injury to particular tissues, which causes tissue cells to release their enzymes.

17.26  The CK(MB) band originates in the leakage of this isoenzyme only from damaged heart muscle.

17.27  Of the five LD isoenzymes, $LD_1$ normally is less concentrated than $LD_2$.  The "flip" is the reversal of this relationship.  $LD_1$ shows up as *more* concentrated than $LD_2$.  This flip os observed in patients who have suffered a myocardial infarction.

17.28  If glucose is present, it is acted upon by glucose oxidase in the test strip.  This produces hydrogen peroxide, and the enzyme peroxidase (also in the strip) catalyzes a reaction between hydrogen peroxide and an aromatic compound in the test strip.  The product is a dye whose appearance and intensity of color signals that glucose is present and its approximate concentration.

17.29  Of fibrin, which exists as soluble fibrinogen first.

17.30  A tissue plasminogen activator, tPA, acts on plasminogen when the latter has become absorbed onto a developing clot.  One of three enzymes, streptokinase, APSAC, or (genetically engineered) tPA, can be administered to help start the dissolution of a clot.

17.31  By reducing the numbers of calcium ions that enter heart muscle cells to activate the muscle contraction process.

17.32  They are primary chemical messengers.

17.33   When activated by a primary chemical messenger, it catalyzes the formation of cyclic AMP (from ATP), which then activates an enzyme inside the target cell.

17.34   It is an enzyme activator.

17.35   The cyclic AMP becomes AMP.

17.36   (a)  Endocrine glands.

       (b)  Axon ends of presynaptic nerve cells.

17.37   It explains how molecules of hormones or neurotransmitters recognize their particular target cells.

17.38   (1)  Activating an enzyme, such as the work of epinephrine.

       (2)  Activating a gene, such as the work of sex hormones.

       (3)  Altering the permeability of a cell membrane toward a specific substance, as in the work of insulin.

17.39   It is hydrolyzed back to acetic acid and choline.  The enzyme is cholinesterase. Nerve poisons inactivate this enzyme.

17.40   It inactivates the receptor protein for acetylcholine.

17.41   It blocks the receptor protein for acetylcholine.

17.42   They catalyze the deactivation of neurotransmitters such as norepinephrine and thus reduce the level of signal-sending activity that depends on such neurotransmitters.

17.43   Iproniazid inhibits the monoamine oxidases and thus lets norepinephrine work at a high level of activity.

17.44   They inhibit the reabsorption of norepinephrine by the presynaptic neuron and thus reduce the rate of its deactivation by the monoamine oxidases.

17.45   Norepinephrine, acting as a hormone, serves as a backup to its acting as a neurotransmitter should some injury disrupt the latter action.

17.46   Dopamine.

17.47   They inhibit the action of dopamine.

17.48   They accelerate the release of dopamine from the presynaptic neuron.

17.49   Degenerated neurons can use L-DOPA to make dopamine.

17.50   GABA (gamma-aminobutyric acid), whose signal-inhibiting activity is enhanced by Valium® and Librium®.

17.51   Enkephalin molecules enter pain-signalling neurons and inhibit the release of substance P, a neurotransmitter that helps to send pain signals.  Thus enkephalin, like an opium-drug, inhibits pain.

17.52   By being used to make a slightly altered bacterial enzyme that then will not work.

17.53   By inhibiting the enzyme needed to complete the formation of the bacterium's cell wall.

17.54    Separation of proteins.

17.55    Polypeptides with different masses and different net charges are induced to migrate to electrodes under the influence of an electrical current. Because they migrate at different rates, they become separated and show up at different zones on the support plate.

## Review Exercises, Chapter 18

18.1    Interstitial fluid and blood.

18.2    Saliva, gastric juice, pancreatic juice, and intestinal juice.

18.3    (a)  α-Amylase.

   (b)  Pepsinogen and gastric lipase.

   (c)  α-Amylase, lipase, nuclease, trypsinogen, chymotrypsinogen, procarboxypeptidase, and proelastase

   (d)  None

   (e)  Amylase, aminopeptidase, sucrase, lactase, maltase, lipase, nucleases, enteropeptidase.

18.4    (a)  Pepsin from its zymogen in gastric juice; trypsin, chymotrypsin, and elastin from zymogens in pancreatic juice.

   (b)  Lipases provided in gastric juice, pancreatic juice, and intestinal juice.

   (c)  Amylases in saliva, pancreatic juice, and intestinal juice.

   (d)  Sucrase in intestinal juice.

   (e)  Carboxypeptidase from its zymogen in pancreatic juice; aminopeptidase from its zymogen in intestinal juice.

   (f)  Nucleases in pancreatic juice and intestinal juice.

18.5    (a)  Amino acids.

   (b)  Glucose, fructose, and galactose.

   (c)  Glycerol and fatty acids.

18.6    (a)  Peptide (amide) bonds in proteins.

   (b)  Acetal systems in carbohydrates.

   (c)  Ester groups in triacylglycerols.

18.7    It catalyzes the conversion of trypsinogen to trypsin. Then trypsin catalyzes the conversion of other zymogens to chymotrypsin and carboxypeptidase. Thus enteropeptidase turns on enzyme activity for three major protein-digesting enzymes.

18.8    They would catalyze the digestion of proteins that make up part of the pancreas to the serious harm of this organ.

18.9    They are surface active agents that help to break up lipid globules, wash lipids from the particles of food, and aid in the absorption of fat-soluble vitamins.

18.10  (a)  Lubricates the food.

(b)  Protects the stomach lining from gastric acid and pepsin.

18.11  (a)  HCl          (b)   Enteropeptidase

(c)  Trypsin     (d)   Trypsin

18.12  It helps to coagulate the protein in milk so that this protein stays longer in the stomach where it can be digested with the aid of pepsin.

18.13  This enzyme is inactive at the high acidity of the digesting mixture in the adult stomach, but the acidity of this mixture in the infant's stomach is less.

18.14  Pancreatic juice delivers its zymogens and enzymes into the duodenum, whereas the enzymes of the intestinal juice work within cells of the intestinal wall.

18.15  Dilute sodium bicarbonate released from the pancreas. This raises the pH of the chyme to the optimum pH for the action of the enzymes that will function in the duodenum.

18.16  They recombine to molecules of triacylglycerol during their migration from the intestinal tract toward the lymph ducts.

18.17  The flow of bile normally delivers colored breakdown products from hemoglobin in the blood that give the normal color to feces. When no bile flows, no colored products are available to the feces.

18.18  These lipids are less in need of the surfactant activity of bile salts as an aid to their digestion.

18.19  The concentration of soluble proteins is greater in blood.

18.20  The serum-soluble proteins (albumins, mostly).

18.21  Fibrinogen is a protein in blood that is changed to fibrin, the insoluble protein of a blood clot, by the clotting mechanism.

18.22  Albumin molecules transport hydrophobic molecules such as fatty acids and cholesterol, and albumins contribute as much as 75-80% of the osmotic effect of the blood.

18.23  It protects the body against infections.

18.24  $Na^+$ is in blood plasma and other extracellular fluids; $K^+$ is chiefly in intracellular fluids.

18.25  Blood pressure that tends to force blood fluids out and osmotic pressure that tends to force fluids back. The return of fluids to the blood from the interstitial compartment is overbalanced by the blood pressure so that the net effect is a diffusion of fluids from the blood.

18.26  Blood pressure and osmotic pressure. The natural return of fluids to the blood from the interstitial compartment is not balanced by the now reduced blood pressure, so fluids return to the blood from which they left on the arterial side.

18.27  Serum proteins are lost from the blood, which upsets the osmotic pressure of the blood. Water leaves the blood for the interstitial compartment, and the blood volume drops.  Loss of blood delivery to the brain leads to the symptoms of shock.

18.28  (a)  Blood proteins leak out which allows water to leave the blood and enter interstitial spaces throughout various tissues.

(b)  Blood proteins are lost to the blood by being consumed which also leads to the loss of water from the blood and its appearance in interstitial compartments.

(c)  Capillaries are blocked at the injured site reducing the return of blood in the veins, so fluids accumulate at the site.

18.29  Oxygen and carbon dioxide.

18.30  Hemoglobin.

18.31  The first oxygen molecule to bind changes the shapes of other parts of the hemoglobin molecule and makes it much easier for the remaining three oxygen molecules to bind.  This ensures that all four oxygen-binding sites of each hemoglobin molecule will leave the lungs fully loaded with oxygen.

18.32  $HHb + O_2 \rightleftharpoons HbO_2^- + H^+$

(a)  To the left.          (b)  To the left.

(c)  To the right.         (d)  To the left.

(e)  To the left.          (f)  To the right.

18.33  (a)  $HHB + O_2 \longleftarrow HbO_2^- + H^+$

$H_2CO_3 \longrightarrow HCO_3^- + H^+$

Isohydric shift in active tissue

(b)  $HHb + O_2 \longrightarrow HbO_2^- + H^+$

$H_2CO_3 \longleftarrow HCO_3^- + H^+$

Isohydric shift in an alveolus

18.34  It generates $H^+$ needed to convert $HCO_3^-$ to $CO_2$ and $H_2O$ and to convert $HbCO_2^-$ to HHb and $CO_2$.

18.35  It combines with water to give $HCO_3^-$ and the $H^+$ needed to react with $HbO_2^-$ to form HHb and release $O_2$.

18.36  It helps to shift the following equilibrium to the left:

$HHb + O_2 \rightleftharpoons HbO_2^- + H^+$

18.37  It is found in red cells.

(a)  It catalyzes the conversion of $HCO_3^-$ and $H^+$ to $CO_2$ and $H_2O$.

(b)  It catalyzes the conversion of $CO_2$ and $H_2O$ to $HCO_3^-$ and $H^+$.  It can do both because it accelerates *both* the forward and the reverse reactions in the equilibrium:

$CO_2 + H_2O \rightleftharpoons H^+ + HCO_3^-$

Other factors, such as the value of the partial pressure of carbon dioxide, determine whether the forward or the reverse reaction is favored.

18.38 As $HCO_3^-$ in the serum and as $HbCO_2^-$ (carbaminohemoglobin) in red cells.

18.39 The exchange of a chloride ion for a bicarbonate ion between a red blood cell and blood serum. This brings $Cl^-$ inside the red cell when it is needed to help deoxygenate $HbO_2^-$.

18.40 Myoglobin can take oxygen from oxyhemoglobin and thus ensure that the oxygen needs of myoglobin-containing tissue are met.

18.41 These animals can store more oxygen in heart muscle, which helps them to go longer without breathing.

18.42 Fetal hemoglobin can take oxygen from the oxyhemoglobin of the mother's blood and thus ensure that the fetus gets needed oxygen.

18.43

| Condition | pH | $pCO_2$ | $[HCO_3^-]$ |
|---|---|---|---|
| Normal | 7.35-7.45 | 35-45 mm Hg | 19-24 meq/L |
| Metabolic acidosis | ↓7.20 | ↓30 | ↓14 |
| Metabolic alkalosis | ↑7.45 | ↑>45 | ↑>29 |
| Respiratory acidosis | ↓7.10 | ↑68 | ↑40 |
| Respiratory alkalosis | ↑7.54 | ↓32 | ↓20 |

18.44 The pH of the blood decreases in both but both $pCO_2$ and $[HCO_3^-]$ increase in respiratory acidosis and both decrease in metabolic acidosis.

18.45 Hyperventilation is observed in metabolic acidosis, and $HCO_3^-$ can be given intravenously to neutralize excess acid. Hyperventilation is also observed in respiratory alkalosis (because the patient can't help hyperventilating), and $CO_2$ is given (by rebreathing one's own air) to keep up the supply of $H_2CO_3$, which can neutralize excess base.

18.46 Hypoventilation is observed in metabolic alkalosis, and isotonic ammonium chloride can be given to neutralize the excess base. Involuntary hypoventilation is observed in respiratory acidosis, and isotonic sodium bicarbonate might be given to neutralize excess acid.

18.47 In metabolic acidosis, because it helps to blow out $CO_2$ and thereby to reduce the level of $H_2CO_3$ in the blood and simultaneously raise the pH.

18.48 In respiratory alkalosis. The involuntary loss of $CO_2$ reduces the level of $H_2CO_3$ in the blood and so reduces the level of $H^+$.

18.49 In metabolic alkalosis.

18.50 In respiratory acidosis.

18.51 The kidneys work to remove acids from the blood, but to remove them they must also remove water from the blood. If too much water is taken in this way, then the blood obtains more water by taking it from interstitial and intracellular compartments.

18.52 (a) Respiratory alkalosis     (b) Metabolic alkalosis
      (c) Respiratory acidosis     (d) Respiratory acidosis
      (e) Metabolic acidosis     (f) Respiratory alkalosis
      (g) Metabolic acidosis     (h) Metabolic alkalosis
      (i) Respiratory acidosis     (j) Respiratory acidosis

18.53 (a) Hyperventilation     (b) Hypoventilation
      (c) Hypoventilation     (d) Hypoventilation
      (e) Hyperventilation     (f) Hyperventilation
      (g) Hyperventilation     (h) Hypoventilation
      (i) Hypoventilation     (j) Hypoventilation

18.54 $CO_2$ is removed at an excessive rate, which removes carbonic acid, so the blood becomes more alkaline and the pH of the blood rises (alkalosis).

18.55 Hypoventilation in emphysema lets the blood retain carbonic acid, and the pH decreases.

18.56 The loss of acid with the loss of the stomach contents results in a loss of acid from the blood, which means a rise in the blood's pH-(alkalosis).

18.57 The loss of alkaline fluids from the duodenum and lower intestinal tract leads to a loss of base from the bloodstream, too. The result is a decrease in the blood's pH-(acidosis).

18.58 The blood has become more concentrated in solutes.

18.59 It acts to prevent the loss of water via the urine by letting the hypophysis secrete vasopressin whose target cells are in the kidneys. This helps to keep the blood's osmotic pressure from rising further. The thirst mechanism is also activated, which leads to bringing in more water to dilute the blood.

18.60 Aldosterone is secreted from the adrenal cortex, and it instructs the kidneys to retain sodium ion in the blood.

18.61 The rate of diuresis increases.

18.62 They transfer hydrogen ions into the urine and put bicarbonate ions into the bloodstream.

## Review Exercises, Chapter 19

19.1

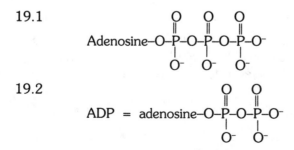

19.2

$$ADP = adenosine-O-\overset{\overset{O}{\|}}{P}-O-\overset{\overset{O}{\|}}{P}-O^-$$

$$AMP = \text{adenosine } -O-\overset{\overset{\displaystyle O}{\|}}{\underset{\underset{\displaystyle O^-}{\|}}{P}}-O^-$$

19.3   The singly and doubly ionized forms of phosphoric acid, $H_2PO_4^- + HPO_4^{2-}$.

19.4   To resynthesize high–energy phosphates.

19.5   (a)  The aerobic synthesis of ATP.

      (b)  The synthesis of ATP when a tissue operates anaerobically.

      (c)  The supply of metabolites for the respiratory chain.

      (d)  The supply of metabolites for the respiratory chain and for the citric acid cycle.

19.6   The disappearance of ATP by some energy-demanding process and the simultaneous appearance of ADP + $P_i$.

19.7   The citric acid cycle.

19.8   (a)  4 2 5 1 3

      (b)  2 4 3 1

19.9   It starts with glycolysis and since this is aerobic as stated, it ends with the respiratory chain.

19.10  A pair of electrons on the left of the arrow.

19.11  Respiratory chain.

19.12
$$\underset{\overset{\displaystyle |}{^-O_2CCH_2}}{\overset{\displaystyle CH_2CO_2^-}{}} + \ FAD \longrightarrow \underset{^-O_2CCH}{\overset{CHCO_2^-}{\|}} + \ FADH_2$$

      (a)  $^-O_2CCH_2CH_2CO_2^-$ is oxidized.

      (b)  FAD is reduced.

19.13
$$\underset{\underset{CH_3CCO_2^-}{\overset{O}{\|}}}{\overset{\overset{OH}{|}}{CH_3CHCO_2^-}} \underset{\nwarrow NADH + H^+}{\overset{\searrow NAD^+}{\bowtie}}$$

19.14  NAD$^+$ FMN FeS–P  Cyt $a$

19.15  It catalyzes the reduction of oxygen to water.

19.16  It is a riboflavin-containing coenzyme that in its reduced form, $FADH_2$, passes electrons and $H^+$ into the respiratory chain.

19.17  Across the inner membrane of the mitochondrion.  The value of $[H^+]$ is higher on the outer side of this inner membrane.

19.18  The flow of protons across the inner mitochondrial membrane.

**19.19** If the membrane is broken, then the simple process of diffusion defeats any mitochondrial effort to set up a gradient of $H^+$ ions across the membrane, but the chain itself can still operate.

**19.20** $MH_2 + nH^+$ from inside $+ 1/2\ O_2 \longrightarrow M + nH^+$ now on the $+ H_2O$
the inner                                                    outside of the
membrane                                                  inner membrane

The value of $n$ ranges from about 9 to 12.

**19.21** The protons that flow back through the membrane cause a configurational change in ATP synthase that causes it to release an enzyme-bound molecule of ATP and so activate the enzyme to make another.

**19.22** When the respiratory chain starts up, the citric acid cycle must start up to keep the chain going.

**19.23** The acetyl unit, $CH_3\overset{\overset{\displaystyle O}{\|}}{C}-$, of acetyl coenzyme A.

**19.24** Two

**19.25** Two

**19.26** (a) $CH_3-\overset{\overset{\displaystyle O}{\|}}{C}-H$, acetaldehyde

(b) $CH_3-\overset{\overset{\displaystyle O}{\|}}{C}-OH$, acetic acid

**19.27** Pyruvate ion.

**19.28** ADP, because when ADP appears in the cell it is time to make ATP, an outcome of the action of this enzyme.

**19.29** $C_6H_{12}O_6 + 2ADP + 2P_i \longrightarrow 2C_3H_5O_3^- + 2H^+ + 2ATP$

**19.30** Glycolysis can operate and make ATP even when the oxygen supply is low, so a tissue in oxygen debt can continue to function.

**19.31** (a) It undergoes oxidative decarboxylation and becomes the acetyl group in acetyl CoA.

(b) Its keto group is reduced to a 2° alcohol group in lactate, which enables NADH to be reoxidized to $NAD^+$ and then reused for more glycolysis.

**19.32** It is reoxidized to pyruvate, which then undergoes oxidative decarboxylation to the acetyl group in acetyl CoA. This enters the citric acid cycle.

**19.33** NADPH forms, and the body uses it as a reducing agent to make fatty acids.

**19.34** They are joined to coenzyme A as fatty acyl CoA.

19.35  (a)

$$CH_3CH_2CH_2CH_2CH_2\overset{\overset{\displaystyle O}{\|}}{C}\text{-SCoA} + FAD \longrightarrow$$

$$CH_3CH_2CH_2CH=CH\overset{\overset{\displaystyle O}{\|}}{C}\text{-SCoA} + FADH_2$$

(b)

$$CH_3CH_2CH_2CH=CH\overset{\overset{\displaystyle O}{\|}}{C}\text{-SCoA} + H_2O \longrightarrow CH_3CH_2CH_2\overset{\overset{\displaystyle OH}{|}}{C}HCH_2\overset{\overset{\displaystyle O}{\|}}{C}\text{-SCoA}$$

(c)

$$CH_3CH_2CH_2\overset{\overset{\displaystyle OH}{|}}{C}HCH_2\overset{\overset{\displaystyle O}{\|}}{C}\text{-SCoA} + NAD^+ \longrightarrow$$

$$CH_3CH_2CH_2\overset{\overset{\displaystyle O}{\|}}{C}CH_2\overset{\overset{\displaystyle O}{\|}}{C}\text{-SCoA} + NADH + H^+$$

(d)

$$CH_3CH_2CH_2\overset{\overset{\displaystyle O}{\|}}{C}CH_2\overset{\overset{\displaystyle O}{\|}}{C}\text{-SCoA} + CoASH \longrightarrow$$

$$CH_3CH_2CH_2\overset{\overset{\displaystyle O}{\|}}{C}\text{-SCoA} + CH_3\overset{\overset{\displaystyle O}{\|}}{C}\text{-SCoA}$$

19.36

$$CH_3CH_2CH_2\overset{\overset{\displaystyle O}{\|}}{C}\text{-SCoA} + FAD \longrightarrow CH_3CH=CH\overset{\overset{\displaystyle O}{\|}}{C}\text{-SCoA} + FADH_2$$

$$CH_3CH=CH\overset{\overset{\displaystyle O}{\|}}{C}\text{-SCoA} + H_2O \longrightarrow CH_3\overset{\overset{\displaystyle OH}{|}}{C}HCH_2\overset{\overset{\displaystyle O}{\|}}{C}\text{-SCoA}$$

$$CH_3\overset{\overset{\displaystyle OH}{|}}{C}HCH_2\overset{\overset{\displaystyle O}{\|}}{C}\text{-SCoA} + NAD^+ \longrightarrow CH_3\overset{\overset{\displaystyle O}{\|}}{C}CH_2\overset{\overset{\displaystyle O}{\|}}{C}\text{-SCoA} + NADH + H^+$$

$$CH_3\overset{\overset{\displaystyle O}{\|}}{C}CH_2\overset{\overset{\displaystyle O}{\|}}{C}\text{-SCoA} + CoASH \longrightarrow 2\ CH_3\overset{\overset{\displaystyle O}{\|}}{C}\text{-SCoA}$$

No more turns of the fatty acid cycle are possible.

19.37  FADH$_2$ passes its hydrogen into the respiratory chain and is changed back to FAD.

19.38  NADH passes its hydrogen into the respiratory chain and is changed back to NAD$^+$.

19.39  Steps 1 – 3 succeed in oxidizing the beta position of the fatty acyl unit to a ketone group.

19.40  (a)  7

(b)  6

(c)  6

(d)

| Intermediate | Maximum No. of ATP from each | Total number of ATP possible from each as acetyl CoA forms |
|---|---|---|
| 6 FADH$_2$ | 2 | 12 |
| 6 NADH | 3 | 18 |
| $\overset{O}{\overset{\|}{7 \; CH_3CSCoA}}$ | 12 | 84 |

Sum = 114

Deduct 2                                    −2

Net ATP produced                   112

19.41  The acetyl units in the acetyl CoA produced by the fatty acid cycle are accepted by the citric acid cycle and used to make raw materials for the respiratory chain and the synthesis of ATP.

19.42  The entire collection of nitrogen compounds found anywhere in the body.

19.43  To synthesize protein.

To synthesize nonprotein compounds of nitrogen.

To synthesize nonessential amino acids.

To contribute to the synthesis of ATP.

19.44  Infancy.

19.45  They are catabolized.  Some are converted to fatty acids.

19.46  To make glucose via gluconeogenesis, and to generate ATP.

19.47      $\overset{O}{\overset{\|}{C_6H_5CH_2CCO_2H}}$

19.48  $\overset{O}{\overset{\|}{(CH_3)_2CHCCO_2H}}$

19.49

$$\underset{CH_3CHCO_2^-}{\overset{NH_3^+}{|}} + \; ^-O_2CCH_2CH_2\overset{O}{\overset{\|}{C}}CO_2^- \longrightarrow CH_3\overset{O}{\overset{\|}{C}}CO_2^- \; +$$

$$^-O_2CCH_2CH_2\underset{CHCO_2^-}{\overset{NH_3^+}{|}}$$

Then:

$$^-O_2CCH_2CH_2\underset{CHCO_2^-}{\overset{NH_3^+}{|}} + NAD^+ + H_2O \longrightarrow \; ^-O_2CCH_2CH_2\overset{O}{\overset{\|}{C}}CO_2^-$$

$$+ \; NADH + H^+ + NH_4^+$$

19.50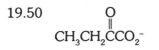

19.51  Fatty acids.

19.52  Acetyl CoA.

19.53  Acetoacetic acid, $CH_3CCH_2CO_2H$ (with C=O above central carbon)

19.54  Acetoacetate:  $CH_3CCH_2CO_2^-$ (with C=O)

β-Hydroxybutyrate:  $CH_3CHCH_2CO_2^-$ (with OH on the CH)

Acetone:  $CH_3CCH_3$ (with C=O)

19.55  An above normal concentration of the ketone bodies in the blood.

19.56  An above normal concentration of the ketone bodies in the urine.

19.57  Enough acetone vapor in exhaled air to be detected by its odor.

19.58  Ketonemia + ketonuria + acetone breath.

19.59  Metabolic acidosis brought on by a rise in the level of the ketone bodies in the blood.

19.60  Acetoacetic acid.

19.61  Diuresis is accelerated to remove ketone bodies from the blood, and their removal requires the simultaneous removal of water, so the urine volume rises.

19.62  Their *over*-production leads to acidosis.

19.63  Amino acids are catabolized at a faster than normal rate to participate in gluconeogenesis, and their nitrogen is excreted largely as urea.

19.64  Each $Na^+$ ion that leaves corresponds to the loss of one $HCO_3^-$ ion, the true base, because $HCO_3^-$ neutralizes acid generated as the ketone bodies are made.  And for every negative ion that leaves with the urine a positive ion, mostly $Na^+$, has to leave to ensure electrical neutrality.

19.65  3 2 1 4 5 7 6

19.66  It imparts a yellowish coloration to the skin.

19.67  (a)  A too rapid breakup of red cells.

(b)  Liver infection or cirrhosis.

19.68  This inhibits the export of birubin to the intestinal tract (in bile), so bilirubin backs up into the blood stream.

## Review Exercises, Chapter 20

20.1    Glucose, fructose, and galactose.

20.2    After a few steps, the metabolic pathways of galactose and fructose merge with the pathway of glucose.

20.3    The concentration of reducing monosaccharides, chiefly glucose, in the blood is called the blood sugar level. The normal fasting level is the blood sugar level after several hours of fasting.

20.4    65–95 mg/dL. (Note. Various references give slightly different ranges of values, e.g., 70–110 mg/dL.)

20.5    (a)  Glucose in urine.

(b)  A low blood sugar level.

(c)  A high blood sugar level.

(d)  The conversion of glycogen to glucose.

(e)  The synthesis of glucose from smaller molecules.

(f)  The synthesis of glycogen from glucose.

20.6    The lack of glucose means the lack of the one nutrient most needed by the brain.

20.7    It rises.

20.8    Muscle tissue.

20.9    Glucose 1-phosphate is the end product and phosphoglucomutase catalyzes its change to glucose 6-phosphate.

20.10   Liver, but not muscles, has the enzyme glucose 6-phosphatase that catalyzes the hydrolysis of glucose 6-phosphate. This frees glucose for release from the liver to the bloodstream.

20.11   It is a polypeptide hormone made in the alpha cells of the pancreas and released into circulation when the blood sugar level drops. At the liver it activates adenylate cyclase, which leads to glycogenolysis and the release of glucose into circulation.

20.12   Glucagon, because it works better at the liver than epinephrine in initiating glycogenolysis, and when glycogenolysis occurs at the liver there is a mechanism for releasing glucose into circulation.

20.13   It stimulates the release of glucagon, which leads to the release of glucose into circulation.

20.14   It is a polypeptide hormone released from the beta cells of the pancreas in response to a rise in the blood sugar level, and it acts most effectively at adipose tissue.

20.15   A rise in the blood sugar level.

20.16   Too much insulin leads to a sharp drop in the blood sugar level and therefore a drop in the supply of the chief nutrient for the brain.

20.17  Somatostatin is a polypeptide hormone released by the hypothalamus, and it acts at the pancreas to inhibit the release of glucagon and slow down the release of insulin.

20.18  The body's ability to manage dietary glucose without letting the blood sugar level swing too widely from its normal fasting level.

20.19  To test for the possibility of diabetes mellitus. An adult patient is given a drink that has 75 g of glucose. For children, 1.75 g of glucose per kilogram of body weight is given. Then the blood sugar level is measured at regular intervals.

20.20  (a)  The blood sugar level initially rises rapidly, but then drops sharply and slowly levels back to normal.

(b)  The blood sugar level, already high to start, rises much higher and never sharply drops. It only very slowly comes back down.

20.21  An over-release of epinephrine (as in a stressful situation) that induces an over-release of glucose.

20.22  It makes glucose out of smaller molecules obtained by the catabolism of fatty acids and amino acids.

20.23  All are catabolized, and parts of some of their molecules are used to make fatty acids and, thence, fat.

20.24  (a)  Alanine      (b)  Aspartic acid

20.25  Fatty acids and glycerol.

20.26  They become reconstituted into triacylglycerols.

20.27  (a)  Inside mitochondria.

(b)  Cytosol.

20.28  The pentose pathway of glucose catabolism.

20.29  Cholesterol inhibits the synthesis of an enzyme needed to make additional cholesterol.

20.30  The body cannot make them, so it is essential that they be in the diet.

20.31  The body can make alanine from the pieces of other amino acids and other molecules.

20.32
$$^-O_2CCH_2CH_2\overset{O}{\overset{\|}{C}}CO_2^- + NH_4^+ + NADPH + H^+ \longrightarrow$$
$$^-O_2CCH_2CH_2\overset{NH_3^+}{\overset{|}{C}}HCO_2^- + NADP^+ + H_2O$$

20.33
$$C_6H_5CH_2\overset{O}{\overset{\|}{C}}CO_2H$$

20.34
$$CH_3\overset{CH_3}{\overset{|}{C}}H-\overset{O}{\overset{\|}{C}}CO_2H$$

20.35  (a)  Glucose 6-phosphatase.  Glucose cannot leave the liver.  Liver enlarges as glycogen reserves become high.  Blood sugar level falls.  Blood levels of pyruvate and lactate increase.

  (b)  An enzyme for breaking 1,6-glycosidic bonds.  Not as much glucogen can be used.  Symptoms like those of Von Gierke's disease, but milder.

  (c)  Phosphorylase.  Cannot obtain glucose 1-phosphate from glycogen.  Reduced physical activity follows.

  (d)  Enzymes for making the branches in glycogen.  Liver failure.

20.36  Type I diabetics cannot make insulin.

20.37  Type I

20.38  Type II

20.39  1.  Presence of genetic defects.

  2.  Some triggering incident occurs, like a viral infection.

  3.  Particular antibodies associated with diabetes appear in the blood.

  4.  Gradual loss of ability to secrete insulin.

  5.  Full-fledged diabetes.  Hyperglycemia.

  6.  Destruction of β-cells is complete.

20.40  The β-cells of the pancreas.

20.41  Changes occur in body molecules that the immune system reads as new antigens.  The antibodies then made destroy body tissue.

20.42  The basement membrane of capillaries.

20.43  Its aldehyde group reacts with amino groups on proteins and genes to form C=N systems by means of which the glucose units are tied to other molecules.

20.44  It decreases.  The equilibrium involving glucose, an amino group, and glycosylated hemoglobin shifts back to regenerate glucose.

20.45  The change to more permanent Amadori compounds.  They are more permanent changes in cell molecules.

20.46  The sorbitol made by the hydrogenation of glucose stays inside the eye lens where it draws water osmotically.  This produces a swelling of the lens, pressure, glaucoma, and eventually blindness.

20.47  They transport lipids received from the digestive tract to the liver.

20.48  They unload some of their triacylglycerol.

20.49  They are absorbed.

20.50  Some cholesterol has originated in the diet and some has been synthesized in the liver.

20.51  (a)  Very low density lipoprotein complex.

  (b)  Intermediate density lipoprotein complex.

  (c)  Low density lipoprotein complex.

  (d)  High density lipoprotein complex.

20.52   Triacylglycerol.
20.53   Triacylglycerol.
20.54   The liver.
20.55   Cholesterol.
20.56   The synthesis of steroids and the fabrication of cell membranes.
20.57   IDL and LDL.
20.58   When the receptor proteins are reduced in number, the liver cannot remove cholesterol from the blood, so the blood cholesterol level increases.
20.59   Return to the liver any cholesterol that extrahepatic tissue cannot use.
20.60   A low level of HDL means a low ability to carry cholesterol from extrahepatic tissue back to the liver for export via the bile.

## Practice Exercises, Chapter 21

1.      (a)  Proline                    (b)  Arginine
        (c)  Glutamic acid             (d)  Lysine
2.      (a)  Serine                     (b)  CT (chain termination)
        (c)  Glutamic acid             (d)  Isoleucine

## Review Exercises, Chapter 21

21.1    Nucleic acid and histones.
21.2    Deoxyribonucleic acid (DNA).
21.3    Chromosomes are discrete, microscopically visible bodies made of chromatin, and they become visible because of the thickening of chromatin during its replication.
21.4    Replication.
21.5    The cell nucleus.
21.6    Nucleic acids.
21.7    Nucleotides.
21.8    Ribose and deoxyribose.
21.9    (a)  Adenine, A. Thymine, T. Guanine, G. Cytosine, C.
        (b)  Adenine, A. Uracil, U. Guanine, G. Cytosine, C.
21.10   The main chains all have the same phosphate-ribose-phosphate-ribose repeating system.
21.11   In the sequence of bases attached to the deoxyribose units of the main chain.
21.12   The main chains all have the same phosphate-deoxyribose-phosphate-deoxyribose repeating system.

21.13  DNA occurs as a double helix, and it alone has the base thymine (T). RNA alone has the base uracil (U). (The remaining three bases, A, G, and C, are the same in both DNA and RNA.) DNA molecules have one less –OH group per pentose unit then RNA molecules.

21.14  A and T pair to each other, so they must be in a 1:1 ratio regardless of the species. Similarly, G and C pair to each other and must be in a 1:1 ratio.

21.15  Hydrogen bond.

21.16  Bases that project from the twin spirals of DNA chains fit to each other on opposite strands by hydrogen bonds. The geometries and functional groups are such that only A and T can pair (or only A and U can pair when RNA is involved), and only G and C can pair.

21.17  Original strand (given):    AGTCGGA  5' $\longrightarrow$ 3'

$$\begin{matrix} \cdot & \cdot & \cdot & \cdot & \cdot & \cdot & \cdot \\ \cdot & \cdot & \cdot & \cdot & \cdot & \cdot & \cdot \\ \cdot & \cdot & \cdot & \cdot & \cdot & \cdot & \cdot \\ \cdot & \cdot & \cdot & \cdot & \cdot & \cdot & \cdot \end{matrix}$$

Opposite strand:    TCAGCCT  3' $\longleftarrow$ 5'

21.18  The synthesis of two new DNA double helices under the direction of and identical with an original DNA double helix.

21.19  The base pairings of A with T and G with C.

21.20  In cells of higher animals, a series of segments of a DNA molecule comprise one gene, each segment called an *exon* and each separated by DNA segments called *introns*.

21.21  The introns are b, d, and f, because they are the longer segments.

21.22  A sequence of base triplets that comprise a gene corresponds to a specific sequence of amino acid residues of a polypeptide.

21.23  It consists of several proteins plus *r*RNA, and it serves as the assembly area for the synthesis of polypeptides.

21.24  It is heterogeneous nuclear RNA (sometime called primary transcript RNA). Its sequence of bases is complementary to a sequence of bases on DNA — those of both the exons and introns. It is processed to make *m*RNA whose base sequence is complementary only to the exons of the DNA.

21.25  A codon is a specific triplet of bases that corresponds to a specific amino acid residue in a polypeptide, and *m*RNA consists of a continuous sequence of codons.

21.26  (a)  5'                                    3'
           UUUCUUAUAGAGUCCCCAACAGAU

        (b)  5'                    3'
           UUUUCCACAGAU

21.27  A triplet of bases found on *t*RNA and which is complementary to a codon found on *m*RNA.

21.28   ATA cannot be a codon because T does not occur in any type of RNA.

21.29   (a)  Phenylalanine          (b)  Serine

       (c)  Threonine              (d)  Aspartic acid

21.30   Writing them in the 5' to 3' direction:

       (a) AAA        (b) GGA        (c) UGU        (d) AUC

21.31   Translation is the *mRNA*-directed synthesis of a polypeptide.  Transcription is the DNA-directed synthesis of *mRNA*.

21.32   (a)  A large number of sequences are possible because three of the specific amino acid residues are coded by more than one codon.  The possibilities are indicated by:

Met · Ala · Try · Ser · Tyr

| AUG | GCU | UGG | UCU | UAU | (5' ⟶ 3') |
|-----|-----|-----|-----|-----|-----------|
|     | GCC |     | UCC | UAC |           |
|     | GCA |     | UCA |     |           |
|     | GCG |     | UCG |     |           |

       (b)  CAU (5' ⟶ 3')

21.33   Some block the expression of a gene by interfering with DNA-directed polypeptide synthesis.

21.34   The four-letter DNA language and the twenty-letter amino acid language.

21.35   The same codons are used for the same amino acids in virtually all species, plants and animals.

21.36   All viruses have nucleic acid, and many also contain a protein.

21.37   The protein part of a virus catalyzes the digestion of part of the cell's membrane.  This opens a hole for the virus particle or its nucleic acid to enter the cell.

21.38   They synthesize new virus particles (and some become silent genes.)

21.39   That retroviruses are able to use RNA information to make DNA, the reverse of the normal direction of information flow.

21.40   Human immunodeficiency virus.

21.41   The T4 lymphocyte.  This host is part of one's immune systems, so damage to it makes an individual vulnerable to infectious diseases.

21.42   Circular molecule of super-coiled DNA found in bacteria.

21.43   One source is the plasmid and another is new material added to the bacterium or yeast.

21.44   Polypeptides of critical value to human medicine or technology.

21.45   The technology whereby genes are cloned and they or their products are used, for example, in medicine.

21.46   A defect in a gene.

21.47   An alteration in DNA or the absence of a gene.

21.48   A gene needed for the metabolism of phenylalanine is defective. This leads to an increase in the level of phenylpyruvic acid in the blood, which causes brain damage. A diet very low in phenylalanine is prescribed.

21.49   Every cell from every tissue has the full complement of DNA and genes.

21.50   Minisatellites (or core segments). They are allowed to bind to specially-made radioactively labeled DNA. The combinations affect a photographic film to produce a pattern of paralled dark zones or lines unique for each person.

## Practice Exercises, Chapter 22

1.      $^{131}_{53}I \longrightarrow \ ^{131}_{54}Xe + \ ^{0}_{-1}e + \ ^{0}_{0}\gamma$

2.      $^{239}_{92}Pu \longrightarrow \ ^{235}_{92}U + \ ^{4}_{2}He + \ ^{0}_{0}\gamma$

3.      10,000 units

4.      8.5 m

## Review Exercises, Chapter 22

22.1    It emits radiations such as alpha radiations, beta radiation, or gamma radiation (and sometimes two of these).

22.2    Transmutation

22.3    Gamma radiation

22.4    (a) $^{4}_{2}He$        (b) $^{0}_{-1}e$        (c) $^{0}_{0}\gamma$

22.5    It is the most massive of the particles and it carries the largest charge. Therefore, it collides very quickly with a molecule in the air or other matter that it enters.

22.6    The mass number is reduced by 4 units and the atomic number is reduced by 2 units.

22.7    There is no change in mass number, because the mass number of the beta particle is 0. The atomic number increases by one because the loss of $^{0}_{-1}e$ from a neutron creates an additional proton.

22.8    No change in either occurs.

22.9    No. No transmutation occurs if only gamma radiation is emitted.

22.10   A neutron changes into a proton as an electron is ejected.

22.11   (a) $^{241}_{94}Pu$        (b) $^{22}_{10}Ne$

22.12   (a) $^{216}_{84}Po$        (b) $^{140}_{57}La$

22.13   (a) $^{252}_{99}Es \longrightarrow \ ^{4}_{2}He + \ ^{248}_{97}Bk$

        (b) $^{28}_{12}Mg \longrightarrow \ ^{0}_{-1}e + \ ^{28}_{13}Al$

        (c) $^{20}_{8}O \longrightarrow \ ^{0}_{-1}e + \ ^{20}_{9}F$

(d) $^{251}_{98}\text{Cf} \longrightarrow ^4_2\text{He} + ^{247}_{96}\text{Cm} + ^0_0\gamma$

22.14  (a) $^{211}_{83}\text{Bi} \longrightarrow ^0_{-1}e + ^{211}_{84}\text{Po}$

(b) $^{242}_{94}\text{Pu} \longrightarrow ^4_2\text{He} + ^{238}_{92}\text{U} + ^0_0\gamma$

(c) $^{30}_{13}\text{Al} \longrightarrow ^0_{-1}e + ^{30}_{14}\text{Si}$

(d) $^{243}_{96}\text{Cm} \longrightarrow ^4_2\text{He} + ^{239}_{94}\text{Pu}$

22.15  Lead-214 forms by successive decays of U-238. The quantity of lead-214 in the sample diminishes by half each 19.7 minutes.

22.16  The beta and gamma emitter, because its radiations are more penetrating.

22.17  0.750 ng

22.18  2.20 x 10$^{-3}$ ng

22.19  The ions that radiations produce are strange, unstable, highly reactive ions that initiate undesirable reactions in the body.

22.20  Any particle with an unpaired electron. The particles lacks an outer octet and so is reactive.

22.21  They can cause birth defects.

22.22  Its radiations can cause cancer.

22.23  Their intensities diminish with the square of the distance; and they can be blocked by dense absorbing materials such as lead.

22.24  There is no level of exposure below which no damage is possible.

22.25  In low doses over a long period, radiation can initiate cancer. In well-focused, massive doses over a short period, radiations can kill cancer cells.

22.26  4

22.27  They move in straight lines and spread out in the way that light from a light bulb spreads out.

22.28  Radionuclides in natural materials such as the soil, fallout from nuclear testing, cosmic rays, medical X rays, radioactive wastes released from nuclear power plants, television tubes, and medical radionuclides.

22.29  Cosmic rays are more intense at higher altitudes.

22.30  Sr-90 is a bone seeker. I-131 is taken up by the thyroid gland. Cs-137 gets as widely distributed as the sodium ion.

22.31  In decay, only a tiny part of the atom breaks away. In fission, the whole atom splits roughly in half.

22.32  Each fission event produces more neutron initiators than were needed to cause the fission event.

22.33  The rods absorb neutrons and prevent them from causing fissions.

22.34  15.1 m

22.35  4000 millirad

22.36  The curie

22.37   The sample is undergoing $1.5 \times (3.7 \times 10^{10}) \times 10^{-3} = 5.6 \times 10^7$ disintegrations per second.

22.38   The roentgen

22.39   The rad (The SI unit is the gray.)

22.40   650 rad

22.41   They are basically equivalent.

22.42   Different kinds of tissue have different responses to the same quantity of rads.

22.43   The rad doses are adjusted for the kinds of tissues and radiations and expressed as rems.

22.44   295 mrem

22.45   The electron-volt

22.46   100 keV or less

22.47   Radiations can cause cancer, so for diagnosis the lowest usable energies are in order.

22.48   Radiations fog the film and the degree of fogging is proportional to the exposure.

22.49   Alpha particles can't penetrate the tube's window.

22.50   Iron-55. $^{55}_{25}\text{Mn} + {}^{1}_{1}\text{H} \longrightarrow {}^{1}_{0}n + {}^{55}_{26}\text{Fe}$

22.51   $^{109}_{47}\text{Ag} + {}^{4}_{2}\text{He} \longrightarrow {}^{113}_{49}\text{In}$

22.52   $^{113}_{49}\text{In} \longrightarrow {}^{111}_{49}\text{In} + 2\,{}^{1}_{0}n$

22.53   $^{66}_{30}\text{Zn} + {}^{1}_{1}\text{H} \longrightarrow {}^{67}_{31}\text{Ga}$

22.54   $^{67}_{31}\text{Ga} + {}^{0}_{-1}e \xrightarrow{\text{electron capture}} {}^{67}_{30}\text{Zn}$

22.55   $^{19}_{9}\text{F} + {}^{4}_{2}\text{He} \longrightarrow {}^{23}_{11}\text{Na} \longrightarrow {}^{22}_{11}\text{Na} + {}^{1}_{0}n$

22.56   $^{10}_{5}\text{B} + {}^{4}_{2}\text{He} \longrightarrow {}^{13}_{7}\text{N} + {}^{1}_{0}n$

22.57   $^{14}_{7}\text{N} + {}^{2}_{1}\text{H} \longrightarrow {}^{16}_{8}\text{O} \longrightarrow {}^{15}_{8}\text{O} + {}^{1}_{0}n$

22.58   $^{27}_{13}\text{Al} + {}^{6}_{3}\text{Li} \longrightarrow {}^{32}_{15}\text{P} + {}^{1}_{1}\text{H}$

22.59   $^{32}_{16}\text{S} + {}^{1}_{0}n \longrightarrow {}^{32}_{15}\text{P} + {}^{1}_{1}\text{H}$

22.60   The shorter the half-life, the more active is the radionuclide and hence the smaller is the dose that is needed to get results.

22.61   We can get by with a smaller quantity.

22.62   Gamma radiation can penetrate the entire body and reach a detector and thereby serve a diagnostic purpose. Alpha and beta radiation would be absorbed in the body, causing harm, without giving diagnostic value.

22.63   It emits only gamma radiation, and it has a shorter half-life.

22.64   It has a more intense radiation and it involves only gamma radiation.

22.65   As phosphate ion, because bone contains this ion.

22.66   As a product in the uranium-238 disintegration series.

22.67    Alpha and gamma

22.68    Its decay products, which are not gases, but are also radioactive stay in the lungs.

22.69    4 picocuries/L

22.70    Neutron bombardment of molybdenum-98 in the device changes some of it to molybdenum-99, which they decays to technetium-99m.

22.71    From molybdenum-99

22.72    Initially as the pertechnitate ion, $TcO_4^-$, but this can be changed into a number of other chemical forms to suit a particular radiological purpose.

22.73    By bombarding a metal surface with high-energy electrons, which are able to expel first level electrons from the metal atoms.  As higher level electrons drop into the lower level 'holes', X rays are emitted.

22.74    It consists of a large array of X-ray generators that can all be focused onto an area of the body.  Brief pulses are sent through the body, and the emerging radiation is processed into an image on a film.

22.75    Same masses, but the charge on the positron is 1+, not 1– as on the electron.

22.76    A nuclear proton breaks into a neutron and a positron:

$$_1^1p \longrightarrow \ _0^1n + \ _1^0e$$

22.77    In any collision with an electron it is annihilated, and electrons are abundant in matter.

22.78    Gamma radiation (annihilation radiation).

22.79    The radionuclide is incorporated into molecules of some substance, like glucose, that can be taken up by the target cell of the tissue.

22.80    Its ability to identify extremely small regions of the brain that are in their early stages of breakdown.

22.81    No ionizing radiation or alien chemicals are used.

22.82    Bones do not obscure the MRI scan.  Soft, water-abundant tissues are imaged easily.

22.83    The inorganic ions in bone, like calcium ion, do not have magnetic nuclei that give off signals that would confuse signals from water molecules or others with hydrogen nuclei.